MW00814606

Springer Proceedings in Physics

Volume 173

More information about this series at http://www.springer.com/series/361

Mustapha Tlidi · Marcel G. Clerc

Editors

Nonlinear Dynamics: Materials, Theory and Experiments

Selected Lectures, 3rd Dynamics Days South America, Valparaiso 3–7 November 2014

 Springer

Editors

Mustapha Tlidi
Physique des systèmes dynamiques
Université Libre de Bruxelles
Brussels
Belgium

Marcel G. Clerc
Facultad de Ciencias Físicas y Matemáticas
Universidad de Chile
Santiago
Chile

ISSN 0930-8989 ISSN 1867-4941 (electronic)
Springer Proceedings in Physics
ISBN 978-3-319-24869-1 ISBN 978-3-319-24871-4 (eBook)
DOI 10.1007/978-3-319-24871-4

Library of Congress Control Number: 2015951395

Springer Cham Heidelberg New York Dordrecht London

© Springer International Publishing Switzerland 2016
This work is subject to copyright. All rights are reserved by the Publisher, whether the whole or part of the material is concerned, specifically the rights of translation, reprinting, reuse of illustrations, recitation, broadcasting, reproduction on microfilms or in any other physical way, and transmission or information storage and retrieval, electronic adaptation, computer software, or by similar or dissimilar methodology now known or hereafter developed.
The use of general descriptive names, registered names, trademarks, service marks, etc. in this publication does not imply, even in the absence of a specific statement, that such names are exempt from the relevant protective laws and regulations and therefore free for general use.
The publisher, the authors and the editors are safe to assume that the advice and information in this book are believed to be true and accurate at the date of publication. Neither the publisher nor the authors or the editors give a warranty, express or implied, with respect to the material contained herein or for any errors or omissions that may have been made.

Printed on acid-free paper

Springer International Publishing AG Switzerland is part of Springer Science+Business Media
(www.springer.com)

Preface

Nonlinear dynamics in out equilibrium system has been appealing a great deal of attention for several decades due to their tremendous influence in many physical, chemical, biological, economical, ecological, and social processes. Many efforts in the past have focused on the understanding of complex behavior of low-dimensional system, characterization of primary, spatial, or temporal bifurcations, which generate dissipative structures that can be either periodic or localized in space or/and in time. In two-dimensional setting, regular patterns like stripes, hexagons, squares, and super lattices, and the understanding of the macroscopic particle type solutions. Prigogine and Lefever to describe the spontaneous appearance of periodic structures as a result of irreversible processes maintained by exchanges of matter and/or energy with a non-equilibrium environment have introduced the term "dissipative structures." Dissipative structures generally evolve on macroscopic scales and can only be maintained by continuous application of a non-equilibrium constraint. The spontaneous emergence of dissipative structures arises from a principle of self-organization that can be either in space or in time. A classic example of spatial self-organization is provided in the context of chemical reaction diffusion system referred to as Turing instability. This instability is characterized by a symmetry breaking that leads spontaneously to the formation of periodic structures with an intrinsic wavelength. In this case the wavelength is determined solely by the dynamical parameters such as diffusion coefficient and the inverse characteristic time associated with chemical kinetics. However, because of theoretical and numerical advances, and experimental observations are still offering many fundamental questions, which present the challenge of the nonlinear community. Such as characterization and emergence of complex spatiotemporal behavior, self-organization of life matter and nanoscale structure, and geometric characterization of localized structures. To make progress in this problematic, it is necessary to develop new nonlinear concepts, methods, and instruments.

This book contains the selected lectures given at the third Dynamics Days South America 2014, which took place at Valparaiso November 3–7, 2014. Since its beginning, a key goal of Dynamics Days South America conference has been to promote the participation and cross-discipline interaction of young scientists together with more established experts from a wide variety of fields. The lectures were devoted to various aspects of pattern formation, coarsening dynamics, driven systems, front propagation, stochastic process, nonlinear Anderson localization, interface dynamics, defects dynamics, phase transition and nonlinear dynamics in magnetic, fluid, optical, mechanical, and social systems.

We want to thank, here, our sponsors and supporters whose interest and help were essential for the success of the meeting. However, the physical realization of the workshop would have not been possible without the decisive collaboration of the Universidad de Chile, Pontificia Universidad Católica de Valparaiso and Center for mathematical modeling.

Brussels, Belgium   Mustapha Tlidi
Santiago, Chile   Marcel G. Clerc

Contents

Part III   Robust Phenomena

Contributors

W.W. Ahmed Departament de Fisica i Enginyeria Nuclear, Universitat Politècnica de Catalunya (UPC), Terrassa, Barcelona, Spain

Cristóbal Arratia Departamento de Física FCFM, Universidad de Chile, Santiago, Chile; LadHyX, CNRS-École Polytechnique, Palaiseau, France

Juan A. Asenjo Centre for Biochemical Engineering and Biotechnology, Centre for Biotechnology and Bioengineering, University of Chile, Santiago, Chile

Sébastien Aumaître Service de Physique de l'Etat Condensé, DSM, CEA-Saclay, Gif-sur-Yvette, France; Laboratoire de Physique, ENS de Lyon, Lyon, France

Raouf Barboza Departamento de Física, Facultad de Ciencias Físicas Y Matemáticas, Universidad de Chile, Santiago, Chile

Cédric Beaume Department of Aeronautics, Imperial College London, London, UK

D. Becerra-Alonso Universidad Loyola Andalucía, Córdoba, Spain

Ignacio Bordeu Departamento de Física, Facultad de Ciencias, Universidad de Chile, Santiago, Chile

Umberto Bortolozzo Institut Nonlinéaire de Nice, Université de Nice-Sophia Antipolis, CNRS, Valbonne, France

M. Botey Departament de Fisica i Enginyeria Nuclear, Universitat Politècnica de Catalunya (UPC), Terrassa, Barcelona, Spain

R. Bustos-Guajardo Departamento de Física Aplicada, CINVESTAV del IPN, Mérida, Yucatán México

Thomas Butler Centre for Advanced Photonics and Process Analysis, Cork Institute of Technology and Tyndall National Institute, Cork, Ireland

Iberê L. Caldas Instituto de Física, Universidade de São Paulo, São Paulo, Brazil

Jean-Marc Chomaz LadHyX, CNRS-École Polytechnique, Palaiseau, France

Marcel G. Clerc Facultad de Ciencias Físicas y Matemáticas, Departamento de Física, Universidad de Chile, Santiago, Chile

Carlos Conca Department of Mathematical Engineering (DIM), Centre for Mathematical Modelling (CMM) (UMI CNRS 2807), and Centre for Biotechnology and Bioengineering (CeBiB), University of Chile, Santiago, Chile

Vasco Cortez Facultad de Ingeniería y Ciencias, Universidad Adolfo Ibáñez, Peñalolén, Santiago, Chile

S. Coulibaly PhLAM, Université de Lille 1, Bât. P5-bis; UMR CNRS/USTL 8523, Villeneuve D'ascq, France

Patricio Cumsille Group of Applied Mathematics, Department of Basic Sciences, University of Bío-Bío, Chillán, Chile; Centre for Biotechnology and Bioengineering, University of Chile, Santiago, Chile

Daniel Escaff Facultad de Ingeniería y Ciencias Aplicadas, Complex Systems Group, Universidad de Los Andes, Las Condes, Santiago, Chile

C. Falcón Facultad de Ciencias Físicas y Matemáticas, Departamento de Física, Universidad de Chile, Santiago, Chile

Sergej Flach Center for Theoretical Physics of Complex Systems, Institute for Basic Science, Daejeon, Korea; New Zealand Institute for Advanced Study, Centre for Theoretical Chemistry and Physics, Massey University, Auckland, New Zealand

Punit Gandhi Department of Physics, University of California, Berkeley, CA, USA

Mónica A. García-Ñustes Instituto de Física, Pontificia Universidad Católica de Valparaíso, Valparaíso, Casilla, Chile

I. Giden Department of Electrical and Electronics Engineering, TOBB University of Economics and Technology, Ankara, Turkey

Gregorio González-Cortés Departamento de Física, Facultad de Ciencias, Universidad de Chile, Santiago, Chile

David Goulding Centre for Advanced Photonics and Process Analysis, Cork Institute of Technology and Tyndall National Institute, Cork, Ireland

Pablo Gutiérrez Facultad de Ciencias Físicas y Matemáticas, Departamento de Física, Universidad de Chile, Santiago, Chile; Service de Physique de l'Etat Condensé, DSM, CEA-Saclay, Gif-sur-Yvette, France

Stephen P. Hegarty Centre for Advanced Photonics and Process Analysis, Cork Institute of Technology and Tyndall National Institute, Cork, Ireland

R. Herrero Departament de Fisica i Enginyeria Nuclear, Universitat Politècnica de Catalunya (UPC), Terrassa, Barcelona, Spain

Guillaume Huyet Centre for Advanced Photonics and Process Analysis, Cork Institute of Technology, Cork, Ireland; National Research University of Information Technologies, Mechanics and Optics, Saint Petersburg, Russia and Tyndall National Institute, Cork, Ireland

Bryan Kelleher Department of Physics, University College Cork and Tyndall National Institute, Cork, Ireland

Edgar Knobloch Department of Physics, University of California, Berkeley, CA, USA

A.V. Kovalev ITMO University, St. Petersburg, Russia

Tiago Kroetz Universidade Tecnológica Federal do Paraná, Pato Branco, Paraná, Brazil

S. Kumar Departament de Fisica i Enginyeria Nuclear, Universitat Politècnica de Catalunya (UPC), Terrassa, Barcelona, Spain

H. Kurt Department of Electrical and Electronics Engineering, TOBB University of Economics and Technology, Ankara, Turkey

D. Laroze Instituto de Alta de Investigación, Universidad de Tarapacá, Casilla 7D, Arica, Chile; SUPA School of Physics and Astronomy, University of Glasgow, Glasgow, UK

René Lefever Faculté des Sciences, Université Libre de Bruxelles (ULB), Bruxelles, Belgium

F. Leo Photonics Research Group, Department of Information Technology, Ghent University-IMEC, Ghent, Belgium

Edson D. Leonel Departamento de Física, Universidade Estadual Paulista, São Paulo, Brazil

Alejandro O. León Facultad de Ciencias Físicas y Matemáticas, Departamento de Física, Universidad de Chile, Santiago, Chile

Z. Liu PhLAM, Université de Lille 1, Bât. P5-bis; UMR CNRS/USTL 8523, Villeneuve D'ascq, France

André L.P. Livorati Departamento de Física, Universidade Estadual Paulista, São Paulo, Brazil

E. Louvergneaux Laboratoire de Physique des Lasers, Atomes Et Molécules, CNRS UMR8523, Université de Lille 1, Villeneuve d'Ascq, France

J.E. Macías Facultad de Ciencias Físicas y Matemáticas, Departamento de Física, Universidad de Chile, Santiago, Chile

A. Makhoute Physique du Rayonnemet et des Interactions Laser-Matiere, Faculté des Sciences, Département de Physique, Université Moulay Ismail, Meknès, Morocco

Boris A. Malomed Faculty of Engineering, Department of Physical Electronics, School of Electrical Engineering, Tel Aviv University, Tel Aviv, Israel

H.L. Mancini Departamento de Física y Matemática Aplicada, Universidad de Navarra, Pamplona, Spain

Fernando Mora Facultad de Ingeniería y Ciencias, Universidad Adolfo Ibáñez, Peñalolén, Santiago, Chile

Cristian F. Moukarzel Departamento de Física Aplicada, CINVESTAV del IPN, Mérida, Yucatán México

V. Odent Laboratoire de Physique des Lasers, Atomes Et Molécules, CNRS UMR8523, Université Lille 1, Villeneuve d'Ascq, France

Sabine Ortiz LadHyX, CNRS-École Polytechnique, Palaiseau, France; UME/DFA, ENSTA, Palaiseau, France

Vincent Padilla Service de Physique de l'Etat Condensé, DSM, CEA-Saclay, Gif-sur-Yvette, France

Krassimir Panajotov Brussels Photonics Team (B-Phot), Department of Applied Physics and Photonics (TONA), Vrije Universiteit Brussel, Brussel, Belgium; Institute of Solid State Physics, Sofia, Bulgaria

L.M. Pérez Departamento de Física y Matemática Aplicada, Universidad de Navarra, Pamplona, Spain

M. Radziunas Weierstrass Institute for Applied Analysis and Stochastics, Leibniz Institute in Forschungsverbund Berlin e.V., Berlin, Germany

Stefania Residori Institut Nonlineaire de Nice, Université de Nice-Sophia Antipolis, CNRS, Valbonne, France

Sergio Rica Facultad de Ingeniería y Ciencias, Universidad Adolfo Ibáñez, Peñalolén, Santiago, Chile

Marc Sciamanna OPTEL Research Group, LMOPS EA-4423, CentraleSupelec - Universite de Lorraine, Belin, France

Svetlana Slepneva Centre for Advanced Photonics and Process Analysis, Cork Institute of Technology and Tyndall National Institute, Cork, Ireland

N. Slimani Physique du Rayonnement et des Interactions Laser-Matière, Faculté des Sciences, Université Moulay Ismail, Meknès, Morocco

K. Staliunas Departament de Fisica i Enginyeria Nuclear, Universitat Politècnica de Catalunya (UPC), Terrassa, Barcelona, Spain; Institució Catalana de Recerca i Estudis Avançats (ICREA), Barcelona, Spain

M. Taki PhLAM, Université de Lille 1, Bât. P5-bis; UMR CNRS/USTL 8523, Villeneuve D'ascq, France

Hugo Thienpont Brussels Photonics Team (B-Phot), Department of Applied Physics and Photonics (TONA), Vrije Universiteit Brussel, Brussel, Belgium

Mustapha Tlidi Faculté des Sciences, Université Libre de Bruxelles (ULB), Bruxelles, Belgium

M. Turduev Department of Electrical and Electronics Engineering, TED University, Ankara, Turkey

Felipe Urbina Facultad de Ingeniería y Ciencias, Universidad Adolfo Ibáñez, Peñalolén, Santiago, Chile

D. Urzagasti Instituto de Investigaciones Físicas, UMSA, La Paz, Bolivia

Lautaro Vergara Departamento de Física, Universidad de Santiago de Chile, USACH, Santiago 2, Chile

E.A. Viktorov ITMO University, St Petersburg, Russia

Martin Virte Brussels Photonics Team (B-Phot), Department of Applied Physics and Photonics (TONA), Vrije Universiteit Brussel, Brussel, Belgium

Mario Wilson DFI, Facultad de Ciencias Físicas y Matemáticas, Universidad de Chile, Santiago, Chile

Valeska Zambra Departamento de Física, Facultad de Ciencias, Universidad de Chile, Santiago, Chile

Part I
Optics

Visualisation of the Intensity and Phase Dynamics of Semiconductor Lasers via Electric Field Reconstructions

David Goulding, Thomas Butler, Bryan Kelleher, Svetlana Slepneva, Stephen P. Hegarty and Guillaume Huyet

Abstract In any physical dynamical system an understanding of both the intensity and phase dynamics can provide an unique insight into the dynamical behaviour observed. A thorough picture of the phase and intensity dynamics of an optical system can greatly aid in understanding the device dynamics from the perspective of fundamental physics as well as providing a deeper insight into device performance. This work details a novel interferometric analysis technique which allows the measurement and visualisation of the complex electric field of a laser source in a time-resolved, single shot format. To demonstrate the effectiveness of this technique, it has been applied to the dynamically rich system of a semiconductor laser undergoing external optical injection, allowing for the first time a direct study of the phase trajectories of the various non-linear dynamical responses observed in such a system. Furthermore a full electric field reconstruction of a fast frequency swept source has also been performed. A time-resolved analysis of this wide bandwidth laser allows direct measurement of many fast time-scale dynamics and a precise calculation of many device parameters which were previously inaccessible. This novel technique allows for previously unobtainable insights into the underlying dynamics of laser systems with many potential applications for this technique in studying future light sources and in gaining further understanding of fundamental laser dynamics.

D. Goulding (✉) · T. Butler · S. Slepneva · S.P. Hegarty
Centre for Advanced Photonics and Process Analysis, Cork Institute of Technology
and Tyndall National Institute, Lee Maltings, Cork, Ireland
e-mail: david.goulding@tyndall.ie

B. Kelleher
Department of Physics, University College Cork
and Tyndall National Institute, Lee Maltings, Cork, Ireland

G. Huyet
Centre for Advanced Photonics and Process Analysis,
Cork Institute of Technology, Cork, Ireland

G. Huyet
National Research University of Information Technologies,
Mechanics and Optics, Saint Petersburg, Russia and Tyndall National Institute,
Lee Maltings, Cork, Ireland

© Springer International Publishing Switzerland 2016
M. Tlidi and M.G. Clerc (eds.), *Nonlinear Dynamics: Materials,
Theory and Experiments*, Springer Proceedings in Physics 173,
DOI 10.1007/978-3-319-24871-4_1

1 Introduction

In order to gain an insight into the behaviour of any dynamical system, an understanding of both the phase and intensity evolution is desired. Typically access to the intensity dynamics is achievable in many experimental systems but the phase dynamics can often remain hidden. From cases such as rotating pendula to Van der Pol oscillators, the phase of the system plays a key role in allowing one to gain fundamental understanding of the various non-linear dynamical behaviours which can occur. In optical experiments, typical tools available to experimenters include optical and radio frequency spectra and intensity time series. On the other hand, numerical simulations provide excellent opportunities for identifying specific features associated with the system using phasor plots obtained from the components of the laser electric field. Identifying certain dynamical features experimentally is more difficult in the absence of such phasor plots, although various criteria for recognising certain features have been found [32, 41]. For example the approach taken in [41] to reconstruct the complete phase space trajectory from scalar or few-variable measurements is time-delayed embedding of a single measured variable. While successful in reconstructing a topologically equivalent trajectory to that of the real phase space trajectory this approach does not allow accurate association of the phase space trajectories with physical variables of the system. So while such techniques give a good indication of the behaviour of the system they do not provide a full understanding or exposure of the full dynamics that take place in the system. As a result, direct experimental phasor plots are extremely desirable and will provide an unique opportunity to gain a deeper understanding of the non-linear system. Achieving this will allow one to for example trace the position of separatrices in non-linear systems as well as accurately follow both local and global bifurcations. This work discusses a novel technique by which experimental phase space trajectories may be obtained, in particular focusing on phase plane reconstructions in semiconductor laser systems.

Semiconductor lasers are among the most useful and widely used lasers in the world at present. Their various applications in telecommunications, data processing and medical imaging, for example, have marked them out as technologically important devices. In addition to their technical applications and advantages, semiconductor lasers have allowed the straightforward construction of non-linear systems in which it is possible to make quantitative comparisons between experimental and theoretical results. One such system of great interest is the injection locked laser system [26, 33] in which a wealth of non-linear dynamical features have been predicted [46].

This work considers the case of an optically injected single mode quantum dot (QD) laser. Experiments using quantum dot (QD) semiconductor lasers have shown them to demonstrate both low linewidth enhancement (α) factors and strongly damped relaxation oscillations. These two parameters have been shown to play a key role in the improved stability of QD lasers undergoing both feedback and injection. These parameters also play a role in the appearance of certain regimes of non-linear dynamical behaviour in the injection locked configuration and motivate the use of QD lasers in the work described here on injection locked semiconductor lasers. Due to their unique properties, many dynamical features previously predicted by the theory but unobserved experimentally have now been observed in injection locked QD lasers, with excitable pulsing and bistability being two such features. Typically in experimental works on injected lasers, authors produce figures of optical spectra and intensity for comparison with theoretical work. This chapter however describes a novel interferometric method which allows direct comparison between not only optical spectra and intensity plots but also for the first time the phase-space trajectories of the slave laser. This will provide additional opportunities to allow qualitative comparison between predicted theoretical dynamics and experimentally obtained results to be made. From the experimental data collected, phase space trajectories of an optically injected quantum dot laser while undergoing complicated dynamical behaviour are reconstructed. Features such as excitable pulsing, limit cycle switching and chaotic intermittency and their corresponding phase trajectories are all clearly demonstrated.

In addition to studying the phase dynamics of optically injected semiconductor lasers, the technique is also applied in the study of fast frequency swept lasers, used in optical coherence tomography (OCT). OCT is now a well-known imaging technique used to obtain real time two or three dimensional cross-sectional images of biological tissue and non-biological materials. The light source examined in this work has an optical bandwidth of approximately 20 THz with a fast tuning frequency ($\sim$50 kHz). The performance of modern day OCT systems, in terms of imaging depth, speed and resolution, is governed by the coherence properties of the light source used. Typical measurements of the coherence properties of swept sources rely on a time and resource consuming averaging of numerous repeated measurements of the light source. Clearly as the coherence properties of a light source are defined by the properties of the complex electric field, a knowledge of same would provide rapid measurement of the device properties.

This work therefore also presents a novel time-delayed self-referencing single-shot experimental technique, based on the phase reconstruction method, which has been developed to fully characterise and reconstruct the dynamic complex electric field of a fast frequency swept source. This technique allows a direct, instantaneous examination of the coherence and linewidth properties of the source, providing a method to rapidly characterise and guide further improvement of next generation light sources in medical imaging.

2 Overview of Phase Reconstruction Technique

The interferometric technique discussed in this chapter is based upon the mixing of signals in a 3×3 coupler. The coupler has three inputs E_n^{in} and outputs E_n^{out}, for $n = 1, 2, 3$. Each output consists of a contribution from its associated input, along with phase shifted contributions from the other two inputs. The relative magnitude of the associated phase shift, in each case, is $\frac{2\pi}{3}$. For example, the output of port 1 can be expressed as:

$$E_1^{out} = \frac{1}{\sqrt{3}} \left(E_1^{in} + e^{\frac{2\pi}{3}} \left(E_2^{in} + E_3^{in} \right) \right). \tag{1}$$

Each experimental arrangement considered in this chapter is based upon the mixing of two signals in a passive optical coupler and as such can be represented by the schematic presented in Fig. 1. Here the electric fields E_1^{in} and E_2^{in} are passed to the first two inputs of the 3×3 coupler, while the third arm remains unused ($E_3^{in} = 0$).

The input electric fields can be represented by $E_n^{in} = R_n e^{i\phi_n}$, for $n = 1, 2$, where R_n and ϕ_n are the complex amplitude and phase of the respective input signals. The full set of equations for the output electric fields of the 3×3 coupler can, in this situation, be further reduced to read:

$$E_1^{out} = \frac{1}{\sqrt{3}} \left(R_1 e^{i\phi_1} + R_2 e^{i\left(\phi_2 + \frac{2\pi}{3}\right)} \right), \tag{2}$$

$$E_2^{out} = \frac{1}{\sqrt{3}} \left(R_1 e^{i\left(\phi_1 + \frac{2\pi}{3}\right)} + R_2 e^{i\phi_2} \right), \tag{3}$$

$$E_3^{out} = \frac{1}{\sqrt{3}} \left(R_1 e^{i\left(\phi_1 + \frac{2\pi}{3}\right)} + R_2 e^{i\left(\phi_2 + \frac{2\pi}{3}\right)} \right). \tag{4}$$

Clearly the output intensities measured on the oscilloscope, $I_n = |E_n^{out}|^2$, can then be expressed in terms of the complex amplitude and phase of the input signals as:

$$I_1 = \frac{1}{3} \left(R_1^2 + R_2^2 + 2R_1 R_2 \cos\left(\phi_1 - \phi_2 - \frac{2\pi}{3}\right) \right), \tag{5}$$

$$I_2 = \frac{1}{3} \left(R_1^2 + R_2^2 + 2R_1 R_2 \cos\left(\phi_1 - \phi_2 + \frac{2\pi}{3}\right) \right), \tag{6}$$

$$I_3 = \frac{1}{3} \left(R_1^2 + R_2^2 + 2R_1 R_2 \cos\left(\phi_1 - \phi_2\right) \right). \tag{7}$$

Fig. 1 Schematic overview of the measurement technique used for phase reconstruction

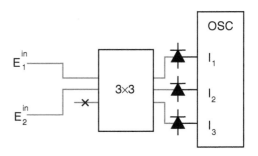

From (5)–(7), straight forward algebraic manipulations will allow the relative phase difference between the two input signals to be isolated. To proceed, it is worthwhile to consider the following linear combinations of the output intensities measured on the oscilloscope:

$$I_1 - I_2 = \frac{2R_1R_2}{\sqrt{3}} \, \text{Sin} \, (\phi_1 - \phi_2), \tag{8}$$

$$I_1 + I_2 = \frac{2}{3} \left(R_1^2 + R_2^2 - R_1R_2 \, \text{Cos} \, (\phi_1 - \phi_2) \right), \tag{9}$$

$$I_3 - \frac{1}{2} (I_1 + I_2) = R_1R_2 \, \text{Cos} \, (\phi_1 - \phi_2). \tag{10}$$

Then combining (8)–(10), the relative phase difference can then be calculated as:

$$\phi_1 - \phi_2 = \text{atan} \left(\frac{\sqrt{3}}{2} \left(\frac{I_1 - I_2}{I_3 - \frac{1}{2} (I_1 + I_2)} \right) \right). \tag{11}$$

Equation (11) provides a direct measurement of the phase difference between two input signals in terms of the three separate intensity measurements. It should also be noted that summing the three output intensities (5)–(7) gives the sum of the input intensities of the input fields E_1^{in} and E_2^{in}. From this, provided that an independent measurement of one of the intensities is taken, it is then possible to measure both the intensity and phase dynamics in a single measurement.

To demonstrate the power of the technique, the work that follows examines implementations of the technique in two differing situations. Firstly, the technique is applied in the study of semiconductor lasers under external optical injection. The technique is used to examine the, heretofore hidden, experimental phase trajectories corresponding to various non-linear dynamical responses of semiconductor lasers to external injection. Secondly a novel time-resolved interferometric technique is presented, based on the phase reconstruction technique that allows the reconstruction of the complex electric field output of a fast frequency swept laser in a single-shot measurement.

3 Optical Injection of Semiconductor Lasers

Optical injection is a process in which light from one laser (deemed the master) is injected into the cavity of another laser (called the slave). Provided that the fields of both lasers are sufficiently close in frequency, the master field forces the slave field to oscillate at the same frequency and at a fixed phase relative to the phase of the master field. Such synchronisation of oscillators is a familiar process in many areas of science and engineering and so-called master-slave configurations have long been studied. The configuration provides a tractable, non-linear system with effects including multi-stability [27], chaos [37, 38, 44] and excitability [12, 21, 43, 45]. In addition to the wealth of non-linear dynamical behaviour possible in such systems [32], there are also a number of significant engineering benefits in optical injection locking of lasers. These benefits include, for example, spectral narrowing [11], relative intensity noise (RIN) reduction [17] and a modulation bandwidth enhancement [29].

Recent work on quantum dot (QD) lasers under optical injection have demonstrated many of the dynamical features predicted by theoretical considerations, the first time many of these features have been observed in semiconductor lasers under external optical injection [9, 12, 19, 20, 22]. As originally conceived QD lasers were predicted to have a low to zero linewidth enhancement (α) factor [30] and strongly damped relaxation oscillations (ROs) [24]. Experimental work on QD lasers in optical configurations such as optical injection [12, 19], optical feedback [31] and mutual coupling [13] have shown improved performance in direct comparison to their quantum well and bulk counterparts [36, 42]. The ability of QD lasers to stably lock under external optical injection for any injection levels has been shown to be due to the relatively low α-factor and strongly damped relaxation oscillations. It is as a direct result of their unique properties that many dynamical features previously predicted by the theory have now been observed in injection locked QD lasers.

Using the method discussed in the previous section it is possible to obtain a phase resolved measurement of the phase dynamics of the slave laser under external optical injection. Using the technique it is then possible to visualise the trajectory of the slave in the phase plane (E_x, E_y), the real and imaginary parts of the electric field.

3.1 Quantum Dot Lasers Under Optical Injection

The experiments detailed in this section were carried out on several distributed feedback (DFB) InAs devices of similar construction to those described in detail in [28]. The devices are a five-QD-layer structure which had been grown by solid-source Molecular Beam Epitaxy (MBE). It consisted of 2.4 InAs monolayers topped with 5 nm GaInAs, stacked in a 200 nm thick optical cavity. A 35 nm GaAs spacer was used between the QD layers. Optical confinement is provided by $Al_{0.85}Ga_{0.15}As$ cladding layers. The single-mode ridge waveguide lasers were approximately 2 μm wide. The master laser was a commercially available external cavity tunable device with linewidth <100 kHz and tunable in steps of 0.1 pm. In each experiment carried out, the three control parameters were the power of the master laser, S_m, the slave laser injection current, J, and the angular frequency detuning between the master and slave lasers, Δ.

Figure 2 shows the evolution of the power spectrum of the slave DFB laser as the detuning is varied at a fixed relatively low injection strength. In this case, both the power of the master and the slave injection current were fixed and the detuning was systematically varied by changing the wavelength of the master laser. At each value of the master wavelength an RF spectrum was recorded and Fig. 2 shows a false color plot of the spectra over a wide range of detuning. Clear from this figure are

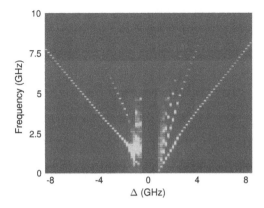

Fig. 2 Experimental power spectrum mask showing fast dynamics occurring near both the positive and negative detuning unlocking boundaries and a stable locking region at zero detuning. The master laser wavelength is swept across resonance with the slave laser at a fixed injection strength

the regions of beating between the master and slave lasers far from zero detuning and a stable locked region in the center. A noteworthy feature is the large area of stable locked operation close to zero detuning, which was found for each injection strength tested. In particular and in stark contrast to optically injected QW lasers, stable phase-locked operation was found at zero detuning for each injection strength considered with the QD lasers.

In [12] the dynamical response of a multi-mode QD laser to external optical injection was reported. There it was shown that for low to moderate injection strengths, fast dynamics can be observed close to the locking boundary for negative detuning. However in the case of a single-mode QD laser [19] dynamical regimes appear close to the locking boundaries for both positive and negative detuning as shown in Fig. 2. The nature of these regions depends on the injection strength. The injection strength is defined to be the ratio of the intensity of the light injected into the lasing cavity to the intensity of the light in the cavity when free running. At low injection strengths and for negative detuning, intensity pulses such as those in Fig. 3a were observed. While in the case of positive detuning, intensity pulses such as those seen in Fig. 3b were observed. For low absolute values of the detuning, these intensity pulsations were very rare and apparently randomly spaced initially while displaying a broad power spectrum. As the magnitude of the detuning was increased towards unlocking, the pulsations became more frequent, eventually becoming a periodic train of pulses. This periodic train transitions into the unlocked limit cycle corresponding to beating between the master and slave lasers as the magnitude of the detuning is further increased. The shape of the pulses differs depending on the sign of the detuning since there is a non-zero α-factor. For the case of negatively detuned pulses, the intensity drops before rising above the phase-locked intensity and then returns to the steady-state value. For the positively detuned side the intensity increases, then drops below the steady-state value, and eventually again recovers to the steady-state value.

These intensity pulsations can be viewed as excitable pulses resulting from 2π phase rotations of the slave electric field [19]. In the external optical injection system examined here, these excitable pulses arise close to a saddle-node (SN) bifurcation

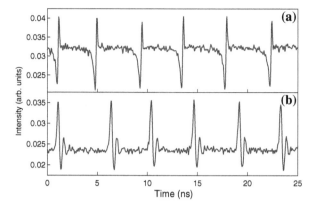

Fig. 3 Experimentally obtained intensity pulses in the case of **a** negative and **b** positive detuning

and their behaviour is similar to the phase slips observed in the classical Adler model [1]. The logic behind the consideration of these events as excitable pulses is as follows: at very low absolute values of the detuning, the stable and unstable points are sufficiently separated to avoid any noise-induced pulsations (that is the noise in the system is not sufficient to push the laser away from the stable point and beyond the unstable threshold point) and as a result quiet regions of locking are observed. However as the magnitude of the detuning is increased, the stable and unstable fixed points become progressively closer until eventually the noise level is sufficient to push the system past the unstable point resulting in a 2π rotation of the electric field and a consequent intensity pulsation. As the magnitude of the detuning is increased even further the pulses occur more and more frequently until eventually the two fixed points collide and annihilate each other leaving a ghost at which point the time series of the slave's output represents a periodic train (modulo noise) with the characteristic shape of the pulses. Using the phase measurement method detailed in the previous section, it should be possible to directly examine and observe the phase rotations corresponding to the two different intensity pulses observed in Fig. 3.

3.2 Phase Space Reconstructions of QD Lasers Under Optical Injection

The experimental set-up used in obtaining the phase resolved measurement is shown in Fig. 4. The master output is initially split in a 50/50 passive coupler. One part of the output of the 50/50 is injected into the slave laser using a lensed fibre via a set of polarisation controllers. In order to prevent undesired feedback, optical isolators with isolation greater than 40 dB were used at both the input and output sides of the slave laser. The second output of the 50/50 passive coupler serves as a reference

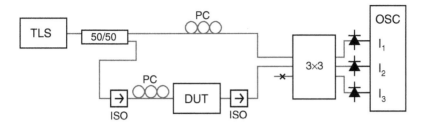

Fig. 4 Experimental arrangement used to measure the phase space trajectory of a semiconductor laser under external optical injection. Light from the master tunable laser source (TLS) is split via a 50/50 passive coupler with one part injected into the slave quantum dot device under test (DUT) via an isolator (ISO) and a set of polarisation controllers. The other half of the TLS acts as a reference beam for the phase measurement. The output intensities of the 3×3 coupler are directly measured on a 12 GHz, 40 GaS/s real-time oscilloscope

for phase measurement and is connected to port 1 of the 3×3 optical coupler. The output of the slave laser is connected to port 2 of the 3×3 optical coupler, the outputs of which are connected to three 12 GHz photodiodes and to a 40 GSa/s real time oscilloscope via equal lengths of fibre. Prior to commencing the experiment, a signal with an identifiable feature was chosen, and the optical paths and oscilloscope channel time skews were carefully adjusted to ensure simultaneous reading of the 3×3 outputs I_1, I_2 and I_3.

As mentioned previously, in order to observe the intensity pulsations a low injection strength is used and the master laser is swept across resonance with the lasing mode of the slave laser and at each value of the detuning the intensities I_n are measured and recorded on the oscilloscope. Figure 5 shows an example of the three intensities recorded on the oscilloscope for the case of a single isolated pulse observed for negative detuning. For each of the experimental measurements carried out in this section, it is assumed that the phase of the master laser is constant since the measurements are carried out in the frame of the master laser. Following the analysis described

Fig. 5 Measured output intensities of the 3×3 coupler for the case of a single pulsation observed for negative detuning. The three output intensities are of course, by (5)–(7), phase shifted by $2\pi/3$

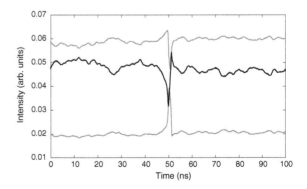

previously leading to (5)–(7), it is possible to calculate the phase difference between the master and slave lasers as:

$$\phi_S - \phi_M = \text{atan}\left(\frac{\sqrt{3}}{2}\left(\frac{I_2 - I_1}{I_3 - \frac{1}{2}(I_1 + I_2)}\right)\right), \tag{12}$$

where ϕ_S and ϕ_M are the phase of the slave and master laser respectively. (It should be noted here that (12) differs slightly from (11). In essence since it is the phase difference $\phi_2 - \phi_1$ that is required in this situation, the roles of I_1 and I_2 are effectively swapped in the equations.) As ϕ_M is considered constant throughout the experiment, it can be set equal to zero and therefore (12) gives the value of the phase of the slave laser at any one instance.

Figure 6 shows a time resolved plot of the evolution of the phasor of the slave laser during a single intensity pulsation for the case of negative (Fig. 6a) and positive (Fig. 6b) values of the detuning. In both cases, the time evolution of the phasor is colour coded as a transition from dark blue to green to red. For example in the case of the negatively (positively) detuned pulse, it can be seen that modulo noise the slave laser initially has a fixed phase, corresponding to continuous wave steady state emission while during the pulse the phase rapidly rotates anticlockwise (clockwise) before again returning to a steady state value. It is clear that in the case of the negatively (positively) detuned pulse that the phase change corresponds to a positive (negative) 2π phase rotation and this is further confirmed by plotting the corresponding phase angle ϕ in Fig. 7. This clear demonstration that the pulsing observed for both positive and negative detuning is caused by slips of the phase of the slave electric field confirms the excitable nature of these phase rotations. These excitable phase slips result in intensity pulses over which the phase of the electric field either increases or decreases by 2π (hence returning to the same point physically).

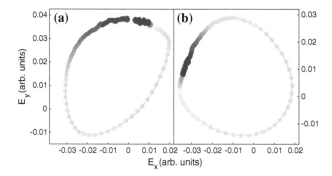

Fig. 6 Experimental, time-resolved reconstructed phase trajectory in the case of an **a** negatively and **b** positively detuned intensity pulsation. The colour coded plots evolve in time from the dark blue colour to *green* to *red*. As can be seen in both cases, the system evolves from a stable state and undergoes a large excursion in the phase space before returning to the stable state, with differing directions of rotation for negative and positive detuning

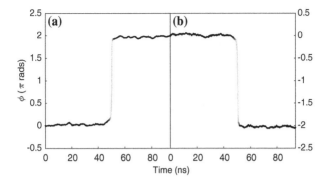

Fig. 7 Experimental, time-resolved evolution of the phase, $\phi(t)$, of the slave laser corresponding to the cases of the **a** negatively and **b** positively detuned intensity pulses from Fig. 6. It should be noted that the phase slowly drifts away from its steady state value before undergoing the rapid large scale 2π rotation corresponding to the intensity pulsation

As mentioned previously there is a wealth of non-linear dynamical features present in the optically injected semiconductor laser system. As the injection strength is increased, a broadening of the stable locking region is observed concurrent with a change in the dynamical behaviour of the slave laser output at both the unlocking boundaries. For example features such as switching between stable states and limit cycles and chaotic intermittency can be observed by altering the injection strength and detuning.

Figure 8 shows both the (a) intensity and (b) reconstructed phase space for the case of the slave laser switching from a stable state to a higher average power limit cycle (the master laser is positively detuned from the slave laser in this case). Again it is worthwhile to recall here that the slave laser intensity can be calculated by summing the three intensity outputs on the oscilloscope, recalling that the master laser intensity is constant. Figure 8b clearly demonstrates the effectiveness of the phase reconstruction technique and it is seen that the laser switches from a stable state (modulo noise) to a period one limit cycle oscillation.

The technique is also capable of examining more complicated non-linear responses and as an example a chaotic intermittency (negative detuning) is plotted in Fig. 9 showing the laser switching between a chaotic output and a distorted limit cycle behaviour. Again in the situation in which the master laser is negatively detuned from the slave laser, Fig. 10 presents a time-resolved plot of both the (a) intensity and (b) reconstructed phase space for the case of the slave laser switching from a stable state to a lower average power period two limit cycle. As can be seen from the figure, there are two separate rotations visible in the phase space, corresponding to the two separate cycles within the period two limit cycle. However one of the limitations of the phase measurement technique is that in both of these cases it is not possible to directly verify that the cycles in phase space do not cross each other. In order to achieve this, the experimental method used must be extended in order to include a third dimension to fully visualise the phase space trajectory.

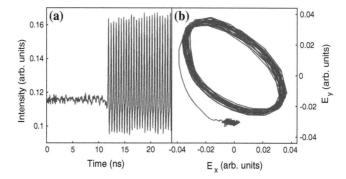

Fig. 8 Experimentally measured **a** intensity and **b** phase space trajectory for a switch from a stable state to a higher average power period one limit cycle

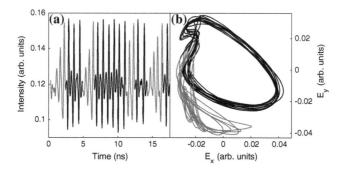

Fig. 9 Experimentally measured **a** intensity and **b** phase space trajectory for a chaotic intermittency for the case of negative detuning, showing intermittency between a distorted limit cycle behaviour and a chaotic output. The phase space trajectory clearly indicates two distinct behaviours

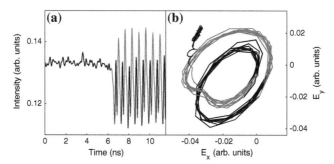

Fig. 10 Experimentally measured **a** intensity and **b** phase space trajectory for a noise induced switch from a stable state to a lower average power period two limit cycle. The phase space trajectory clearly indicates the presence of two different limit cycles in the system

3.3 Full Three Dimensional Phase Space Reconstruction of Optically Injected QD Lasers

One of the most frequently used representations of the phase space of a laser is the three dimensional space (E_x, E_y, N), where E_x, E_y are the real and imaginary parts of the complex electric field and N is the carrier density. The technique described above allows the reconstruction of two dimensional projections of the phase space trajectories on to the plane of the electric field. As direct access to the carrier density of the slave is not viable in the experimental arrangement of Fig. 4, it is possible that by sampling the slave's gain dynamics the experimental phase trajectories can be extended from two dimensions to full three-dimensional reconstructions. This will therefore allow, for example, a fuller examination of the phase trajectory reconstructions and in addition allow qualitative comparison between experimentally obtained trajectories and those computed numerically using standard rate equations to be carried out.

In order to sample the carrier gain dynamics at the same time as the phase dynamics, a modification to the two dimensional set-up of Fig. 4 is required. The full three dimensional reconstruction is achieved by inserting a low power detuned probe laser, set to the wavelength of a non-lasing side mode of the slave laser, into the experimental set-up. This detuned laser is then used to probe the modal gain dynamics far from the DFB peak. Since this detuned peak is located away from the DFB peak it is can be concluded that any variations observed in this signal at the probe wavelength will have resulted from gain variations in the non-lasing side mode of the DFB laser.

The experimental set-up used to achieve this is shown in Fig. 11. The master output is again initially split in a 50/50 passive coupler with one output of the couple again used as a reference for the phase measurement. The second output of this coupler

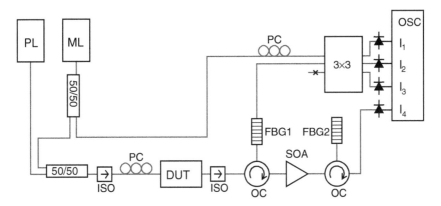

Fig. 11 Experimental set-up used to sample both the phase and gain dynamics of an injection locked QD DFB laser, PL and ML are the probe and master lasers, respectively. FBG1 and FBG2 are two fibre bragg gratings tuned to the off-resonance probe wavelength and OC are optical circulators. The rest of the experimental set-up is as described in Fig. 4

is combined with the low power detuned probe laser and both signals are injected into the slave laser using lensed fibre (again via a set of polarisation controllers). The output of the slave laser is collected at port 1 of a passive optical circulator, with port 2 connected to a Fibre Bragg Grating (FBG). The transmitted light from the first FBG is combined in the 3 × 3 passive coupler for the phase measurement as previously discussed. The temperature (or current) of the slave laser must be set so that a non-lasing side mode of the slave laser coincides exactly with the band gap of the FBGs used in the set-up and these gratings are then used to isolate the response at the probe wavelength. The reflected beam of the first FBG is output at port three of the circulator and amplified through a commercial semiconductor optical amplifier and passed to a second optical circulator. The second port of this circulator is connected to another FBG at the probe wavelength. The reflected light from this second FBG is collected and coupled to a 12 GHz photodiode connected to the real-time oscilloscope. This collected data gives a clear indication of the behaviour of the modal gain as the master laser is varied with respect to the slave laser. The intensity fluctuations observed at the detuned probe wavelength and the calculated phase changes can then be simultaneously plotted to reconstruct the desired three dimensional trajectory.

Figure 12 shows the optical spectra at three different locations within the experimental set-up of Fig. 11. Firstly, Fig. 12a shows the output of the slave laser with both the master and probe lasers on and injected into the slave. From this figure it should be noted that the DFB peak and the probe peak are well separated by approximately 3 nm and that the injected DFB peak and the probe peak powers differ by approximately 20 dB. This optical signal is then passed to a circulator and through

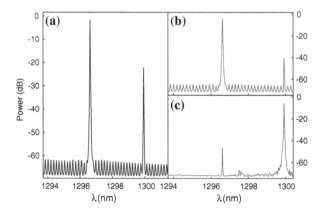

Fig. 12 Optical spectra measured at three different points in the experimental set-up of Fig. 11. The three locations correspond to **a** output of the slave laser prior to entering FBG1, **b** transmission of FBG1 (this is the input to port 2 of the 3 × 3 coupler) and **c** reflection of FBG2 (this is the field measured on channel 4 of the oscilloscope). The DFB peak is at ∼1296.7 nm, while the off-resonance probe wavelength is ∼1299.7 nm

the first of two FBGs. Examining the transmission of the first FBG (Fig. 12b), it can be seen that the DFB peak now exceeds the probe peak by roughly 35 dB. It is of great importance that the DFB peak far exceeds the probe peak so that the influence of the probe laser may be ignored for the two dimensional reconstruction of the phase as described previously. Examining the reflected output of the second FBG (Fig. 12c), it is clear that the probe peak power is larger by approximately 40 dB in comparison to the peak observed at the DFB peak wavelength. This output is used to sample the gain dynamics of the slave laser and in order for the assumption that the changes observed are due to gain variations, it is vital that the probe peak be significantly larger than the main DFB peak. By separating out the response at the DFB and probe wavelengths it is possible to isolate and examine both the phase and gain dynamics of the slave laser.

In Fig. 13, the experimentally reconstructed full three-dimensional phase trajectory for the chaotic intermittency shown in Fig. 9 is plotted. From the plot it is clear to see that the phase space trajectory has two distinct and separated parts, one corresponding to a distorted sine curve with the fast/slow characteristic motion of the pulses and the other to the chaotic slips (both these features are easily observed in Fig. 9a). Similarly shown in Fig. 14 is the three-dimensional phase trajectory for the noise induced switch between a stable state and a period two limit cycle corresponding to Fig. 10. The three-dimensional reconstruction in this case clearly shows that the trajectories of the two different limit cycles, as expected, do not cross in phase space.

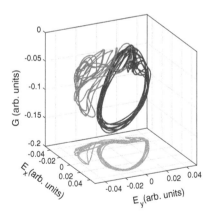

Fig. 13 Reconstructed three-dimensional phase space trajectory for the chaotic intermittency of Fig. 9. The *blue* and *red* sections of the trajectory correspond to the two separate trajectories present in the system at this injection strength and detuning. The two-dimensional projection in *grey* corresponds to the result plotted in Fig. 9

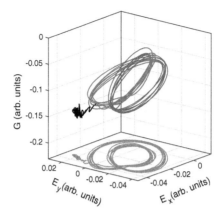

Fig. 14 Reconstructed three-dimensional phase space trajectory for the period two limit cycle switch of Fig. 10. The *blue* and *red* sections of the trajectory are used to identify the two separate limit cycle trajectories present in the system at this injection strength and detuning. The noise induced transition from the stable state into the period two limit cycle is plotted in *black*. The two-dimensional projection in *grey* corresponds to the result plotted in Fig. 10

4 Frequency Swept Lasers in Optical Coherence Tomography

Optical coherence tomography (OCT) is a well-known imaging technique used to obtain real time two- or three-dimensional cross-sectional images of biological and non-biological scattering media, semi-transparent to near infrared radiation [7]. Swept source OCT (SS-OCT) is currently the fastest and most sensitive of all OCT implementation techniques [8, 48]. The main elements of a swept source laser are a gain medium (typically a semiconductor optical amplifier (SOA) is used) and a tunable filter element. Two of the key control parameters for these systems are the full-sweep spectral width and the filter tuning rate or sweep-speed. In order to achieve high-resolution, real-time images both wide spectra and fast tuning rates are required. The semiconductor gain medium will determine the maximum tuning range of the device, while the sweep speed is mainly limited by the properties of the tunable filter and fundamental device physics. For example consider the behaviour of a laser under going mode-hopping, clearly this device will have its speed limited by the switch-on time of the laser modes. While in contrast coherent effects such as those employed in sliding frequency mode-locking [2, 10, 25] and Fourier domain mode-locking (FDML) [15] can admit much higher tuning rates. Both sliding frequency mode-locked lasers and FDML lasers are currently widely used in OCT imaging systems, and while both employ coherent effects, they nonetheless operate via very different lasing mechanisms, the basic descriptions of which have been recently reported in [39, 40]. The goal of any frequency swept source for OCT is to provide a wide tuning bandwidth of ∼100 nm or more in order to ensure a good axial resolution, while also providing a fast tuning frequency ($\sim$100 s of kHz) to lower image acquisition times.

The key requirements of modern day OCT systems include real time measurements with micron-scale resolution and sufficient depth penetration. Over the past decade, OCT technology has evolved from time-domain OCT to spectral or Fourier-domain OCT to swept-source OCT and in each case the advancement of OCT technology has progressed with advancements of the broadband light source used in each case.

This light source is the key technological element of all OCT systems and the characteristics of the laser or light source directly define the performance in terms of both the imaging speed and resolution. For example in the case of SS-OCT, it is the linewidth of the swept source that defines the imaging depth of the system and is most often measured using an unbalanced Michelson interferometer. In this situation, the output power oscillates at a frequency proportional to the path difference between the arms, and the coherence typically decreases with increasing optical path difference, thereby leading to a decrease of the amplitude of the intensity modulation. This method, commonly referred to as the roll-off measurement technique [4, 14, 23] gives the coherence properties of the laser averaged over the duration of at least one full frequency sweep and thus is not well-suited to measure the intra-sweep temporal evolution of the coherence of novel swept sources such as the compact short cavity laser [25] or FDML laser [15] where both the phase and amplitude can vary strongly within a single frequency sweep.

Current methods used to measure the instantaneous linewidth of a frequency swept source in a time-resolved manner rely on the repetitive analysis of the optical spectrum of independent sections of the laser sweep. The two most common techniques used involve examining either the reflection from a fibre Bragg grating [47] or implementing a time-gating measurement using a modulator [5].

The critical feature of the technique introduced in Sect. 2 is that it will allow for the recovery of the full phase of the electric field as a function of time with a single shot measurement across the full sweep. If a concurrent measurement of the laser intensity is available, it is then possible to reconstruct the full electric field of the swept source. Access to the full field allows properties such as the coherence roll-off and instantaneous linewidth to be calculated directly.

4.1 Short Cavity Frequency Swept Laser

The study in this section is concerned with examining the operation of a commercially available Axsun SS-OCT laser [2, 10]. This short-cavity frequency swept source is an external cavity laser that is swept by a reflective micro-electro-mechanical (MEMS) tunable Fabry-Pérot bandpass filter, that can be rapidly tuned over the 100 nm bandwidth of the SOA at a rate of 50 kHz. Gain in the 1300 nm region is provided by a quantum well based SOA, while the Fabry-Pérot filter and a further output coupling mirror are used as the cavity reflectors. Previous theoretical and

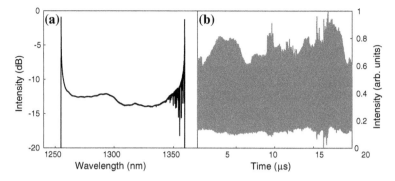

Fig. 15 **a** Time averaged optical spectrum of the AXSUN laser obtained with a grating spectrometer, showing a spectral width of ~100 nm. **b** Laser intensity measured over one sweep period, showing the complex dynamics present during the sweep

experimental works on this type of laser [3, 39] have shown the different intensity dynamical regimes that are observed during different parts of the sweep cycle.

Figure 15a shows the full optical spectrum of the device, with an optical sweep between 1260 and 1360 nm. Also shown in Fig. 15b is the recorded optical intensity of the laser over a single sweep period. Due to the fast tuning rate of the filter, the laser exhibits much dynamical intensity behaviour. Three main dynamical regimes can be seen in the intensity output, recorded with a 40 GSa/s real-time oscilloscope and a 12 GHz bandwidth photodiode. Figure 16 shows a zoom of these regimes on a nanosecond time-scale. The three regions observed are periodic pulses, deterministic chaos, and constant intensity regions separated by transients. Of particular interest is the region of periodic pulsing which is reminiscent of mode-locking operation. This region is generally referred to as the forward sweep of the laser and in fact is the region most often used in the acquisition of OCT images. As can be seen from these two figures, the output of the swept source is non-trivial and alternative methods to analyse and characterise the output will provide deeper insight into both device performance and optimisation.

4.2 Electric Field Visualisation of Frequency Swept Lasers

The experimental set-up used to characterise the fast frequency swept source is shown in Fig. 17 and can be classified as a time delayed, self-referencing interferometric technique. The experimental arrangement is similar in nature to that of Fig. 4 in Sect. 3. The laser's output is first split by a 50/50 coupler. One of the outputs serves as an input for the phase detection set-up while the other one allows a simultaneous measurement of the laser intensity. The phase detection section of the set-up consists of another 50/50 coupler, one output of which is connected to a polarisation controller to optimise the interference signal while the second output is connected to a single

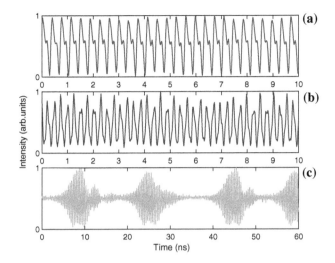

Fig. 16 Three representative plots of the output intensity, on a nanosecond time-scale, of the frequency swept laser, taken at three different parts of the laser sweep of Fig. 15b. **a** Pulsations reminiscent of mode locking observed at ∼5.5 μs, **b** chaotic output observed at ∼16.5 μs and **c** mode-hopping between single mode continuous wave solutions separated by complex transients at ∼9.5 μs

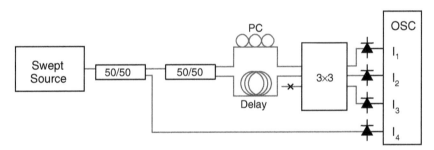

Fig. 17 Experimental set up used to measure the phase of the swept laser output. Again the 3 × 3 denotes the passive fibre coupler used to mix the laser signal with a delayed version of itself. PC represents a polarisation controller placed in the 3 × 3 fibre path, while a time delay τ is introduced by increasing the length of the second fibre arm. OSC: 40 GSa/s real time oscilloscope recording the 3 × 3 and laser intensities on four identically skewed channels

mode optical fibre that provides a short time-delay τ between the two input signals. These two signals are then recombined by the 3 × 3 coupler while the third input remains empty. Again prior to the experiment, a signal with an identifiable feature was chosen, and the optical paths and oscilloscope channel time skews were carefully adjusted to ensure simultaneous reading of the 3 × 3 outputs I_1, I_2 and I_3, and the laser intensity I_4. (In this case, the laser intensity I_4 is measured separately as due to the time delayed self-referencing nature of this experiment it is not possible to use the intensity reconstruction described in Sect. 3.) The intensity outputs of the 3 × 3 coupler are again analysed using four identical photodiodes and a 40 GSa/s real time oscilloscope.

The phase difference $\eta(t)$ between the two input signals, that is the difference between the phase of the laser $\phi(t)$ and the time delayed laser phase $\phi(t - \tau)$, is again calculated using (11) as:

$$\eta(t) = \phi(t) - \phi(t - \tau) = \text{atan}\left(\frac{\sqrt{3}}{2}\left(\frac{I_1 - I_2}{I_3 - \frac{1}{2}(I_1 + I_2)}\right)\right). \tag{13}$$

Provided that the time delay τ is sufficiently small [16], an expansion for the derivative of the phase can be written as:

$$\frac{d\phi(t)}{dt} = \sum_{n=0}^{\infty} a_n \tau^{n-1} \eta^{(n)}(t), \tag{14}$$

where $\eta^{(n)}(t)$ is the nth derivative of $\eta(t)$ and a_n are coefficients given by:

$$a_0 = 1, \quad a_1 = \frac{1}{2}, \quad a_{2n} = \frac{(-1)^n B_n}{(2n)!}, \quad a_{2n+1} = 0, \tag{15}$$

where B_n are the Bernoulli numbers. Under the condition that the time delay is small, it is sufficient to truncate (14) and retain only the leading order terms to obtain:

$$\frac{d\phi(t)}{dt} \approx \frac{\eta(t)}{\tau} + \frac{1}{2}\frac{d\eta(t)}{dt}. \tag{16}$$

The merit of (16) is that it will allow for the calculation of the instantaneous phase of the laser rather than merely the time delayed phase difference $\eta(t)$ and with a simple rescaling of (16), the instantaneous frequency, $f(t)$, of the swept source is obtained via:

$$f(t) = \frac{1}{2\pi}\frac{d\phi(t)}{dt} \approx \frac{1}{2\pi}\left(\frac{\eta(t)}{\tau} + \frac{1}{2}\frac{d\eta(t)}{dt}\right). \tag{17}$$

The instantaneous phase of the electric field is then calculated by integrating (16) (ignoring the arbitrary constant phase offset) and combining this with the intensity measurement $I_4(t)$, the full electric field can be reconstructed via:

$$E(t) = \sqrt{I(t)}e^{i\phi(t)}. \tag{18}$$

For convenience of comparison with a direct measurement of the optical spectrum, the instantaneous frequency can be straight forwardly converted to instantaneous wavelength. Figure 18, therefore, shows the resulting measurement of the instantaneous wavelength superimposed on the laser intensity during a full single sweep of the laser calculated using the phase reconstruction technique. The labels (a)–(c)

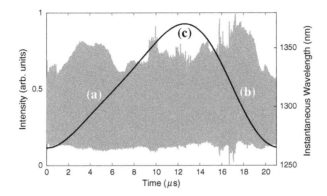

Fig. 18 Plot of the reconstructed instantaneous wavelength (*black*) superimposed on the output intensity of the frequency swept laser (*grey*). The labels **a**, **b** and **c** correspond to the location of the three different intensity features observed in Fig. 16

on Fig. 18 are included to represent the locations at which the intensity measurements from Fig. 16 were taken. As described in [39] sliding frequency mode-locking (Fig. 18a) is observed during the forward sweep of the laser, chaotic output (Fig. 18b) is observed during the backward sweep while mode-hopping between single mode frequency solutions (Fig. 18c) is observed close to the turning points of the laser sweep. From Fig. 18 it can be inferred that the instantaneous wavelength follows the sweep of the filter. As prescribed by (18), the time-resolved instantaneous wavelength sweep combined with the intensity trace (I_4) can be used to reconstruct the optical spectrum of the swept source laser. This is accomplished by calculating the fast Fourier transform of the entire electric field sweep. Figure 19 shows the reconstructed optical spectrum averaged over the two sweep periods, compared to the optical spectrum as measured with a standard grating optical spectrum analyser. The good agreement between the spectra indicates that the time-delayed self-referencing interferometric method provides a highly accurate measurement of the electric field of the laser.

4.3 Coherence Properties of Frequency Swept Lasers

As mentioned previously, for use in OCT systems, two of the key device performance measurements are the roll-off and the device linewidth. The advantage of using the phase measurement technique to characterise fast frequency swept sources is that knowledge of the full laser electric field allows a single shot calculation of the roll-off. In the short cavity device used in this work, it is the forward (increasing wavelength) sweep that is used in the acquisition of OCT images. Therefore in this study, when calculating both the roll-off and instantaneous linewidth, only the forward sweep is considered. As can be seen in Fig. 18, the full 20 μs wavelength

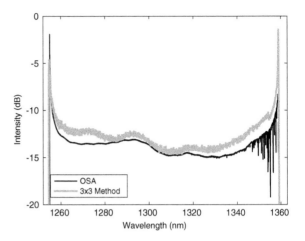

Fig. 19 Optical spectra measured with a grating optical spectrum analyser (*black*) and calculated from the full-field 3 × 3 phase measurement (*grey*)

sweep is asymmetric with an approximately 12 μs long forward sweep. The reconstructed electric field is numerically delayed in order to simulate the Michelson interferometer measurement, which is commonly used to determine the roll-off. The Fourier transform peak of the combined fields, $E(t)$ and $E(t - \tau)$, is then calculated for a range of delays, corresponding to variations in the optical path length from 0 to 30 mm (corresponding to a variation of 0–15 mm in imaging depth). Figure 20 shows the resulting single-shot roll-off measurement. The 3 dB roll-off depth of ~5 mm is in excellent agreement with the figures of merit for similar lasers produced by the manufacturer [18, 25, 34].

In addition to providing a measurement of the roll-off, the results of the 3 × 3 interferometric technique can also be used to obtain a single-shot, time-resolved measurement of the dynamic linewidth of the laser. Calculation of the instantaneous dynamic linewidth of the swept source is achieved by transforming the optical field into the frame of the moving tunable filter (recall that from Fig. 18 the instantaneous

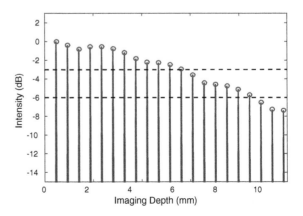

Fig. 20 Imaging roll-off calculated using the single-shot phase measurement technique, showing a 3-dB coherence imaging depth in air of 5 mm

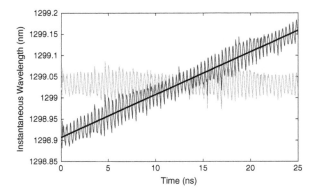

Fig. 21 Instantaneous wavelength (*dark grey*) and quasi-linear filter sweep (*black*) in the lab frame. The background (*grey*) shows the difference between the two representing the instantaneous wavelength in the filter frame. The zero on the wavelength axis is chosen to match the central wavelength of the forward sweep of the laser

frequency/wavelength of the optical field follows the sweep of the filter). Again focusing attention on the forward sweep of the laser, by removing the quasi-linear filter sweep from the instantaneous frequency measurement the instantaneous frequency can be transformed to a stationary trajectory, allowing a direct calculation of the dynamic linewidth to be made. Figure 21 shows a section of the instantaneous frequency before and after this transformation. With the field transformed into the filter frame, the power spectral density of the field can be calculated as [35]:

$$S(\nu) = \left| \int_{-\infty}^{+\infty} E(t)e^{-2\pi i \nu t} dt \right|^2 , \tag{19}$$

which is the amplitude of the Fourier transform of the field. The average power spectrum of the forward sweep calculated from the reconstructed field is shown in Fig. 22. The linewidth $\Delta\nu$ can then be calculated explicitly via:

$$\Delta\nu = \frac{\left(\int_0^\infty S(\nu)d\nu \right)^2}{\int_0^\infty S(\nu)^2 d\nu} . \tag{20}$$

By splitting the sweep region into small sections (in this case approximately 250 ns), using (20) the linewidth can be calculated for each section, yielding a single-shot time-resolved measurement of the dynamic linewidth over the course of a single sweep. Figure 23 shows the evolution of the linewidth versus the intensity and the instantaneous wavelength. Again the linewidth is calculated only over the linear part of the sweep. An average value of 5.6 GHz is found for the instantaneous

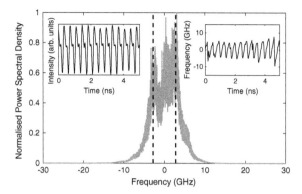

Fig. 22 Average power spectral density of the forward sweep field in the filter frame, from 1260 to 1350 nm. Zero frequency is defined as the center of the moving optical filter. The dashed lines indicate the frequency range from $-0.5\Delta\nu$ to $0.5\Delta\nu$, calculated with (20). The insets show sample 5 ns long outputs of the laser intensity (*left*) and the transformed instantaneous frequency (*right*) corresponding to the region over which the averaged power spectral density is calculated

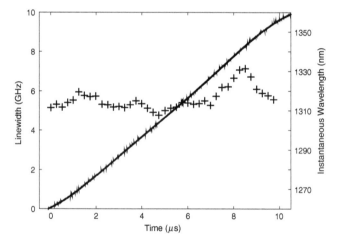

Fig. 23 The instantaneous linewidth is shown in the *black crosses* versus the time. The *solid black curve* shows the reconstructed instantaneous wavelength of the forward sweep. The corresponding output intensity can be seen in Fig. 18

linewidth. In the early part of the sweep the linewidth is approximately constant. While it varies slightly across the sweep, the linewidth is noticeably higher (by approximately 1 GHz) in the later part of the sweep around 8 µs. This change in the dynamic linewidth motivates a closer analysis of the behaviour in this region. Examining the intensity trace during this region, the quasi-periodic intensity pulsing behaviour (see Fig. 16a) that is observed during the majority of the forward sweep is lost, with a chaotic like intensity section being observed instead. Thus, this single-shot measurement technique also helps to reveal changes in the intra sweep dynamical behaviour of the source.

The linewidth as calculated from the reconstructed full field 3×3 technique can be compared to that obtained from theoretical calculations obtained using a noiseless delay-differential equation model, previously shown to reproduce the experimental intensity observations [6, 39]. The calculation follows exactly the same form as for the experiment: the electric field is numerically obtained and then (19) and (20) are used to find the linewidth in an identical manner as above. In this case an average linewidth of approximately 5.4 GHz is found. This result from the noise-free model is in excellent agreement with the experimental findings, clearly indicating that the main contribution to the linewidth is from deterministic multimode effects rather than stochastic effects.

5 Conclusions

In conclusion a novel time-resolved interferometric phase reconstruction technique has been introduced. The power of this technique has been illustrated by examining the use of this technique to study semiconductor lasers in two distinct experimental arrangements, namely semiconductor lasers under optical injection and fast frequency swept lasers for use in OCT.

The vastly improved stability of QD lasers under optical injection has provided an opportunity to systematically study the various bifurcations and non-linear dynamics observed in the injected semiconductor laser system. The evolution of the phase of the electric field of the slave is of immense interest in such systems, as the phase plays a key role in determining the dynamical behaviour observed and this work presents a technique capable of reconstructing the phase of the electric field of a QD laser undergoing optical injection. Optical injection experiments on a single mode QD laser have been presented and complicated dynamical behaviours including excitable pulsing, limit cycle switching and chaotic intermittency have been demonstrated. The phase reconstruction technique describe in this chapter is used to measure full phase space trajectories of these dynamical events. The results clearly demonstrate the both the validity and power of the technique in terms of examining non-linear dynamical features in optical systems.

In the case of frequency swept sources, the phase reconstruction technique was examined using a commercially available short cavity swept source laser. The phase and intensity of the laser can be recovered during an arbitrary time window in a single-shot measurement, ranging from sub-sweep period to multi-sweep period time-scales, limited only by the memory depth of the real time oscilloscope. The resulting information about the complex electric field of the laser can be used to calculate the optical spectrum, roll-off properties and time-resolved dynamic linewidth. These results are in excellent agreement with other measurement techniques and theoretical simulations, as well as previously published figures of merit. Unlike previous experimental techniques used in the study of the dynamics of swept sources, the technique presented here is not limited by the direct detection bandwidth. As long as the instantaneous power spectral density of the source remains smaller than the detection bandwidth, this interferometric technique can recover the phase of a swept source of arbitrary spectral width. This scalability and applicability to swept sources ensures that this electric field reconstruction technique can provide valuable insights into the dynamics and design of future swept sources for OCT and other applications.

Acknowledgments The authors gratefully acknowledge the support of Science Foundation Ireland under Contracts No. 11/PI/1152 and No. 12/RC/2276 and the EU FP7 Marie Curie Action FP7-PEOPLE-2010-ITN through the PROPHET project, Grant No. 264687.

References

1. R. Adler, Proc. IEEE **61**, 1380 (1973)
2. W. Atia, M. Kuznetsov, D. Flanders, US Patent App. 12/027710 (2009)
3. E. Avrutin, L. Zhang, IEEE Int. Conf. Trans. Opt. Net. **14** (2012)
4. B.R. Biedermann, W. Wieser, C.M. Eigenwillig, T. Klein, R. Huber, Opt. Express **17**, 9947 (2009)
5. B.R. Biedermann, W. Wieser, C.M. Eigenwillig, T. Klein, R. Huber, Opt. Lett. **35**, 3733 (2010)
6. T. Butler, S. Slepneva, B. OShaughnessy, B. Kelleher, D. Goulding, S.P. Hegarty, H.-C. Lyu, K. Karnowski, M. Wojtkowski, G. Huyet. Opt. Lett. **40**(10), 2277–2280 (2015)
7. S.R. Chinn, E.A. Swanson, J.G. Fujimoto, Opt. Lett. **22**, 340 (1997)
8. M.A. Choma, M.V. Sarunic, C. Yang, J.A. Izatt, Opt. Express **11**, 2183 (2003)
9. T. Erneux, E.A. Viktorov, B. Kelleher, D. Goulding, S.P. Hegarty, G. Huyet, Opt. Lett. **35**(7), 937 (2010)
10. D. Flanders, M. Kuznetsov, W. Atia, US Patent App. 11/158617 (2008)
11. P. Gallion, H. Nakajima, G. Debarge, C. Chabran, Electron. Lett. **22**, 626 (1985)
12. D. Goulding, S.P. Hegarty, O. Rasskazov, S. Melnik, M. Hartnett, G. Greene, J.G. McInerney, D. Rachinskii, G. Huyet, Phys. Rev. Lett. **98**, 153903 (2007)
13. S.P. Hegarty, D. Goulding, B. Kelleher, G. Huyet, M.T. Todaro, A. Salhi, A. Passaseo, M. De Vittorio, Opt. Lett. **32**, 3245 (2007)
14. R. Huber, M. Wojtkowski, K. Taira, J.G. Fujimoto, Opt. Express **13**, 3513 (2005)
15. R. Huber, M. Wojtkowski, J.G. Fujimoto, Opt. Express **14**, 3225 (2006)
16. G. Huyet, P.A. Porta, S.P. Hegarty, J.G. McInerney, F. Holland, Opt. Comm. **180**, 339 (2000)
17. K. Iwashita, K. Nakagawa, IEEE J. Quantum Electron **18**, 1669 (1982)
18. J. Jing, J. Zhang, A.C. Loy, B.J.F. Wong, Z. Chena, J. Biomed. Opt. **17**, 110507 (2012)
19. B. Kelleher, D. Goulding, S.P. Hegarty, G. Huyet, Ding-Yi Cong, A. Martinez, A. Lemaitre, A. Ramdane, M. Fischer, F. Gerschutz, J. Koeth. Opt. Lett. **34**, 440 (2009)
20. B. Kelleher, D. Goulding, B. Baselga Pascual, S.P. Hegarty, G. Huyet, Eur. Phys. D **58**(2), 175–179 (2010)
21. B. Kelleher, D. Goulding, S.P. Hegarty, G. Huyet, E.A. Viktorov, T. Erneux, in *Quantum Dot Devices*, ed. by Z.M. Wang (Springer, New York, 2012)
22. B. Kelleher, D. Goulding, B. Baselga Pascual, S.P. Hegarty, G. Huyet, Phys. Rev. E **85**, 046212 (2012)
23. T. Klein, W. Wieser, L. Reznicek, A. Neubauer, A. Kampik, R. Huber, Biomed. Opt. Express **4**, 1890 (2013)
24. M. Kuntz, N.N. Ledentsov, D. Bimberg, A.R. Kovsh, V.M. Ustinov, A.E. Zhukov, Y.M. Shernyakov, Appl. Phys. Lett. **81**(20), 3846 (2002)

25. M. Kuznetsov, W. Atia, B. Johnson, D. Flanders, Proc. SPIE **7554**, 75541F (2010)

26. R. Lang, IEEE, J. Quant. Elect. **18**(6), 976 (1982)

27. L.A. Lugiato, Prog. Opt. **21**, 69 (1984)

28. A. Martinez, A. Lematre, K. Merghem, L. Ferlazzo, C. Dupuis, A. Ramdane, J.G. Provost, B. Dagens, O. Le Gouezigou, O. Gauthier-Lafaye, Appl. Phys. Lett. **86**, 211115 (2005)

29. X. Meng, T. Chau, M.C. Wu, Electron. Lett. **34**(21), 2031 (1998)

30. T.C. Newell, D.J. Bossert, A. Stintz, B. Fuchs, K.J. Malloy, L.F. Lester, IEEE Photonics Tech. Lett. **11**(12), 1527 (1999)

31. D. O'Brien, S.P. Hegarty, G. Huyet, A.V. Uskov, Opt. Lett. **29**(10), 1072 (2004)

32. J. Ohtsubo, in *Semiconductor Lasers*, ed. by W.T. Rhodes (Springer, Heidelberg, 2013)

33. I. Petitbon, P. Gallion, G. Debarge, C. Chabran, IEEE J. Quant. Elect. **24**(2), 148 (1988)

34. B. Potsaid, B. Baumann, D. Huang, S. Barry, A.E. Cable, J.S. Schuman, J.S. Duker, J.G. Fujimoto, Opt. Express **83**, 20029 (2010)

35. B.E.A. Saleh, M.C. Teich, *Fundamentals of Photonics* (Wiley, New York, 1991)

36. T. Sano, Phys. Rev. A **50**, 2719 (1994)

37. T.B. Simpson, J.M. Liu, A. Gavrielides, V. Kovanis, P.M. Alsing, Phys. Rev. A **51**(5), 4181 (1995)

38. T.B. Simpson, J.M. Liu, K.F. Huang, K. Tai, Quant. Semiclass. Opt. **9**, 765 (1997)

39. S. Slepneva, B. O'Shaughnessy, B. Kelleher, S.P. Hegarty, A. Vladimirov, H.-C. Lyu, K. Karnowski, M. Wojtkowski, G. Huyet, Opt. Express **22**, 18177 (2014)

40. S. Slepneva, B. Kelleher, B. O'Shaughnessy, S.P. Hegarty, A. Vladimirov, G. Huyet, Opt. Express **21**, 19240 (2013)

41. F. Takens, in *Dynamical Systems and Turbulence*, ed. by A. Rand, L.S. Young (Springer, Berlin, 1981)

42. G.H.M. Van Tartwijk, A.M. Levine, D. Lenstra, J. Sel. Top. Quant. Electron **1**(2), 466 (1995)

43. O. Vaudel, N. Péraud, P. Besnard, Proc. SPIE **6997**, 69970F (2008)

44. S. Wieczorek, B. Krauskopf, D. Lenstra, Phys. Rev. E **64**, 056204 (2001)

45. S. Wieczorek, B. Krauskopf, D. Lenstra, Phys. Rev. Lett. **88**, 063901 (2002)

46. S. Wieczorek, B. Krauskopf, T.B. Simpson, D. Lenstra, Phys. Rep. **416**(1), 1 (2005)

47. S.H. Yun, D.J. Richardson, D.O. Culverhouse, B.Y. Kim, IEEE J. Sel. Topics Quant. Electron **3**, 1087 (2010)

48. M. Wotjkowski, Appl. Opt. **49**, D30 (2010)

Coarsening Dynamics of Umbilical Defects in Inhomogeneous Medium

Raouf Barboza, Umberto Bortolozzo, Marcel G. Clerc, Stefania Residori and Valeska Zambra

Abstract Non-equilibrium systems with coexistence of equilibria exhibit a rich and complex defects dynamics in order to reach a more stable configuration. Nematic liquid crystals layer with negative dielectric constant and homeotropic anchoring under the influence of a voltage are the ideal context for studying the interaction of gas of topological vortices. The number of vortices decreases with time. Experimentally, we show that the presence of imperfections drastically changes this coarsening law. Imperfections are achieved by considering glass beads inside the nematic liquid crystal sample. Depending on the disorder of these imperfections, the system exhibits different statistical evolution of the number of umbilical defects. The coarsening dynamics is persistent and is characterized by power laws with different exponents.

1 Introduction

Macroscopic systems under the influence of injection and dissipation of energy and momenta exhibit instabilities leading to spontaneous symmetry breaking and pattern

R. Barboza (✉) · M.G. Clerc
Departamento de Física, Facultad de Ciencias Físicas Y Matemáticas,
Universidad de Chile, Casilla 487-3, Santiago, Chile
e-mail: raouf.barboza@ing.uchile.cl

M.G. Clerc
e-mail: marcel@dfi.uchile.cl

U. Bortolozzo · S. Residori
Institut Nonlinéaire de Nice, Université de Nice-Sophia Antipolis, CNRS,
1361 Route des Lucioles, 06560 Valbonne, France
e-mail: umberto.bortolozzo@inln.cnrs.fr

S. Residori
e-mail: stefania.residori@inln.cnrs.fr

V. Zambra
Departamento de Física, Facultad de Ciencias, Universidad de Chile, Santiago, Chile
e-mail: valesk.za@gmail.com

© Springer International Publishing Switzerland 2016
M. Tlidi and M.G. Clerc (eds.), *Nonlinear Dynamics: Materials,
Theory and Experiments*, Springer Proceedings in Physics 173,
DOI 10.1007/978-3-319-24871-4_2

formation [1]. Due to the inherent fluctuations of these macroscopic systems, different organizations may emerge in distinct regions of the same sample; hence, these spatial structures are usually characterized by domains, separated by interfaces, as grain boundaries, and defects or dislocations [2, 3]. Among others, defects in rotationally invariant two dimensional systems, i.e. vortices, attract a great deal of attention because of their universal character and intriguing topological properties. These defects have been observed in different systems such as fluids, superfluids, superconductors, liquid crystals, fluidized anisotropic granular matter, magnetic media, and optical dielectrics, to mention a few [4]. Vortices occur in complex fields and can be identified as topological defects, that is, point-like singularities which locally break rotational symmetry. They exhibit a zero intensity at the singular point with a phase spiraling around it. The topological charge is assigned by counting the number of spiral arms in the phase distribution, while the sign is given by the sense of the spiral rotation.

Nematic liquid crystals with negative anisotropic dielectric constant and homeotropic anchoring are a natural physical context where dissipative vortices can be observed and analyzed [5, 6]. In this context, the dissipative vortices are usually called umbilical defects. These defects in nematic liquid crystals have long been reported in the literature (see textbooks [5–7] and reference therein). Two types of stable vortices with opposite charges are observed, which are characterized by being attracted to (repelled by) the opposite (identical) topological charge. The nematic liquid crystal phase is characterized by rod-shaped molecules that have no positional order but tend to point in the same direction [5–7]. Then, the description of the nematic liquid crystal is given by a vector—the director **n**—which accounts for the molecular orientational order. The direction of this vector is irrelevant, only the orientation of **n** has a physical meaning. Note that the defects observed in this context are similar to those observed in magnetic systems, superfluids, superconductors, and Bose-Einstein condensates. However, these vortices exhibit a entirely different dynamic evolution due to the strongly dissipative nature of liquid crystals.

The vortex-like defects have accompanied liquid crystals since their discovery in 1889 by Lehmann [8], who called these structures *kernel*. Later, they were observed in a similar experimental setup by Friedel, who called them *noyaux* [9]. Moreover, he also resolved their detailed topological structure. From the theory of elasticity of nematic liquid crystals, Frank calculated the detailed structure of these defects [10]. Due to the fact that these defects break the orientational order and by analogy with dislocations in crystals of condensed matter, Frank called these defects *disclinations*. Despite the different names given to the observed vortices in this context, none of them were adopted by the community of liquid crystals. There the most widely used name for these defects is *nematic umbilical defects*. The term umbilics was coined by Rapini [11] and refers to the topological structure of the defect which corresponds to a string-like object in three dimensions. Because of the complex elasticity theory associated with nematic liquid crystals, characterized by three types of deformation (bend, twist and splay), the theoretical dynamic study of defects is a thorny task [5–7].

Based on weak nonlinear analysis, valid close to the orientational instability of the molecules (Fréedericksz transition [5, 6]), the dynamics of the director can be reduced at main order to the Ginzburg-Landau equation with real coefficients [12–14]. This amplitude equation allows to understand the emergence of different orientational domains, two types of stable vortices and their respective dynamics. Since the vortices have a $\pm 2\pi$ azimuthal phase jump (winding number), usually they are referred to as vortex "+" and "−", respectively. In this approach, both defects are indistinguishable in their amplitude and, as a result of the phase invariance of the Ginzburg-Landau equation, they account for a continuous family of solutions, characterized by a phase parameter [4]. From this model one can characterize the interaction of vortex pairs [4], which shows a good agreement with experimental observations [15]. From the interaction of defect pairs and through the use of self similarity statements, one can infer the law of number of defects as a function of time [16, 17]. This type of self-similar behavior is well-known as *coarsening process*, which is equivalent to the growth process of domains in phase separations transitions observed in metallic alloys [18]. Using the law of vortices interaction, one shows that the number of defects decreases inversely proportional to time, which it has been experimentally observed in nematic liquid crystal samples [19, 20]. Similarly, using XY phase model, one obtains the same decay law for the vortices number [19].

The aim of this manuscript is to investigate experimentally the persistence and coarsening law when inhomogeneities are considered in a liquid crystal sample. The inhomogeneities are achieved by considering glass beads inside a nematic liquid crystals sample with negative dielectric constant and homeotropic anchoring. Depending on the disorder of these glass beads, the system exhibits different statistical temporal evolution of number of umbilical defects. This evolution is found to exhibit power laws with different exponents.

2 Experimental Setup

Let us consider an interaction geometry in which, a uniform thin layer of nematic liquid crystal has the molecular director constrained to be normal to the two parallel bounding plates, direction which we later denote by z. Due to inherent elastic forces between the molecules, the alignment in the bulk will be uniform and parallel to z, this in order to minimize the elastic energy. When a low frequency ($\approx$100 kHz in our case) electric field is applied in the z direction, if the dielectric anisotropy of the liquid crystal is negative, the resulting electric torque will try to rotate the molecules away from the z-axis. Only over a critical threshold voltage, called Fréedericksz transition voltage [6], the molecules effectively tilt away from their equilibrium position. Due to the 2π degeneracy in the possible direction of orientation, defects called umbilics will be generated in the nematic layer [6, 11].

The observation of these umbilical defects and their dynamics was done by using two different types of liquid crystal cells about the same thickness. The first cell, uniform, is made of two ITO (Indium Tin Oxide, transparent conductor) coated glass

Fig. 1 Sketch of the
experimental setup of a
nematic liquid crystal layer
with negative dielectric
constant anisotropy and
homeotropic anchoring
under the influence of a
voltage. The essential parts
of the setup are emphasized.
Crossed polarizers, either
linear or circular are used to
analyze the texture of the
liquid crystal

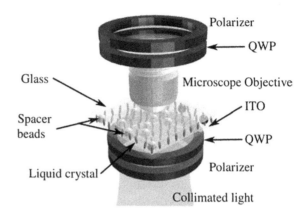

slabs. The glass slabs are treated on the ITO side in order to promote orthogonal
alignment of the liquid crystal molecules. This alignment is termed as homeotropic
alignment or homeotropic anchoring [5–7]. The glass slabs are held together with thin
sheet of polymer spacers such that, the treated faces form a gap in which the liquid
crystal will be infiltrated later. The spacers, which fix the thickness of the gap are
about 15 μm thick. The second cell, non uniform, from Instec Inc. (SB100A150uT180
liquid crystal cell), has the same homeotropic alignment. The spacing gap of the cell
is achieved by sputtering spacer beads made of clear/transparent ceramics or glass
onto the substrate of glass slab before assembly [21]. The diameter of these glass
micro-spheres fixes the cell gap, and, for the chosen cell, it is about 15 μm. The
two cells were filled by capillarity with the MLC-6608 nematic liquid crystal (from
Merck) which has a negative dielectric anisotropy. Both cells are biased with low
frequency sinusoidal voltage. The experimental setup is sketched in the figure Fig. 1.
To achieve maximum resolution, a collimated white light (Köhler illumination) from
a microscope condenser is sent onto the liquid crystal cell, the latter mounted on
a translation stage. The texture of the liquid crystal is imaged on a CCD camera
through a microscope objective and relay lenses.

As the cells contain liquid crystalline materials, which are an intrinsically birefringent in the nematic phase, two crossed polarizers, the first to polarize the illumination
source and the second to analyze the polarization of the light coming from the cell,
are used in order to recover averaged two dimensional texture of the liquid crystal
layer. For simplicity the cell will be considered as a uniform, along the longitudinal
z coordinate, uniaxial birefringent material with optical axis aligned in the xy plane
at angle θ with the x axis, with retardation $\delta = 2\pi L(\tilde{n}_e - n_o)/\lambda$; L represents the
thickness of the cell, λ the operating wavelength, n_o and $\tilde{n}_e$ respectively the ordinary
refractive index and the average extraordinary refractive index over the longitudinal
coordinate. The optical axis can be viewed as the averaged azimuthal direction of
molecules in the xy plane, equivalently their projection onto the xy plane. The averaged extraordinary index $\tilde{n}_e$ is related to the tilt ψ of the molecules with respect the
z axis by the expression

$$\tilde{n}_e = \int_0^L \frac{n_e n_o}{\sqrt{n_e^2 \cos^2 \psi + n_o^2 \sin^2 \psi}} dz. \tag{1}$$

The texture of the liquid crystal layer will vary accordingly with the spatial variation in the xy optical axis at angle θ representing the director orientation in the xy plane and δ the retardation which depends on the average tilt ψ of the molecules with respect to the z axis. Using Jones matrix formalism we can show that the intensity recorded using crossed linear polarizers, the polarizers axis are perpendicular to each other, is given by

$$I(x, y) = I_0 \sin^2 \frac{\delta(x, y)}{2} \sin^2 2\theta(x, y). \tag{2}$$

Likewise, the crossed circular polarizer configuration is achieved when two quarter wave plate (QWP) are inserted in the previous configuration, with the first waveplate at $\pm 45°$ with respect to the axis of the input polarizer, and the fast axis of the second wave-plate is orthogonal to the first one. In this case the recorded intensity writes as follow

$$I(x, y) = I_0 \sin^2 \frac{\delta(x, y)}{2} \tag{3}$$

We used both polarizing microscope imaging, depending on the feature we want to enhance of the umbilical defects dynamics.

3 Results and Discussions

To understand the coarsening dynamics of the vortices in homogeneous cell, we must first establish the vortex pair interaction law and then, by means of self-similarity properties, we can deduce a coarsening law of vortices.

3.1 Vortex-Pair Interaction Law

As we have mention before, close to the orientational instability of the molecules, the dynamics of the director can be reduced at main order to the Ginzburg-Landau equation with real coefficients. This amplitude equation admits stable vortex solutions with topological charge ± 1. The analysis of the vortex interaction law is complex because the energy associated with each vortex diverges logarithmically with the size of the system [5]. Thereby, the interaction between distant vortices has an infinite mobility [4]. However, in the case of considering that the system has a finite size, the mobility is finite and the vortex-pair interaction law can be approximated for long distances by the expression [4]

$$M\dot{r} = \frac{q}{r}, \tag{4}$$

where $r(t)$ is the vortex separation, q is the product of the topological charges of vortices ($q = \pm 1$), then it is positive (negative) when both vortex has the same (different) charge, and M stands for the vortex mobility which depends of the size of the system, the properties of the liquid crystal and the applied voltage. Thus the interaction between vortices is equivalent to overdamped particles with Keplerian type interaction potential. When the distances between the vortices is small enough— the order of the size of the vortex core—the previous dynamics is not valid. But in this case, vortices of opposite charge merge and disappear. In brief, the dynamics of interaction between vortices tries to homogenize the deformations of molecular orientation.

3.2 Vortex Coarsening Law

Considering a gas of n-vortices, the position of the ith-vortex is given by $\mathbf{r}_i$. Hence, the interaction between them is given by

$$M\dot{\mathbf{r}}_i = \sum_{i \neq j} \frac{q_{ij}}{r_{ij}^2} (\mathbf{r}_i - \mathbf{r}_j), \tag{5}$$

where $r_{ij} \equiv \|\mathbf{r}_i - \mathbf{r}_j\|$ is the distance between the ith and jth-vortex, and q_{ij} is the product of the topological charges of vortices. Hence, the dynamics of a gas of n-vortices corresponds to overdamped n-body problem. It is worthy to note the above set of equations is invariant under the self-similarity transformation

$$\mathbf{r}_i \rightarrow \lambda \mathbf{r}_i,$$
$$t \rightarrow \lambda^2 t. \tag{6}$$

Therefore, if one dilates or expands the space and time then the set of (5) are invariant.

Let us introduce $N(t)$, the number of vortices at time t. This number of vortices can be estimated as

$$N(t) = \frac{A}{\langle r \rangle^2}, \tag{7}$$

where A is the area of the sample under study and $\langle r \rangle$ is the average distance between vortices. Because the dynamics of vortices is given by the set of (5), also the average distance $\langle r \rangle$ and $N(t)$ is determined by this dynamics. Then, $\langle r \rangle$ and $N(t)$ should also be self similar with transformation (6). Hence, $N(\lambda^2 t) = A/\lambda\langle r\rangle^2$. From the previous equality, one infers that the only possibility is that

$$N(t) = \frac{\beta}{t},\tag{8}$$

with β a constant. Therefore, the number of defects decreases inversely proportional to time, *coarsening law*.

3.3 Experimental Observation of Coarsening Law in Uniform Cell

To verify the previous law, we have conducted several experimental analysis of the dynamics of vortex gas. This by applying a large enough voltage to the liquid crystal layer between two cross polarizers, which spontaneously generates hundreds umbilical defects in different positions as a result of thermal fluctuations and inhomogeneities in the system. The position of the umbilical defects are recognized by the intersection of four black curves [5]. Subsequently, the defects have a dynamic of attraction and repulsion following the interaction law (5). Figure 2 shows a temporal sequence of snapshots, which emphasizes the characteristic evolution of a gas of umbilical defects. From the temporal sequences and through an appropriate recognition software we can determine the number of vortices and their respective positions. Thus, we acquire the evolution of the number of vortices as a function of time. Figure 3 shows this evolution. From this plot, one concludes that the number of

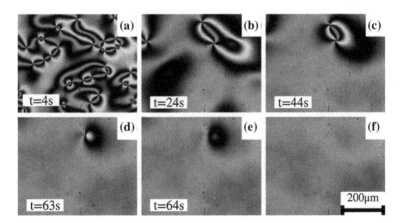

Fig. 2 Annihilation dynamics of umbilical defects in a uniform liquid crystal layer between two crossed linear polarizers. Temporal sequence of snapshots from the *left* to *right* and the *top* to *bottom* (**a–e**). The position of the umbilical defects are given by the intersection of four *black brushes*. Texture of the sample after the anihilation of all the defects (**f**)

Fig. 3 Coarsening dynamics in a uniform cell. Number of umbilical defects as a function of time. The *solid black* and *dashed curve*, respectively, are the experimental evolution of $N(t)$ and the fitting *curve* $N_f(t) = \beta t^{-\alpha} + N_\infty$ with $\alpha = 0.9134 \pm 0.00124$, $\beta = 4.448 \times 10^3 \pm 10$ and $N_\infty = 17 \pm 0.22$

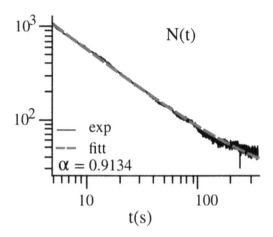

vortices decays as a function of time with a power law. To determine the exponent, we have considered the following fit

$$N_f(t) = \beta t^{-\alpha} + N_\infty, \tag{9}$$

where $\{\beta, \alpha, N_\infty\}$ are fitting parameters, which accounts for the features of the liquid crystal and cell under study. N_∞ stands for the number of imperfections of the system—which trap the vortices in given positions—and the inaccuracy of recognition method. Experimentally we found that in our samples, the exponent $\alpha = 0.9134$ is in reasonable agreement with the simplified description (5).

3.4 Experimental Observation of Coarsening Law in Inhomogeneous Cell

To investigate of the interaction of vortices in inhomogeneous media, we have conducted several experimental analysis of the dynamic of vortex gas in a liquid crystal layer with glass beads between two cross polarizers and applying a large enough voltage. Figure 4 shows a temporal sequence of snapshots with the characteristic evolution of a gas of umbilical defects in an inhomogeneous medium. The glass beads are emphasized by dashed circumferences. Again, the position of the umbilical defects are recognized by the intersection of four black curves. As we have observed in the temporal sequence of snapshots, when one applies a sufficiently large voltage in the liquid crystal layer a large number of vortices appear in different spatial positions, which are determined by the inherent fluctuations and imperfections in the cell. Following the emergence of these umbilical defects, they begin to repel or attract, causing the annihilation process of these defects. This process is characterized by the fact that initially close defects annihilate quickly, and then the more distant umbilical defects annihilate one another, but each time in a slower process, coarsening

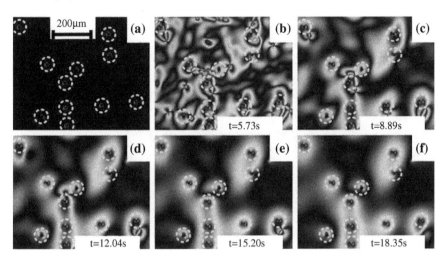

Fig. 4 Umbilical defects annihilation dynamics in a liquid crystal layer with glass beads between two linear crossed polarizers. Temporal sequence of snapshots from the *left* to *right*. The position of the umbilical defects are given by the intersection of four *black curves*

dynamics. The glass beads, as seen in the snapshots, remain motionless. However, the dynamics of vortices is strongly affected by the presence of glass beads. Figure 5 shows a glass bead attracting radially an umbilical defect . Both appear as dark spots as the cell is observed with circular crossed polarizers. In this experimental setup umbilical defects are recognized as small gray circles. Experimentally, this interaction is weaker than the interaction between umbilical defects. It is known that glass beads without surface treatment, generate homeotropic anchoring at their boundaries, that the liquid crystal molecules tend to be oriented normal to the glass beads [21, 22]. In addition, due to the fact that the glass beads are in contact with the glass plates of the sample, one expects a saturn ring like defect loop around each glass inclusion [21, 22]. The trajectory as consequence of the interaction between this defect and the umbilical one is depicted in Fig. 5. Likewise, the interactions between vortices are affected by the presence of the glass beads. Figure 6 illustrates the vortex interaction in presence of a close glass bead in a liquid crystal layer observed with circular crossed polarizers. Clearly from this trajectory, we note that the interaction of the umbilical defects is not a central force as those obtained by (4). Therefore, the interaction of vortices is modified and it is not clear if the process of coarsening is persistent.

Using an appropriate recognition software, based on particle tracking, we can determine the number of vortices and their respective positions. Figures 7c and 8c show the evolution of the number of umbilical defects as a function of time. In both graphs, we observe that the system exhibits coarsening process with power laws. These power laws are obtained by realizing several experiments. Hence, the coarsening dynamics is a persistent phenomenon, though depending on the distribution of the glass beads, we observe different power laws. To characterize the clustering

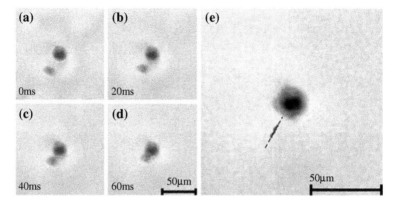

Fig. 5 Interaction between a glass bead and an umbilical defect in a liquid crystal layer with circular crossed polarizers. Temporal sequence of snapshots from (**a**) to (**e**). *Dashed circle* accounts for the glass bead. The small *gray circle* stands for the umbilical defect. In the *right panel*, the *dashed line* sums up the trajectory of the umbilical defect and points are the position of the defect

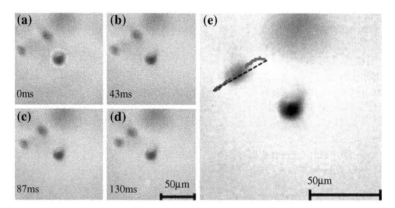

Fig. 6 Vortex interaction in presence of a glass bead in a liquid crystal layer with circular crossed polarizers. Temporal sequence of snapshots from (**a**) to (**e**). *Dashed circle* accounts for a glass bead. The small *gray circle* stands for the umbilical defects. In the *right panel*, the color points account for the trajectories of the umbilical defects, different colors account for the different defects, and the *dashed line* joints initially the defects

and distribution of the glass beads, we have computed the Voronoi diagram of glass beads in the different observed zones (cf. Figs. 7a and 8a) and their histogram of the mutual distance of the glass beads (cf. Figs. 7b and 8b). From these diagrams, we can measure the density of glass beads and we obtain for zone I and III, respectively, 13.630 and 20.803 glass beads per mm^2. Zone III is more ordered than zone I, since its histogram of the mutual distance is closer to a Rayleigh distribution around a length (cf. Fig. 7b) and the other is closer to a uniform distribution without a feature length. Analogously, from this distribution we can compute the Shannon entropy

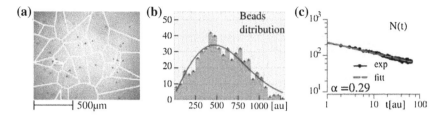

Fig. 7 Coarsening process of umbilical defects in an inhomogeneous medium, zone III. **a** Voronoi diagram of glass beads in the observed zone. **b** Histogram of the mutual distance of the glass beads. The *solid curve* is a fitting curve using a Rayleigh distribution. **c** Corresponding scaling curve of the number of defects vs normalized time. *Black* points stand for experimental observations and the *dashed line* corresponds to a fitting curve of the form $N(t) = \beta/t^{\alpha}$ with $\alpha = 0.29$

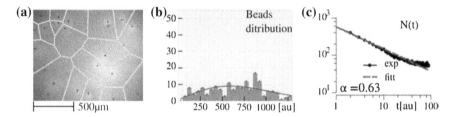

Fig. 8 Coarsening process of umbilical defects in an inhomogeneous medium, zone I. **a** Voronoi diagram of glass beads in the observed zone. **b** Histogram of the mutual distance of the glass beads. The *solid curve* is a *fitting curve* using a Rayleigh distribution. **c** Corresponding scaling *curve* of the number of defects vs normalized time. *Black* points stand for experimental observations and the *dashed line* corresponds to a *fitting curve* of the form $N(t) = \beta/t^{\alpha}$ with $\alpha = 0.63$

S_e and obtain for each zone, respectively $S_e(III) = 0.00767$ and $S_e(I) = 0.021$. Therefore, depending on the different configurations of the glass beads, the evolution of the number of defects as a function of time changes.

To measure the exponent of the coarsening laws, we have considered the following fitting curve $N(t) = \beta t^{-\alpha}$. Table 1 summarizes our results for different zones of our liquid crystal layer. Different zones of the sample exhibit different coarsening laws. However, from this table we are not able to establish a correlation between the

Table 1 Results over an observed area about 1.394 mm^2

Zone	Density (mm^2)	α	β	Entropy
I	13.630	0.63	597.0	0.0210
II	17.217	0.85	1800.0	0.0136
III	20.803	0.29	228.1	0.00767
IV	21.521	0.75	1286.0	0.0091
V	23.673	0.758	1286.0	0.00772
VI	27.977	0.73	187.0	0.0057

density of vortices, Shannon entropy, and spatial distributions with their coarsening exponents found experimentally.

4 Conclusions and Remarks

Far from equilibrium systems with coexistence of equilibria exhibit a rich and complex defects dynamics in order to reach a more stable configuration. This dynamic of defects can generate a rich variety of spatial textures. Defects in rotationally invariant two dimensional systems, attract a great deal of attention because of their universal character and intriguing topological properties. Nematic liquid crystals layer with negative dielectric constant and homeotropic anchoring under the influence of a voltage are the ideal context for studying the interaction of gas of topological vortices with opposite charges.

By considering a uniform sample of nematic liquid crystal layer under the influence of electrical voltage with high frequency, we observe that the number of vortices decrease inversely proportional to time. This coarsening dynamics results when vortices are more close, the interaction between them increases in a self-similarity manner. Experimentally, we show that the presence of imperfections in the liquid crystal layer drastically changes this coarsening process. Imperfections are achieved by considering glass beads inside the nematic liquid crystal sample. We observed that the coarsening process is persistent under the presence of spatial inhomogeneities. Depending on the disorder of these imperfections, the system exhibits different statistical evolution of number of umbilical defects. This evolution is characterized by power laws with different exponents.

From the theoretical point of view, one can model the effect of the glass beads as a screening effect, that is, the law of interaction of pairs of vortices, (4), is modified by considering an effective exponent, which is a function of the properties and distributions of the glass beads. This kind of effective dynamic is self-similar, then this could explain the observed coarsening dynamics.

Acknowledgments M.G.C. acknowledges the support of FONDECYT N 1150507. R.B. acknowledges the support of FONDECYT POSTDOCTORADO N. 3140577.

References

1. G. Nicolis, I. Prigogine, *Self-Organization in Non Equilibrium Systems* (Wiley, New York, 1977)
2. L.M. Pismen, *Patterns and Interfaces in Dissipative Dynamics* (Springer Series in Synergetics, Berlin, 2006)
3. M.C. Cross, P.C. Hohenberg, Pattern formation outside of equilibrium. Rev. Mod. Phys. **65**, 851–1112 (1993)
4. L.M. Pismen, *Vortices in Nonlinear Fields* (Clarendon press, Oxford, 1999)

5. S. Chandrasekhar, Liquid Crystals (Cambridge University Press, Cambridge, 1992)
6. P.G. de Gennes, J. Prost, The physics of Liquid Crystals, 2nd edn. (Oxford Science Publications, Clarendon Press, Oxford, 1993)
7. P. Oswald, P. Pieranski, Nematic and Cholesteric Liquid Crystals (Taylor & Francis Group, Boca Raton, 2005)
8. O. Lehmann, Über fliessende krystalle. Zeitsch Phys. Chem. 4, 462-472 (1889)
9. G. Friedel, The mesomorphic states of matter. Ann. Physique **18**, 273–474 (1922)
10. F.C. Frank, I. Liquid crystals. On the theory of liquid crystals. Discuss. Faraday Soc. 25, 19-28 (1958)
11. A. Rapini, Umbilics: static properties and shear-induced displacements. J. Physique **34**, 629–633 (1973)
12. T. Frisch, S. Rica, P. Coullet, J.M. Gilli, Spiral waves in liquid crystal. Phys. Rev. Lett. **72**, 1471–1474 (1994)
13. T. Frisch, Spiral waves in nematic and cholesteric liquid crystals. Phys. D **84**, 601–614 (1995)
14. M.G. Clerc, E. Vidal-Henriquez, J.D. Davila, M. Kowalczyk, Symmetry breaking of nematic umbilical defects through an amplitude equation. Phys. Rev. E **90**, 012507 (2014)
15. R. Barboza, T. Sauma, U. Bortolozzo, G. Assanto, M.G. Clerc, S. Residori, Characterization of the vortex-pair interaction law and nonlinear mobility effects. New J. Phys. **15**, 013028 (2013)
16. M. Argentina, M.G. Clerc, R. Rojas, E. Tirapegui, Coarsening dynamics of the one-dimensional Cahn-Hilliard model. Phys. Rev. E **71**, 046210 (2005)
17. M.G. Clerc, S. Coulibaly, L. Gordillo, N. Mujica, R. Navarro, Coalescence cascade of dissipative solitons in parametrically driven systems. Phys. Rev. E **84**, 036205 (2011)
18. L. Ratke, P.W. Voorhees, Growth and Coarsening: Ostwald Ripening in Material Processing (Springer Science and Business Media, 2013)
19. A.N. Pargellis, S. Green, B. Yurke, Planar XY-model dynamics in a nematic liquid crystal system. Phys. Rev. E 49, 42504257 (1994)
20. T. Nagaya, H. Orihara, Y. Ishibashi, Coarsening dynamics of +1 and -1 disclinations in two-dimensionally aligned nematics-spatial distribution of disclinations. J. Phys. Soc. Japan, 64, 7885 (1995)
21. M. Hasegawa, Alignment Technology and Applications of Liquid Crystal Devices (CRC Press), pp. 7–54
22. H. Stark, Director field configurations around a spherical particle in a nematic liquid crystal. Eur. Phys. J. B 10, 311321 (1999)

Spreading, Nonergodicity, and Selftrapping: A Puzzle of Interacting Disordered Lattice Waves

Sergej Flach

Abstract Localization of waves by disorder is a fundamental physical problem encompassing a diverse spectrum of theoretical, experimental and numerical studies in the context of metal-insulator transitions, the quantum Hall effect, light propagation in photonic crystals, and dynamics of ultra-cold atoms in optical arrays, to name just a few examples. Large intensity light can induce nonlinear response, ultracold atomic gases can be tuned into an interacting regime, which leads again to nonlinear wave equations on a mean field level. The interplay between disorder and nonlinearity, their localizing and delocalizing effects is currently an intriguing and challenging issue in the field of lattice waves. In particular it leads to the prediction and observation of two different regimes of destruction of Anderson localization—asymptotic weak chaos, and intermediate strong chaos, separated by a crossover condition on densities. On the other side approximate full quantum interacting many body treatments were recently used to predict and obtain a novel many body localization transition, and two distinct phases—a localization phase, and a delocalization phase, both again separated by some typical density scale. We will discuss selftrapping, nonergodicity and nonGibbsean phases which are typical for such discrete models with particle number conservation and their relation to the above crossover and transition physics. We will also discuss potential connections to quantum many body theories.

S. Flach (✉)
Center for Theoretical Physics of Complex Systems,
Institute for Basic Science, Daejeon, Korea
e-mail: sergejflach@googlemail.com

S. Flach
New Zealand Institute for Advanced Study,
Centre for Theoretical Chemistry and Physics,
Massey University, 0745 Auckland, New Zealand

© Springer International Publishing Switzerland 2016
M. Tlidi and M.G. Clerc (eds.), *Nonlinear Dynamics: Materials,*
Theory and Experiments, Springer Proceedings in Physics 173,
DOI 10.1007/978-3-319-24871-4_3

45

1 Introduction

In this contribution we will discuss the regimes of wave packet spreading in nonlinear disordered lattice systems, and its relation to quantum many body localization (MBL). We will consider cases when the corresponding linear (single particle) wave equations show Anderson localization, and the localization length is bounded from above by a finite value.

There are several reasons to analyze such situations. Wave propagation in spatially disordered media has been of practical interest since the early times of studies of conductivity in solids. In particular, it became of much practical interest for the conductance properties of electrons in semiconductor devices more than half a century ago. It was probably these issues which motivated Anderson to perform his groundbreaking lattice wave studies on what is now called Anderson localization [1]. With evolving modern technology, wave propagation became of importance also in photonic and acoustic devices in structured materials [2, 3]. Finally, recent advances in the control of ultracold atoms in optical potentials made it possible to observe Anderson localization there as well [4].

In many if not all cases wave-wave interactions can be of importance, or can even be controlled experimentally. Short range interactions hold for s-wave scattering of atoms. When many quantum particles interact weakly, mean field approximations often lead to effective nonlinear wave equations. Electron-electron interactions in solids and mesoscopic devices are also interesting candidates with the twist of a new statistics of fermions. Nonlinear wave equations in disordered media are of practical importance also because of high intensity light beams propagating through structured optical devices induce a nonlinear response of the medium and subsequent nonlinear contributions to the light wave equations. While electronic excitations often suffer from dephasing due to interactions with other degrees of freedom (e.g. phonons), the level of phase coherence can be controlled in a much better way for ultracold atomic gases and light.

There is a fundamental mathematical interest in the understanding, how Anderson localization is modified in the presence of quantum many body interactions, or classical nonlinear terms in the wave equations. All of the above motivates the choice of corresponding linear (single particle) wave equations with finite upper bounds on the localization length. Then, the corresponding noninteracting quantum systems, as well as linear classical waves, admit no transport. Analyzing transport properties of nonlinear, respectively interacting quantum, disordered wave equations allows to observe and characterize the influence of wave–wave interactions on Anderson localization in a straightforward way.

The chapter is structured in the following way. We will introduce the classical model, discuss its statistical properties, and in particular the nonGibbsean phase, and some of its generalizations. We will then come to wave packet spreading, a self-trapping theorem, and to results of destruction of Anderson localization. Finally, we will discuss the relation of these results to many body localization for the corresponding quantum many body problems.

2 The Model

The Gross-Pitaevskii equation is describing Bose-Einstein condensates (BEC) of interacting ultracold atoms in certain mean-field approximations. It is also known as the nonlinear Schrödinger equation which is integrable in $1 + 1$ dimensions, and has many further applications e.g. in nonlinear optics. Its discretized version— the discrete nonlinear Schrödinger equation (DNLS) or discrete Gross-Pitaevskii equation (DGP)—is typically nonintegrable, and is realized with a BEC loaded onto optical lattices [5]. Similar the discrete nonlinear Schrödinger equation (DNLS) is realized for various one- and two-dimensional networks of interacting optical waveguides [6].

Additional disorder, either due to natural inhomogeneities, or intentionally implanted, finally leads to the disordered discrete Gross-Pitaevsky (dDGP) or disordered discrete nonlinear Schrödinger equation (dDNLS) equation with Hamiltonian

$$\mathcal{H} = \sum_l \varepsilon_l |\psi_l|^2 + \frac{\nu}{2} |\psi_l|^4 - (\psi_{l+1}\psi_l^* + \psi_{l+1}^*\psi_l) \tag{1}$$

with complex variables ψ_l, lattice site indices l and nonlinearity strength $\nu \geq 0$. The (typically uncorrelated) random on-site energies ε_l with zero average $\bar{\varepsilon} = \lim_{N\to\infty} N^{-1}(\sum_{l=1}^{l=N} \varepsilon_l) = 0$ and finite variance $\sigma_\varepsilon = \lim_{N\to\infty} N^{-1}(\sum_{l=1}^{l=N} \varepsilon_l^2)$ are distributed with some probability density distribution (PDF) $\mathcal{P}_\varepsilon$. Here we will use the box distribution $\mathcal{P}_\varepsilon(x) = \frac{1}{W}$ for $|x| \leq \frac{W}{2}$ and $\mathcal{P}_\varepsilon = 0$ otherwise.

The equations of motion are generated by $\dot{\psi}_l = \partial \mathcal{H}_D / \partial(i\psi_l^*)$:

$$i\dot{\psi}_l = \varepsilon_l \psi_l + \nu |\psi_l|^2 \psi_l - \psi_{l+1} - \psi_{l-1}. \tag{2}$$

Equation (2) conserve the energy (1) and the norm $\mathcal{A} = \sum_l |\psi_l|^2$. Note that varying the norm of an initial wave packet is strictly equivalent to varying to varying ν. Note also that the transformation $\psi_l \to (-1)^l \psi_l^*$, $\nu \to -\nu$, $\varepsilon_l \to -\varepsilon_l$ leaves the equations of motion invariant. Therefore the sign of the nonlinear coefficient ν can be fixed without loss of generality to be positive.

3 Gibbsean and NonGibbsean Regimes

The existence of the second—in addition to the energy—conserved quantity $\mathcal{A} \geq 0$, combined with the discreteness-induced bounded kinetic energy part in (1), has a profound impact on the statistical properties of the DGP system. At variance to its space-continuous counterparts, the disorder-free translationally invariant lattice model with $\varepsilon_l = 0$ shows a non-Gibbsean phase [7, 8]. It is separated from the Gibbsean phase by states of infinite temperature. While average densities in the non-Gibbsean phase can be formally described by Gibbs distributions with negative temperature, in truth the dynamics shows a separation of the complex field ψ_l into a two-component one—a first component of high density localized spots and a second component of delocalized wave excitations with infinite temperature [7–11]. The high density localized spots are conceptually very similar to selftrapping and discrete breathers [12].

We will briefly compute the effect of nonvanishing disorder $\varepsilon_l \neq 0$ on the infinite temperature separation line between the Gibbsean and the non-Gibbsean phases. While the final result was listed in [13] in an appendix, no details were provided. We will use the notations of the work by Johansson and Rasmussen [8], to which we refer the reader for further details. We will also generalize to other types of potentials.

We define the local norm per site as $A_l \geq 0$ and the local phase $0 \leq \phi_l \leq 2\pi$ such that

$$\psi_l = \sqrt{A_l}e^{i\phi_l}. \tag{3}$$

The Hamiltonian transforms into

$$\mathscr{H} = \sum_l \varepsilon_l A_l - 2\sqrt{A_l A_{l+1}}\cos(\phi_l - \phi_{l+1}) + \frac{\nu}{2}A_l^2, \tag{4}$$

and the norm simply becomes

$$\mathscr{A} = \sum_l A_l. \tag{5}$$

Assuming a large system of N sites the corresponding average densities become

$$h = \mathscr{H}/N, \quad a = \mathscr{A}/N. \tag{6}$$

From here on we will always implicitly consider the thermodynamic limit $N \to \infty$. We will as well express final results in terms of densities scaled with the nonlinearity parameter ν [13]:

$$y = \nu h, \quad x = \nu a. \tag{7}$$

A number of questions can be posed. First, can any pair of realizable densities $\{x, y\}$ be obtained by assuming a Gibbs distribution

$$\rho_G = \frac{1}{\mathscr{Z}}e^{-\beta(\mathscr{H}+\mu\mathscr{A})} \tag{8}$$

where $\mathscr{Z}$ is the partition function, β the inverse temperature, and μ the chemical potential? And if not, what is the dynamics in the corresponding nonGibbsean phase?

The lowest energy state $E_{\min}$ is evidently given by $\phi_l = const$:

$$E_{\min} = \sum_l (\varepsilon_l - 2)A_l + \frac{\nu}{2}A_l^2, \tag{9}$$

In the absence of any potential $\varepsilon_l = 0$ the lowest energy state is homogeneous: $A_l = \mathscr{A}/n = a$. With the notation of the inverse temperature β we conclude that the zero temperature limit of the scaled energy density at a given value of the scaled norm density is given by

$$y_{\beta \to \infty} = -2x + x^2/2. \tag{10}$$

For a nonzero potential with finite variance the lowest energy density limit will be lowered by a finite value. With the distribution $\mathscr{P}_\varepsilon(x) = \frac{1}{W}$ for $|x| \leq \frac{W}{2}$ we arrive at the upper and lower bound s

$$-(2 + W)x + x^2/2 \leq y_{\beta \to \infty} \leq -2x + x^2/2. \tag{11}$$

At the same time the upper limit for the total energy of a finite system with N sites is obtained by concentrating all the norm on one lattice site which yields $E_{max} = v\mathscr{A}^2/2$ and an energy density $y_{max} = Nx^2/2$. In the thermodynamic limit $N \to \infty$ the upper limit for the energy density is diverging. We conclude that at any given norm density x the energy density is bounded from below by the finite minimum value $y_{\beta \to \infty}$, but is not bounded from above and can take arbitrary large values.

The partition function

$$\mathscr{Z} = \int_0^\infty \int_0^{2\pi} \prod_m d\phi_m dA_m \exp[-\beta(\mathscr{H} + \mu\mathscr{A})] \tag{12}$$

depends on all amplitudes and phases. Integration over the phase variables ϕ_m reduces the symmetrized partition function to

$$\mathscr{Z} = (2\pi)^N \int_0^\infty \prod_m dA_m I_0(2\beta\sqrt{A_m A_{m+1}}) e^{-\beta(\mathscr{H}_0 + \mu\mathscr{A})} \tag{13}$$

with the reduced Hamiltonian

$$\mathscr{H}_0 = \sum_l \varepsilon_l A_l + \frac{1}{2} A_l^2 \tag{14}$$

depending only on the amplitudes, and with I_0 being the Bessel function of 0th order.

The strategy of finding a nonGibbsean phase is simply to find the line of infinite temperature $\beta = 0$ in the control parameter space of the scaled densities. The argument of the Bessel function in (13) vanishes in that limit, turning the Bessel function value to unity—the very argument which is obtained for the absence of any coupling between sites in (1). Therefore the infinite temperature limit corresponds to the case of uncoupled sites, and the results will be valid for any lattice dimension. Note that the infinite temperature limit $\beta \rightarrow 0$ implies $\beta\mu \rightarrow const$. With the notation $\mu_l = \mu + \varepsilon_l$ and after some simple algebra it follows for infinite temperature

$$\ln \mathscr{Z} = N \ln 2\pi - \sum_{l=1}^{N} \left(\ln(\beta\mu_l) + \frac{\beta}{\beta^2 \mu_l^2} \right). \tag{15}$$

With the standard relations $\mathscr{H} = ((\frac{\mu}{\beta} \frac{\partial}{\partial \mu} - \frac{\partial}{\partial \beta}) \ln \mathscr{Z}$ and $\mathscr{A} = -\frac{1}{\beta} \frac{\partial}{\partial \mu} \ln \mathscr{Z}$ we obtain

$$\mathscr{A} = \frac{N}{\beta\mu}, \quad \mathscr{H} = \frac{N}{\beta^2 \mu^2} + \sum_{l=1}^{N} \frac{\varepsilon_l}{\beta\mu_l}. \tag{16}$$

Note that we used $\mu \sim 1/\beta$ in the infinite temperature limit.

The total norm is not affected by the presence of a potential. The energy is affected, however things are different for the densities. The second term in the energy expression in (16) can be expanded as

$$\sum_{l=1}^{N} \frac{\varepsilon_l}{\beta(\mu + \varepsilon_l)} \approx \frac{1}{\beta\mu} \left(\sum_{l=1}^{N} \varepsilon_l - \frac{1}{\mu} \sum_{l=1}^{N} \varepsilon_l^2 \right). \tag{17}$$

The first term on the rhs diverges as $\sqrt{N}$ for any finite variance of ε_l—too slow to contribute to the final relation between the densities at the infinite temperature point (because we have to divide by N). The second term on the rhs is proportional to N/μ—the thermodynamic limit will remain a contribution in the density, however the infinite temperature limit leads to $1/\mu \rightarrow 0$ and therefore both terms vanish. As a result, for any potential with zero average and finite variance, the infinite temperature line in the density control parameter space for any lattice dimension is given by

$$y_{\beta=0} = x^2. \tag{18}$$

Since the energy density is not bounded from above, we conclude that for all densities $y > y_{\beta=0}$ the system will not be described by a Gibbs distribution.

Let us discuss some consequences in the absence of disorder (the presence of disorder will not substantially alter them). First, at a given scaled norm density x, a homogeneous state $A_l = const$ with constant phases $\phi_l = const$ will correspond to the lowest energy state which is in the Gibbsean regime. For staggered phases $\phi_{l+1} = \phi_l + \pi$ the homogeneous state $A_l = const$ yields a scaled energy density $y_{st} = 2x + x^2/2$. Therefore we find that for $x \geq 4$ all homogeneous initial states $A_l = const$, regardless of their phase details, are launched in the Gibbsean regime. However for $x < 4$ a growing set of homogeneous states with nonconstant phases, in particular the staggered case, are located in the nonGibbsean regime. We also stress that for all average scaled norm densities x there exist initial states which are inhomogeneous in the amplitudes such that the state will be located in the nonGibbsean regime.

If an initial state is in the nonGibbsean regime, we can not conclude much about the nature of its dynamics. It could remain to be strongly chaotic, and described by a negative temperature Gibbs distribution. It could be also nonergodic, less chaotic, or non-mixing.

The reported dynamical regimes in the Gibbsean and nonGibbsean are remarkably different [7–12]. While the Gibbsean regime is characterized by a relatively quick decay into a thermal equilibrium on time scales which are presumably inverse proportional to the largest Lyapunov coefficient, the nonGibbsean regime is very different. The dynamics is still chaotic, however the system relaxes into a two-component state—condensed hot spots with concentrated norm in them and corresponding high energy, and cold low energy density regions between them. Some of the results seem to indicate that the system produces as much of a condensate fraction as is needed to keep the remaining noncondensed part in a Gibbsean regime with infinite temperature $\beta = 0$. There is no evident mixing and relaxation in the condensate fraction. The condensed hot spots are similar to discrete breathers which are well known to exist in such models [12].

Interestingly models without norm conservation also allow for discrete breathers [12]. With only one conservation law (energy) and one variational parameter (inverse temperature) the equilibrium state of a Boltzmann distribution is always capable of yielding the prescribed energy density. Still such systems can produce hot spots, or discrete breathers, in thermal equilibrium in certain control parameter domains [12]. The remarkable difference to the above cases is, that the condensed hot spots do have a finite life time, and mixing, ergodicity and thermal equilibrium are obtained after finite times.

4 Selftrapping Theorem

The existence of the second conserved quantity $\mathscr{A}$ has also a nontrivial consequence for the decay of localized initial states or simply wave packets [14]. Consider a compact localized initial state with finite norm $\mathscr{A}$ and energy $\mathscr{H}$. Such a state has nonzero amplitudes inside a finite volume only, and strictly zero amplitudes outside. Note that the theorem can be easily generalized to non-compact localized initial states with properly, e.g. exponentially, decaying tails. The theorem addresses the question whether such a state can spread into an infinite volume and dissolve completely into some homogeneous final states. To measure the inhomogeneity of states we use the participation number (PN)

$$P = \frac{\mathscr{A}^2}{\sum_l A_l^2}. \tag{19}$$

This measure is bounded from below $P \geq 1$ and from above by $P \leq N$ where N is the number of available sites. The lower bound is achieved by concentrating all the available norm onto one single site $A_l = \mathscr{A}\delta_{l,l_0}$. The upper bound is achieved by a completely homogeneous state $A_l = \mathscr{A}/N$. Note that a typical value of the PN in a thermalized state is about $N/2$, due to inavoidable fluctuations in the amplitudes on different sites.

For an infinite system $N \to \infty$, the PN is unbounded from above. A localized initial state has a finite PN. If this state evolves and stays localized, its PN stays localized as well. If it manages to spread into the infinitely large reservoir of the system, and if the densities become on all sites of order $\mathscr{A}/N$, then the PN will be of order N. If the PN stays finite, then a part of the excitation is said to stay localized—either in the area of the initial excitation spot, or in other, possible migrating, locations. Therefore, the participation number turns to be a useful measure of inhomogeneity of a state, including localized distributions on zero or also nonzero delocalized backgrounds. It is the more useful as its inverse, up to a constant, is precisely the anharmonic energy share of the full Hamiltonian (1).

The selftrapping theorem [14] uses the existence of the second integral of motion—the norm. We split the total energy $\mathscr{H} = \langle \psi | \mathbf{L} | \psi \rangle + H_{NL}$ into the sum of its quadratic term of order 2 and its nonlinear terms of order strictly higher than 2. Then, $\mathbf{L}$ is a linear operator which is bounded from above and below. In our specific example, we have $\langle \psi | \mathbf{L} | \psi \rangle \geq \omega_m \langle \psi | \psi \rangle = \omega_m \mathscr{A}$ where $2 + \frac{W}{2} \geq \omega_m \geq -2 - \frac{W}{2}$ and ω_m is an eigenvalue of $\mathbf{L}$.

If the wavepacket amplitudes spread to zero at infinite time, $\lim_{t \to \infty} (\sup_l |\psi_l|) = 0$. Then $\lim_{t \to \infty} (\sum_l |\psi_l|^4) < \lim_{t \to \infty} (\sup_l |\psi_l^2|)(\sum_l |\psi_l|^2) = 0$ since $\mathscr{A} = \sum_l |\psi_l|^2$ is time invariant. Consequently, for $t \to \infty$ we have $\mathscr{H}_{NL} = 0$ and $\mathscr{H} \leq \omega_m \sum_l |\psi_l|^2 = \omega_m \mathscr{A}$. Since $\mathscr{H}$ and $\mathscr{A}$ are both time invariant, this inequality should be fulfilled at all times. However when the initial amplitude $\sqrt{A}$ of the wavepacket is large enough, it cannot be initially fulfilled because the nonlinear energy diverges as A^2 while the total norm diverges as A only. Thus such an initial wavepacket cannot spread to zero amplitudes at infinite time. This proof is valid for any strength of disorder W including the ordered case $W = 0$, and any lattice. Note that the opposite is not true—i.e., if the wavepacket does not fulfill the criterion for selftrapping according to the selftrapping theorem, we can only conclude that it may not selftrap in the course of spreading. We will still coin this regime non-selftrapping.

Let us consider two examples. First, take a single site intial state $\psi_l = \sqrt{A}\delta_{l,0}$. The energy is $\mathscr{H} = \varepsilon_0 A + \frac{\nu}{2}A^2$, and the norm $\mathscr{A} = A$. A zero amplitude final state has an upper energy bound of $(W/2 + 2)A$. For an amplitude $A > A_c$ with $\nu A_c = W + 4 - \varepsilon_0$ the initial state can not spread into a final one with infinite PN. The PN is bounded from above by P_{max} with $P_{max}^{-1} = (2\varepsilon_0 - W - 4)/(\nu A) + 1$.

A second example concerns wave packets excited on many sites. Assume a wave packet of size L with average scaled energy density y_0 and norm density x_0. The selftrapping theorem tells that selftrapping will persist for $y_0 > y_c$ with

$$y_c = \left(\frac{W}{2} + 2 \right) x. \tag{20}$$

Let us discuss the connection between selftrapping and Gibbs-nonGibbs regimes for the second example of a spreading wave packet excited initially on many sites. At any time during its spreading (including the initial time) we can trap it with fixed boundaries at its edges, and address its thermodynamic properties. We also note that if a wave packet spreads, its width L will increase with time, and the densities y and x will correspondingly drop keeping a linear dependence $y = \frac{y_0}{x_0} x$. Then, for $y_0 \leq 0$ the wave packet will stay in the Gibbs regime for all times. Selftrapping does not apply either. However, for positive $y_0 > 0$ things are more complex. A wave packet will be either all the time nonGibbsean, or enter a nonGibbsean phase at some later point in time. With (20) it could also be selftrapped, or not.

We finally note that a spreading wave packet is representing a nonequilibrium process characterized by corresponding time scales. The formation of a nonGibbsean density distribution, as reported in [7–11], takes place on certain time scales as well. It might therefore well happen that a spreading wave packet does not thermalize quickly enough in its core, and therefore never enters a nonGibbsean regime despite the fact that it would do so at equilibrium.

5 Anderson Localization

For $\nu = 0$ with $\psi_l = B_l \exp(-i\lambda t)$ (2) is reduced to the linear eigenvalue problem

$$\lambda B_l = \varepsilon_l B_l - B_{l-1} - B_{l+1}. \tag{21}$$

All eigenstates are exponentially localized, as first shown by Anderson [1]. The normal modes (NM) are characterized by the normalized eigenvectors $B_{\bar{\nu},l}$ ($\sum_l B_{\bar{\nu},l}^2 = 1$). The eigenvalues $\lambda_{\bar{\nu}}$ are the frequencies of the NMs. The width of the eigenfrequency spectrum $\lambda_{\bar{\nu}}$ of (21) is $\Delta = W + 4$ with $\lambda_{\bar{\nu}} \in \left[-2 - \frac{W}{2}, 2 + \frac{W}{2}\right]$. While the usual ordering principle of NMs is with their increasing eigenvalues, here we adopt a spatial ordering with increasing value of the center-of-norm coordinate $X_{\bar{\nu}} = \sum_l l B_{\bar{\nu},l}^2$.

The asymptotic spatial decay of an eigenvector is given by $B_{\bar{\nu},l} \sim e^{-|l|/\xi(\lambda_{\bar{\nu}})}$ where $\xi(\lambda_{\bar{\nu}})$ is the localization length and $\xi(\lambda_{\bar{\nu}}) \approx 24(4 - \lambda_{\bar{\nu}}^2)/W^2$ for weak disorder $W \leq 4$ [15].

The volume occupied by a given eigenstate is denoted by $V_{\bar{\nu}} \sim \xi_{\bar{n}u}$ (for details see [16]). The average spacing d of eigenvalues of neighboring NMs within the range of a localization volume is of the order of $d \approx \Delta/V$, which becomes $d \approx \Delta W^2/300$ for weak disorder. The two scales $d \leq \Delta$ are expected to determine the packet evolution details in the presence of nonlinearity.

Due to the localized character of the NMs, any localized wave packet with size L which is launched into the system for $\nu = 0$, will stay localized for all times. We remind that Anderson localization is relying on the phase coherence of waves. Wave packets which are trapped due to Anderson localization correspond to trajectories in phase space evolving on tori, i.e. they evolve quasi-periodically in time.

Finally, the linear wave equations constitute an integrable system with conserved actions where the dynamics happens to be on quasiperiodic tori in phase space. This can be safely stated for any finite, whatever large, system.

6 Disorder + Nonlinearity

In the presence of nonlinearity the equations of motion of (2) in normal mode space read

$$i\dot{\phi}_{\bar{\nu}} = \lambda_{\bar{\nu}}\phi_{\bar{\nu}} + \nu \sum_{\bar{\nu}_1,\bar{\nu}_2,\bar{\nu}_3} I_{\bar{\nu},\bar{\nu}_1,\bar{\nu}_2,\bar{\nu}_3}\phi_{\bar{\nu}_1}^*\phi_{\bar{\nu}_2}\phi_{\bar{\nu}_3} \tag{22}$$

with the overlap integral

$$I_{\bar{\nu},\bar{\nu}_1,\bar{\nu}_2,\bar{\nu}_3} = \sum_l B_{\bar{\nu},l}B_{\bar{\nu}_1,l}B_{\bar{\nu}_2,l}B_{\bar{\nu}_3,l}. \tag{23}$$

The variables $\phi_{\bar{\nu}}$ determine the complex time-dependent amplitudes of the NMs.

The frequency shift of a single site oscillator induced by the nonlinearity is $\delta_l = \nu|\psi_l|^2 \approx x$. As it follows from (22), nonlinearity induces an interaction between NMs. Since all NMs are exponentially localized in space, each normal mode is effectively coupled to a finite number of neighboring NMs, i.e. the interaction range is finite. However the strength of the coupling is proportional to the norm density $n = |\phi|^2$. Let us assume that a wave packet spreads. In the course of spreading its norm density will become smaller. Therefore the effective coupling strength between NMs decreases as well. At the same time the number of excited NMs grows. One possible outcome would be: (I) that after some time the coupling will be weak enough to be neglected. If neglected, the nonlinear terms are removed, the problem is reduced to an integrable linear wave equation, and we obtain again Anderson localization. That implies that the trajectory happens to be on a quasiperiodic torus—on which it must have been in fact from the beginning. It also implies that the actions of the linear wave equations are not strongly varying in the nonlinear case, and we are observing a kind of anderson localization in action subspace. Another possibility is: (II) that spreading continues for all times. That would imply that the trajectory does not evolve on a quasiperiodic torus, but instead evolves in some chaotic part of phase space. This second possibility (II) can be subdivided further, e.g. assuming that the wave packet will exit, or enter, a Kolmogorov-Arnold-Moser (KAM) regime of mixed phase space, or stay all the time outside such a perturbative KAM regime. In particular if the wave packet dynamics will enter a KAM regime for large times, one might speculate that the trajectory will get trapped between denser and denser torus structures in phase space after some spreading, leading again to localization as an asymptotic outcome, or at least to some very strong slowing down of the spreading process.

Published numerical data [17–22] (we refer to [16, 23] for more original references and details of the theory) show that finite size initial wave packets (i) stay localized if $x \ll d$ and display regular-like (i.e. quasiperiodic as in the KAM regime) dynamics, which is numerically hardly distinguishable from very slow chaotic dynamics with subsequent spreading on unaccessible time scales); (ii) spread subdiffusively if $x \sim d$ with the second moment of the wave packet $m_2 \sim t^\alpha$ and chaotic dynamics inside the wave packet core; segregate into a two component field with a selftrapped component, and a subdiffusively spreading part for $x > \Delta$.

Spreading wave packets reduce their densities x, y in the course of time, and may reach regime (i), without much change in their spreading dynamics. Therefore we can conclude that regime (i) is at best a KAM regime, with a finite probability to launch a wave packet on a KAM torus, and a complementary one to observe spreading [24]. Anderson localization is restored in that probabilistic way, as the probability to stay on a KAM torus will reach value one in the limit of vanishing nonlinearity. Spreading wave packets, when launched in a domain of positive energy densities y, are either from the beginning in the nonGibbsean regime, or have to reach it at some later point in time. Assuming that the wave packet has enough time to develop nonGibbsean structures, one should observe selftrapping. Published numerical studies did not focus on this issue, in particular for parameter values which do not satisfy the seltrapping theorem. The reported numerical selftrapping dynamics is in full accord with the results of the selftrapping theorem.

The most interesting result concerns the spreading dynamics and the exponent α. The assumption of strong chaos—i.e. the dephasing of normal modes on times scales much shorter than the spreading time scales—leads to $\alpha = 1/2$ [13, 22, 25] This exponent can be numerically observed, but only as an intermediate (although potentially extremely long lasting) regime of strong chaos:

$$\alpha = \frac{1}{2} \, , \ x > d \ : \ \text{strong chaos.} \tag{24}$$

In the asymptotic regime of small densities, instead the regime of weak chaos is observed:

$$\alpha = \frac{1}{3} \, , \ x < d \ : \ \text{weak chaos.} \tag{25}$$

Perturbation theories show that in that regime not all normal modes are resonant and chaotic, but only a fraction of them [18, 19, 23, 25]. Correcting the theory of strong chaos with the probability of resonances $\mathscr{P}_r$, the asymptotic value $\alpha = 1/3$ is obtained [22, 25], and a summary of the results reads as follows [23, 25]:

$$D \sim (\mathscr{P}_r(x)x\langle I \rangle)^2, \ \ \mathscr{P}_r = 1 - \mathrm{e}^{-Cx}, \ \ C \sim \frac{\xi^2 \langle I \rangle}{d}. \tag{26}$$

The corresponding nonlinear diffusion equation [26] for the norm density distribution (replacing the lattice by a continuum for simplicity, see also [27]) uses the diffusion coefficient $D(x)$:

$$\partial_t x = \partial_{\bar{v}}(D \partial_{\bar{v}} x).   \tag{27}$$

In the regime of strong chaos it follows $D \sim x^2$, and in the regime of weak chaos—$D \sim x^4$. It is straightforward to show that the exponent α satisfies the relations (24) and (25) respectively [23, 25, 28, 29]. We can also conclude that a large system at equilibrium will show a conductivity which is proportional to D.

7   A Discussion in the Light of Many Body Localization

Let us discuss the expected dynamical regimes of an infinitely large lattice excited to some finite densities. In the nonGibbsean regime $y > x^2$ the dynamics is known to be nonergodic up extremely long time scales. At the same time the Gibbsean phase is characterized by two different regimes—strong chaos at large densities, and weak chaos at small densities. Strong chaos implies that all normal modes are resonant and lead to chaos and mixing. Therefore we could assume that strong chaos is ergodic. For small enough density x we enter the weak chaos regime, where not all normal modes are resonant and lead to chaos and mixing at the same time. Therefore it could be possible that this regime is nonergodic, despite the fact that it characterized by a finite conductivity.

Now we recall the main statements from many body localization. This theory deals with a quantum system of many interacting particles in the presence of disorder. It was initially developed for fermions [30]. The system is considered in a spatially continuous system. At a given particle density, the conductivity is predicted to be zero up to a nonzero critical energy density. Above this transition point, the system exhibits many body states which conduct in a nonergodic (multifractal) fashion. At even larger energy densities the system finally exhibits ergodic metallic states. Later this theory was also developed for bosons, which is the quantum counterpart of our classical model [31]. In that case, again the system is characterized by a finite energy density (at fixed particle density) below which all states are many body localized. Above that critical density states are again extended but nonergodic and fractal.

The cricital many body localization energy density scales to zero in the classical limit. This is consistent with the fact that a wave packet spreads to infinity (assuming that it indeed does so), as the wave packet is characterized by finite densities at any finite time, and approaches zero density in the limit of infinite time. It is therefore tempting to associate the regime of weak chaos with the nonergodic but metallic regime of the quantum theory. The test of this possibility is therefore one of the challenging future tasks to be explored.

References

1. P.W. Anderson, Phys. Rev. **109**, 1492 (1958)
2. T. Schwartz, G. Bartal, S. Fishman, M. Segev, Nature **446**, 52 (2007)
3. Y. Lahini, A. Avidan, F. Pozzi, M. Sorel, R. Morandotti, D.N. Christodoulides, Y. Silberberg, Phys. Rev. Lett. **100**, 013906 (2008)
4. D. Clement, A.F. Varon, J.A. Retter, L. Sanchez-Palencia, A. Aspect, P. Bouyer, New J. Phys. **8**, 165 (2006); L. Sanches-Palencia, D. Clement, P. Lugan, P. Bouyer, G.V. Shlyapnikov, A. Aspect, Phys. Rev. Lett. **98**, 210401 (2007); J. Billy, V. Josse, Z. Zuo, A. Bernard, B. Hambrecht, P. Lugan, D. Clement, L. Sanchez-Palencia, P. Bouyer, A. Aspect, Nature **453**, 891 (2008); G. Roati, C. D'Errico, L. Fallani, M. Fattori, C. Fort, M. Zaccanti, G. Modugno, M. Modugno, M. Inguscio. Nature **453**, 895 (2008)
5. O. Morsch, M. Oberthaler, Rep. Prog. Phys. **78**, 176 (2006)
6. YuS Kivshar, G.P. Agrawal, *Optical Solitons: From Fibers to Photonic Crystals* (Academic Press, Amsterdam, 2003)
7. K.Ø. Rasmussen, T. Cretegny, P.G. Kevrekidis, N. Grønbech-Jensen, Phys. Rev. Lett. **84**, 3740 (2000)
8. M. Johansson, K.Ø. Rasmussen, Phys. Rev. E **70**, 066610 (2004)
9. B. Rumpf, EPL **78**, 26001 (2007)
10. B. Rumpf, Phys. Rev. E **77**, 036606 (2008)
11. B. Rumpf, Phys. D **238**, 2067 (2009)
12. S. Flach, C.R. Willis, Phys. Rep. **295**, 181 (1998); D.K. Campbell, S. Flach, Y.S. Kivshar, Phys. Today **57** (1), 43 (2004); S. Flach, A.V. Gorbach. Phys. Rep. **467**, 1 (2008)
13. D.M. Basko, Phys. Rev. E **89**, 022921 (2014)
14. G. Kopidakis, S. Komineas, S. Flach, S. Aubry, Phys. Rev. Lett. **100**, 084103 (2008)
15. B. Kramer, A. MacKinnon, Rep. Prog. Phys. **56**, 1469 (1993)
16. T.V. Laptyeva, M.V. Ivanchenko, S. Flach, J. Phys. A **47**, 493001 (2014)
17. A.S. Pikovsky, D.L. Shepelyansky, Phys. Rev. Lett. **100**, 094101 (2008)
18. S. Flach, D. Krimer, Ch. Skokos, Phys. Rev. Lett. **102**, 024101 (2009)
19. Ch. Skokos, D.O. Krimer, S. Komineas, S. Flach, Phys. Rev. E **79**, 056211 (2009)
20. M. Johansson, G. Kopidakis, S. Aubry, Europhys. Lett. **91**, 50001 (2010)
21. J. Bodyfelt, T.V. Laptyeva, Ch. Skokos, D. Krimer, S. Flach. Phys. Rev. E **84**, 016205 (2011)
22. T.V. Laptyeva, J.D. Bodyfelt, D.O. Krimer, Ch. Skokos, S. Flach, EPL **91**, 30001 (2010)
23. S. Flach, arxiv:1405.1122
24. M.V. Ivanchenko, T.V. Laptyeva, S. Flach, Phys. Rev. Lett. **107**, 240602 (2011)
25. S. Flach, Chem. Phys. **375**, 548 (2010)
26. Y.B. Zeldovich, Y.P. Raizer, *Physics of Shock Waves and High-Temperature Hydrodynamic Phenomena* (Academic Press, New York, 1966); Y.B. Zeldovich, A. Kompaneets, in *Collected Papers of the 70th Anniversary of the Birth of Academician*, A.F. Ioffe (Moscow, 1950); G.I. Barenblatt. Prikl. Mat. Mekh. **16**, 67 (1952)
27. A.R. Kolovsky, E.A. Gomez, H.J. Korsch, Phys. Rev. A **81**, 025603 (2010)
28. M. Mulansky, A. Pikovsky, EPL **90**, 10015 (2010)
29. T.V. Laptyeva, J.D. Bodyfelt, S. Flach, Phys. D **256–257**, 1 (2013)
30. D.M. Basko, I.L. Aleiner, B.L. Altshuler, Ann. Phys. **321**, 1126 (2006)
31. I.L. Aleiner, B.L. Altshuler, G.V. Shlyapnikov, Nat. Phys. **6**, 900 (2010)

Nonlinear Dynamics of Vertical-Cavity Surface-Emitting Lasers: Deterministic Chaos and Random Number Generation

Martin Virte, Marc Sciamanna, Hugo Thienpont and Krassimir Panajotov

Abstract We report on chaotic polarization fluctuations generated from a free-running vertical-cavity surface-emitting laser, and investigate in details the VCSEL dynamics and its potential application. We provide new experimental evidences of asymmetrical behaviour of VCSEL polarization dynamics, and show that considering a slight misalignment of the anisotropies of the laser cavity allows to reproduce these dynamics numerically. Finally, we propose a simple scheme for random bit generation exploiting polarization chaos, and experimentally demonstrate a 100 Gbps random generation rate.

1 Polarization Switching and Dynamics in Free-Running VCSELs

Vertical-cavity surface-emitting lasers (VCSELs) exhibit several essential advantages over the classic edge-emitting laser structure such as large bandwidth, lower injection current, circular beam and easy on-wafer testing. Yet, the major drawback of these devices is their polarization instabilities [12, 20]. Because of their typical circular geometry combined with the small anisotropies inside the laser cavity, VCSELs typically emit linearly polarized light at threshold, but a variation of the temperature or the injection current can lead to various instabilities including so-called polarization switching (PS) events [1, 8]: in this case the laser emission switches from the linear polarization stable at threshold to the orthogonal polarization mode. This

M. Virte (✉) · H. Thienpont · K. Panajotov
Brussels Photonics Team (B-Phot), Department of Applied Physics
and Photonics (TONA), Vrije Universiteit Brussel, Pleinlaan 2, 1050 Brussel, Belgium
e-mail: mvirte@b-phot.org

M. Sciamanna
OPTEL Research Group, LMOPS EA-4423, CentraleSupelec - Universite de Lorraine,
2 rue Edouard, F-57070 Belin, France

K. Panajotov
Institute of Solid State Physics, 72 Tzarigradsko Blvd., Sofia, Bulgaria

© Springer International Publishing Switzerland 2016
M. Tlidi and M.G. Clerc (eds.), *Nonlinear Dynamics: Materials,
Theory and Experiments*, Springer Proceedings in Physics 173,
DOI 10.1007/978-3-319-24871-4_4

phenomenon can be accompanied by a transition through elliptically polarized states [12], oscillations or noise-induced mode-hopping dynamics between the two orthogonal linearly polarized solutions when the system is close to the switching point [20]. Nevertheless, VCSELs as semiconductor lasers typically behave like damped nonlinear oscillators [11]: after a perturbation the system will slowly oscillate toward a stationary point. As a result, the polarization dynamics observed in VCSELs are typically interpreted as noise-induced phenomena [8, 20].

Recently however, we showed that such view is not entirely accurate and fails to explain the more complex behaviour that can be observed experimentally [6]: thus because of the polarization mode competition taking place, free-running VCSEL can generate chaotic polarization fluctuations without external perturbation [15]. In this contribution, we go one step further and, based on new experimental evidence showing strongly asymmetrical behaviour in VCSELs, demonstrate that a symmetry breaking mechanism also needs to be considered. Finally, we discuss the pertinence of polarization chaos for chaos-based applications and present a simple random bit generator scheme showing competitive performances.

2 From Polarization Mode Competition to Polarization Chaos

As described in [15, 16], the mode competition between two linear and orthogonal polarization states taking place in VCSEL structure can lead to chaotic dynamics. Polarization chaos appears following the bifurcation scenario highlighted in Fig. 1a–d. At threshold, the laser is on a steady-state and emits linearly polarized light: this state is defined as X-LP. When the injection current is increased, X-LP is destabilized by a pitchfork bifurcation which creates two elliptically polarized states (EP) symmetrical with respect to X-LP. These two steady-states are represented by the two crosses in Fig. 1a. When the current is increased further, the two EPs are destabilized by Hopf bifurcations and two limit cycles oscillating around the now unstable EPs are created, see Fig. 1b. A cascade of bifurcations then leads to the emergence of two symmetrical single-scroll chaotic attractors as shown in panel (c). But the two attractors grow in the Stokes parameter phase-space, hence above a critical value of current, they merge to form a double-scroll chaotic attractor. As shown in Fig. 1d, the scrolls of the attractor oscillate around the two unstable EP steady-states.

Experimentally, we cannot observe simultaneously the evolution of the two EP orientation as the system can only settle on one of them at a time. But an analysis of radio-frequency spectra clearly show the apparition of the first limit cycle and then the emergence of a chaotic dynamics. Once the double-scroll chaotic attractor has appeared, we typically observe a random-like hopping dynamics between the two EP orientation as shown in Fig. 1e. In order to observe this dynamics we insert a polarizer between the laser and the detector oriented at 45° with respect to the polarization at threshold to separate the two scrolls of the attractor. Another important feature of

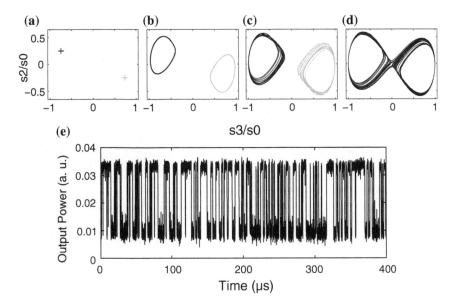

Fig. 1 **a–d** Evolution of the system trajectory in the Stokes parameter space for increasing injection currents. **e** Typical time-series obtained experimentally for a projection at 45° of the polarization at threshold when the laser exhibits polarization chaos dynamics

this dynamics is that the average dwell-time—i.e. the time between two successive jumps between scrolls or between the upper and lower levels of the time-series—shows a exponential decrease along with the increase of the injection current [6]. In fact, the two scrolls of the attractor keep growing in the Stokes parameter space, and thus the switching between the two scroll becomes easier, and the average dwell-time decreases quickly. All these features can be accurately reproduced using the spin-flip model (SFM) for VCSELs [4, 10] as shown in [15].

3 Influence of Anisotropy Misalignement in the Laser Cavity

So far, the theoretical work has been performed assuming that the phase and amplitude anisotropies of the laser cavity are aligned as described in the original SFM framework. As a result, the two EP states and the bifurcation sequence exhibit identical properties apart from their different polarization orientation. Experimentally this is of course an oversimplification: one clue is that the laser always selects the same polarization orientation after the destabilization of X-LP, while in a perfectly symmetrical situation the selection should be random. In this section, we report further experimental evidence of such asymmetry in VCSEL devices which therefore highlights the need for the introduction of symmetry breaking mechanism in

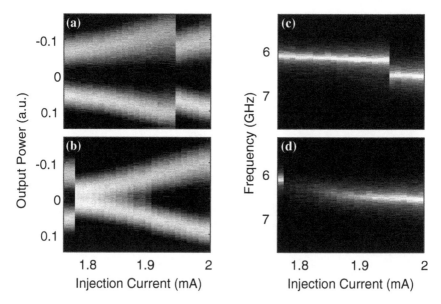

Fig. 2 Experimental bifurcation diagrams (*left*) and radio-frequency spectra evolution (*right*) for increasing (**a, c**) and decreasing injection current (**b, d**). The time-series used to obtain the bifurcation diagrams have been obtained for a polarizer oriented along the same axis as the linear polarization at threshold

the SFM. We then demonstrate that introducing a misalignment between phase and amplitude anisotropy, as suggested in [13] leads to a good qualitative agreement with experimental observations.

First, before the emergence of polarization chaos, we unveiled in some devices a bistability between two limit cycles oscillating around the two EP steady-states shown in Fig. 1a [17]. In Fig. 2, we give the bifurcation diagrams—i.e. the evolution of the histogram of the time-series extrema—and the radio-frequency spectra for increasing and decreasing injection currents. Thus, we clearly observe a hysteresis cycle connecting two periodic solutions with different amplitudes and frequencies: we record a difference of about 300 MHz between the frequencies of the two states. In addition, we can remark that for decreasing currents the system seem to settle on a steady-state just before switching back to the first periodic solution: the amplitude of the oscillation decreases significantly and the peak of the RF-spectrum vanishes. These results therefore clearly indicate that the EP states and their corresponding limit cycles exhibit different properties, and in particular different stability ranges, depending on their orientation.

All these observations are of course not possible in a perfectly symmetrical situation as the two EP states would necessarily exhibit identical properties despite different polarization orientation. However, as shown in [17], introducing a slight misalignment between the phase and amplitude anisotropy in the SFM model allows to reproduce these behaviours numerically. Here, a small misalignment of only

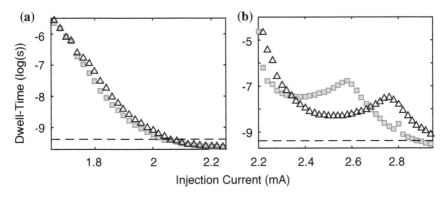

Fig. 3 Experimental measurements of the evolution of the average dwell-time for two different devices **a** and **b**. The *empty triangles* (*filled squares*) show the evolution of the average dwell-time for the upper (*lower*) level of the dynamics

$\theta = -0.0023$, i.e. $\theta \approx 0.13°$, is sufficient to induce an asymmetric evolution of the system leading to a limit cycle bistability: a large hysteresis cycle similar to the one described in Fig. 2 and a frequency difference close to 100 MHz between the two limit cycles. As detailed in [17], this change represent effectively a difference of only about 3 % for the effective anisotropies experienced by the right and left circular polarizations considered in the SFM.

Other signs of asymmetry can be spotted in the statistics of polarization chaos. In Fig. 3 we show the evolution of the average dwell-time for the upper and lower levels of the time-series in two different devices. In panel (a), the evolution is quite close to the symmetrical case as only a small difference—well below one order of magnitude—can be observed. As described in further details in [19], this minor discrepancy can be accurately reproduced in simulation using a small anisotropy misalignment. Panel (b) however shows a more complex picture: a large asymmetry can be seen with dwell-time differences going well above one order of magnitude, two bumps showing an increase of the average dwell-time and clear evidence of an exchange of stability between the two polarization orientations. Again the behaviour reported here obviously requires an asymmetric system but, unlike the previous cases, we were not able to reproduce a similar evolution only considering anisotropy misalignment; in particular no increase of the average dwell-time could be obtained.

In fact, an additional ingredient is required to reproduce such evolution theoretically, and the situation is therefore as follows. In Fig. 4a, we plot a bifurcation diagram of the normalized third Stokes parameter—i.e. the extremas of the trajectory showed in Fig. 1a–d projected on the horizontal axis—for increasing injection current μ. Here, we use a misalignment angle of $\theta = 0.04$—$\theta \approx 2.29°$—other parameters and simulation details can be found in [19]. With such representation we can easily identify the dynamical state of the system and its polarization orientation. Thus we observe that a first single-scroll attractor emerges around $\mu = 1.495$, quickly followed by a switch to the other polarization orientation. Then around $\mu = 1.5$ the

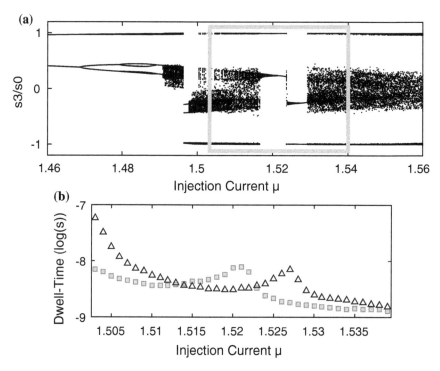

Fig. 4 **a** Bifurcation diagram of the third normalized Stokes parameter for increasing injection current for the case identified in the text without any noise. **b** Simulation of the average dwell-time for the current range delimited by the *grey rectangle* for a spontaneous emission noise of $\beta_{sp} = 4 \times 10^{-9}$

chaotic mode hopping appears as points for positive and negative values of s_3 are recorded. The region of chaotic dynamics extends beyond the $\mu = 1.56$ limit of the diagram and is only interrupted around $\mu = 1.52$ where we observe a restabilization of two limit cycles. Such restabilization is not very surprising as it seems to appear for a broad range of parameters; in particular, such situation can also be observed in the symmetric case, i.e. without any anisotropy misalignment. Hence, breaking the symmetry of the system only leads to different stability ranges depending on the polarization orientation of the limit cycles. The crucial point here however is that these periodic solutions are only marginally stable and co-exist with the chaotic attractor: a small level of noise is therefore sufficient to kick the system out of the basin of attraction of the limit cycle, thus pushing the laser back toward polarization chaos dynamics. This is demonstrated in Fig. 4b where we simulate the evolution of the average-dwell time for injection currents in the range delimited by the gray rectangle and with a spontaneous emission noise level of $\beta_{sp} = 4 \times 10^{-9}$. Around $\mu = 1.52$, we see that indeed the noise is sufficient to kick the system out of the stable limit cycle—as the dwell-time would be infinite otherwise—but we also observe that the existence of stable limit cycles strongly influences the statistics of the chaotic dynamics causing a local increase of the dwell-time. This influence decreases as the noise level is increased suggesting that a certain balance between the noise level and the system asymmetry is required to observe this phenomenon. It is however important to keep in mind that despite the need for some noise to ensure that the laser does not settle on the stable limit cycle, the chaotic dynamics remains fully deterministic. Moreover, the precise level of noise is not that crucial as similar behavior can be observed in a relatively large range of spontaneous emission noise.

4 Application of the Dynamics to Random Bit Generation

Semiconductor lasers show several advantages for industrial applications: they are small, reliable, efficient and electrically pumped. Yet to generate an optical chaos in laser diodes external perturbation, forcing or specific designs are typically required [2, 7, 14, 21]. The ability to generate chaotic fluctuations directly from a free-running device might therefore be an interesting opportunity to simplify the schemes suggested for chaos-based applications. The low-dimension, the limited number of positive Lyapunov exponent and relatively small bandwidth of the dynamics might however cast some doubts about its potential [5, 7, 14]. In this section, we discard these doubts by making a proof-of-concept demonstration of a random bit generator at high-speed based on polarization chaos with performances that are competitive with the latest reports [7, 21].

The scheme we propose is displayed in Fig. 5: the chaotic polarization fluctuation are simply converted to intensity fluctuations using a polarizer oriented at 45° from the linear polarization at threshold and then recorded using a photodiode and the 8-bit analog-to-digital converter of an oscilloscope. For simplicity the optical isolator

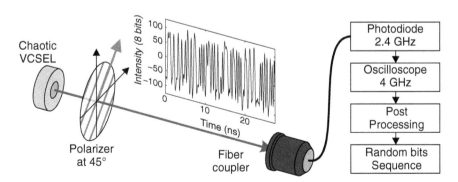

Fig. 5 Scheme of the experimental setup for random bit generation based on polarization chaos. The chaotic polarization fluctuations are first converted in intensity fluctuations by a polarizer. The light beam is coupled into a fiber in order to record the intensity fluctuations using the built-in photodiode of the oscilloscope. The output sequence of random bit is then obtained after a post-processing stage as described in the text

used to avoid any back reflections from the fibre front-facet is not represented. Also a neutral density filter is used to adjust the intensity of the signal received by the photodiode in order to use the full 8-bit range of the ADC. In practice we aimed at a ratio of saturated points between 10^{-4} and 10^{-1} %.

Nevertheless, the 8-bit sequence recorded by the oscilloscope is not random: like in other schemes, a suitable post-processing is required to remove the remaining biases of the sequence, e.g. the uneven occurrence of ones and zeros. As already mentioned in previous works [2, 7, 21], using a multibit generation scheme reduces the experimental constraints compared to a single-bit extraction techniques as used in [14]. With such approach a bit truncation is typically employed: we keep only the least significant bits (LSBs) for each 8-bit sample, in order to remove the highly correlated bits. But here even considering only the least significant bit at a relatively slow sampling rate always led to biased sequences—typically 50.1 % of ones leading to a $P_{value} \approx 4 \times 10^{-12}$. To remove such statistical imperfection, we compare the time-series with a time-shifted version of itself [21]. Thus, using a time-shift of 250 ps corresponding to the autocorrelation decay and then keeping only the 5 LSBs, we obtained sequences passing all the standard statistical tests of randomness for sampling rates up to 20 GSamples/s [3, 9, 18]. The proposed scheme can therefore generate random bit sequences up to 100 Gbps (= 5 bits × 20 GSamples/s). Similarly to what has been done by others [2], adding several comparison stages allows to increase the number of retained bits for the final sequence: with such technique, we were able to keep up to 28 bits when performing 23 comparisons hence leading to a potential bitrate up to 560 GBps. Despite its low-dimension and small bandwidth, the polarization chaos dynamics can therefore efficiently be used for chaos-based applications and exhibit performances competing with the latest reports [7, 21].

To better understand these good performances in spite of dynamical characteristics that seem, at first glance, non-optimal, we investigate the effects of the limited bandwidth of the acquisition electronics and analyze the entropy evolution of the system in different situations.

To evaluate the entropy evolution, we follow the same approach as in [5]: we simulate 1000 time-series with different noise sequences, convert the time-series into a bit sequence using a finely set threshold—to ensure a bias as small as possible between the 0 and 1 s—, and then evaluate the time-dependent entropy

$$H(t) = - \sum_{i=0}^{1} P_i(t) \log_2(P_i(t)) \qquad (1)$$

with $P_{0,1}(t)$ the time-dependent probability of having a 0 or a 1 at time t. Thus $H(t) = 0$ when the system is perfectly predictable $P_0(t) = 1$ or $P_1(t) = 1$, i.e. all the 1000 time-series take the same value at time t, and $H(t) = 1$ when the system is perfectly unpredictable $P_0(t) = P_1(t) = 0.5$, i.e. half of the 1000 time-series take value 0 at time t time while the rest take value 1. In practice, the system is considered unpredictable when $H(t) > 0.995$, and the memory time can be defined as the time the system takes to reach this unpredictability level [5]. Since the evolution of the

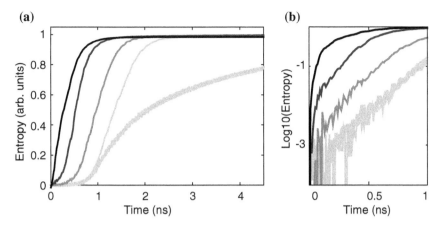

Fig. 6 Simulation of the entropy growth induced by the spontaneous emission noise amplification by the chaotic dynamics for noise levels of –30, –40, –50 and –60 dB from ligth to *dark lines*. The *thin lines* show the evolution of the entropy growth after the low-pass filtering, and the *thick line* give the evolution without filtering for a noise level of –60 dB. **b** shows the same plot but with the entropy in a logarithmic scale

entropy strongly depends on the simulation starting point, we average $H(t)$ over 1000 different starting points. Complete details about the parameters used to qualitatively match experimental observations are given in [18]. We obtain the results displayed in Fig. 6. The tick line shows the entropy evolution for a noise level of –60 dB without any filtering of the time-series: we observe that the increase is relatively slow and the memory time is well beyond 4 ns. However adding a low-pass filter—reproducing the effect of the limited bandwidth of the photodiode—dramatically improves the entropy growth. For very short time scales as shown in Fig. 6b, the effect is quite limited, but after 1 ns, the entropy increase for the filtered case appears to be much steadier than for the unfiltered data, and a memory time of about 3 ns is obtained. In fact this effect is related to the double-scroll structure of the chaotic attractor, as shown in Fig. 1d, where we can identify two time-scales: (1) the fast oscillations around the scroll of the attractor and (2) the slower hopping between the two scrolls. The fast oscillations are actually quite correlated and the randomness mostly originates from the slower hopping behavior, thus filtering out the fast correlated oscillations plays the role of an effective first processing stage. Then as can be expected, increasing the level of noise, leads to a faster increase of the entropy and reduces the memory time down to about 1.5 ns for a noise level of –30 dB. Finally, including the multibit post-processing described previously, we obtained memory times below the 50 ps level coherent with the 20 GSamples/s used experimentally [18].

5 Conclusion

To conclude, starting from the report of the polarization chaos dynamics generated in a free-running VCSEL, we investigate in further details the nonlinear dynamics of VCSELs and the impact of microscopic properties of the laser cavity, and unveil the large potential of the dynamics for chaos-based applications.

We bring new experimental evidence of the asymmetry in the devices hence confirming the need to take into account a symmetry breaking mechanism. We demonstrate that a misalignment between the phase and amplitude anisotropies of the laser cavity explain the experimental observations. In particular, we show that the system asymmetry can have a huge impact on the statistics of the chaotic dynamics.

Then, we make a proof-of-concept demonstration of a random bit generator based on polarization chaos dynamics. Despite some dynamical characteristics which seem non optimal at first glance, we obtain competitive performances comparable with the latest reports. In addition, we identify the limited bandwidth of the acquisition electronics, which plays the role of a low-pass filter for the dynamics, as a crucial stage directly linked with the structure of the chaotic attractor itself.

References

1. K.D. Choquette, R.P. Schneider, K.L. Lear, R.E. Leibenguth, Gain-dependent polarization properties of vertical-cavity lasers. IEEE J. Sel. Top. Quant. Electron **1**, 661 (1995)
2. I. Kanter, Y. Aviad, I. Reidler, E. Cohen, M. Rosenbluh, An optical ultrafast random bit generator. Nat. Photon **4**, 58 (2009)
3. G. Marsaglia, *DieHard: A Battery of Tests of Randomness*. Technical Report (State University, Florida, 1996)
4. J. Martin-Regalado, F. Prati, M. San Miguel, N.B. Abraham, Polarization properties of vertical-cavity surface-emitting lasers. IEEE J. Quant. Electron **33**, 765 (1997)
5. T. Mikami, K. Kanno, K. Aoyama, A. Uchida, T. Ikeguchi, T. Harayama, P. Davis, Estimation of entropy rate in a fast physical random-bit generator using a chaotic semiconductor laser with intrinsic noise. Phys. Rev. E **85**, 016211 (2012)
6. L. Olejniczak, K. Panajotov, H. Thienpont, M. Sciamanna, A. Mutig, F. Hopfer, D. Bimberg, Polarization switching and polarization mode hopping in quantum dot vertical-cavity surface-emitting lasers. Opt. Express **19**, 2476 (2011)
7. N. Oliver, M.C. Soriano, D.W. Sukow, I. Fischer, Fast random bit generation using a chaotic laser: approaching the information theoretic limit. IEEE J. Quant. Electron **49**, 910 (2013)
8. K. Panajotov, F. Prati, *Polarization Dynamics of VCSELs. in VCSELs*, ed. R. Michalzik (2013)
9. A. Rukhin, J. Soto, J. Nechvatal, M. Smid, E. Barker, S. Leigh, M. Levenson, M. Vangel, D. Banks, A. Heckert, J. Dray, S. Vo, *A Statistical Test Suite for Random and Pseudorandom Number Generators for Cryptographic Applications—Special Publication 800–22 Rev1a*, Technical Report (National Institute of Standards and Technology, 2010)
10. M. San Miguel, Q. Feng, J.V. Moloney, Light-polarization dynamics in surface-emitting semiconductor lasers. Phys. Rev. A **52**, 1728 (1995)
11. M. Sciamanna, K.A. Shore, Physics and applications of laser diode chaos. Nat. Photon **9**, 151–162 (2015)
12. M. Sondermann, T. Ackemann, S. Balle, J. Mulet, K. Panajotov, Experimental and theoretical investigations on elliptically polarized dynamical transition states in the polarization switching of vertical-cavity surface-emitting lasers. Opt. Commun. **235**, 421 (2004)

13. M. Travagnin, Linear anisotropies and polarization properties of vertical-cavity surface-emitting semiconductor lasers. Phys. Rev. A **56**, 4094 (1997)
14. A. Uchida, K. Amano, M. Inoue, K. Hirano, S. Naito, H. Someya, I. Oowada, T. Kurashighe, M. Shiki, S. Yoshimori, K. Yoshimura, P. Davis, Fast physical random bit generation with chaotic semiconductor lasers. Nat. Photon **2**, 728 (2008)
15. M. Virte, K. Panajotov, H. Thienpont, M. Sciamanna, Deterministic polarization chaos from a laser diode. Nat. Photon **7**, 6065 (2013)
16. M. Virte, K. Panajotov, M. Sciamanna, Bifurcation to nonlinear polarization dynamics and chaos in vertical-cavity surface-emitting lasers. Phys. Rev. A **87**, 013834 (2013)
17. M. Virte, M. Sciamanna, E. Mercier, K. Panajotov, Bistability of time-periodic polarization dynamics in a free-running VCSEL. Opt. Express **22**, 6772 (2014)
18. M. Virte, E. Mercier, H. Thienpont, K. Panajotov, M. Sciamanna, Physical random bit generation from chaotic solitary laser diode. Opt. Express **22**, 17271 (2014)
19. M. Virte, E. Mirisola, M. Sciamanna, K. Panajotov, Asymmetric dwell-time statistics of polarization chaos from free-running VCSEL. Opt. Lett. **40**, 18651868 (2015)
20. M.B. Willemsen, M.U.F. Khalid, M.P. van Exter, J. Woerdman, P Polarization switching of a vertical-cavity semiconductor laser as a Kramers hopping problem. Phys. Rev. Lett. **82**, 4815 (1999)
21. T. Yamazaki, A. Uchida, Performance of random number generators using noise-based superluminescent diode and chaos-based semiconductor lasers. IEEE J. Sel. Top. Quant. Electron **19**, 0600309 (2013)

Experimental Observation of Front Propagation in Lugiato-Lefever Equation in a Negative Diffractive Regime and Inhomogeneous Kerr Cavity

V. Odent, M. Tlidi, M.G. Clerc and E. Louvergneaux

Abstract A driven resonator with focusing Kerr nonlinearity shows stable localized structures in a region far from modulational instability. The stabilization mechanism is based on front interaction in bistable regime with an inhomogeneous injected field. The experimental setup consist of a focusing Kerr resonator filled with a liquid crystal and operates in negative optical diffraction regime. Engineering diffraction is an appealing challenging topic in relation with left-handed materials. We solve the visible range of current left-handed materials to show that localized structures in a focusing Kerr Fabry-Perot cavity submitted to negative optical feedback are propagating fronts between two stable states. We evidenced analytically, numerically, and experimentally that these fronts stop due to the spatial inhomogeneity induced by the laser Gaussian forcing, which changes spatially the relativity stability between the connected states.

1 Introduction

Localized structures (LS's) often called cavity solitons in dissipative media have been observed in various fields of nonlinear science (see last overview on this issue [1–3]). Localized structures consist of isolated or randomly distributed peaks surrounded by

V. Odent (✉) · E. Louvergneaux
Laboratoire de Physique des Lasers, Atomes Et Molécules, CNRS UMR8523,
Université de Lille 1, 59655 Villeneuve d'Ascq Cedex, France
e-mail: vincent.odent@phlam.univ-lille1.fr

E. Louvergneaux
e-mail: eric.louvergneaux@univ-lille1.fr

M. Tlidi
Faculté des Sciences de l'Université Libre de Bruxelles, Brussel, Belgium
e-mail: mtlidi@ulb.ac.be

M.G. Clerc
Departamento de Física, FCFM, Universidad de Chile, Blanco Encalada 2008,
Santiago, Chile
e-mail: marcel@dfi.uchile.cl

© Springer International Publishing Switzerland 2016
M. Tlidi and M.G. Clerc (eds.), *Nonlinear Dynamics: Materials,*
Theory and Experiments, Springer Proceedings in Physics 173,
DOI 10.1007/978-3-319-24871-4_5

71

regions in the homogeneous steady state. Currently they attract growing interest in optics due to potential applications for all-optical control of light, optical storage, and information processing. They appear in an optical resonator containing a third order nonlinear media such as liquid crystals, and driven coherently by an injected beam in a positive diffraction regime and close to modulational instability [4, 5]. The existence of localized structures and localized patterns, due to the occurrence of a modulational instability, has been abundantly discussed and is by now fairly well understood [6]. In this case, LS's appears in the subcritical modulational instability regime where there is a coexistence between the homogeneous steady state and the spatially periodic pattern. Localized structures consist of isolated or randomly distributed spots surrounded by regions in the uniform state. They may consist of peaks or dips embedded in the homogeneous background.

However, localized structures could be formed in modulationally stable regime [7]. In this case, heterogeneous initial conditions usually caused by the inherent fluctuations generate spatial domains, which are separated by their respective interfaces often called front solutions or interfaces or domain walls [8]. Interfaces between these metastable states appear in the form of propagating fronts and give rise to a rich spatiotemporal dynamics [9, 10]. From the point of view of dynamical system theory at least in one spatial dimension a front is a nonlinear solution that is identified in the comoving frame system as a heteroclinic orbit linking two spatially extended states [11, 12]. The dynamics of the interface depends on the nature of the states that are connected. In the case of a front connecting a stable and an unstable state, it is called as Fisher-Kolmogorov-Petrosvky-Piskunov (FKPP) front [13–16]. One of the characteristic features of these fronts is that the speed is not unique, nonetheless determined by the initial conditions. When the initial condition is bounded, after a transient, two counter propagative fronts with the minimum asymptotic speed emerge [13, 16]. In case that the nonlinearities are weak, this minimum speed is determined by the linear or marginal-stability criterion and fronts are usually referred to as pulled [16]. In the opposite case, the asymptotic speed can only be determined by nonlinear methods and fronts are referred to as pushed [16]. The above scenario changes completely for a front connecting two stable states. In the case of two uniform states, a gradient system tends to develop the most stable state, in order to minimize its energy, so that the front always propagates toward the most energetically favored state [17]. It exists only as one point in parameter space for which the front is motionless, which is usually called the Maxwell point, and is the point for which the two states have exactly the same energy for variational systems [18].

In this chapter, we present analytical, numerical and experimental investigation of front interaction in bistable regime with an inhomogeneous injected beam. Far from any modulational instability, the inhomogeneous injected beam in the form of Gaussian, can lead to the stabilization of LS's. This completes our previous communication of the stability of LS's [19]. We consider an experimental setup which consists of a focusing Kerr resonator filled with a liquid crystal and operates in negative optical diffraction regime. Engineering diffraction is an appealing challenging topic in relation with left-handed materials. We solve the visible range of current left-handed materials [20–23] to show that LS's in a focusing Kerr Fabry-Perot cavity submitted to negative optical feedback are propagating fronts between two stable states. Moreover, we have evidenced that these fronts stop due to the spatial inhomogeneity induced by the laser Gaussian forcing, which changes spatially the relativity stability between the connected states.

This chapter is organized as follows. After an introduction we present the experimental setup and experimental observation in Sect. 2. The analytical and numerical investigations are presented in Sect. 3. The special case with a null diffraction is developed in Sect. 4. We conclude in Sect. 5.

2 Experimental

2.1 Set-Up

The experiments have been carried out using a nonlinear Kerr slice medium inserted in an optical Fabry-Perot resonator. The Kerr focusing medium is a 50 μm-thick layer of E7 nematic liquid crystal homeotropically anchored. Two plane mirrors M_1 and M_2 define the physical cavity but the optical one is delimited by M_1 and M_2' (which is the image of M_2 through the $4f$ lens arrangement) and its optical length is d (Fig. 1). The intensity reflexion coefficients of M_1 and M_2 mirrors are $R_1 = 81.4\%$ and $R_2 = 81.8\%$ respectively so that the cavity finesse is estimated to 15. The experimental recording of the Airy function (blue curve) on Fig. 2 gives a finesse of 11.6 indicating the presence of supplementary losses due to nonlinear medium and lenses transmission coefficients.

The optical cavity length d may be tuned from positive to negative values (positive on Fig. 1a and negative on Fig. 1b). Thus, *for negative optical cavity path ($d < 0$), a beam propagating along this path experiences negative diffraction*. Together with the positive Kerr index, the experimental setup is then equivalent to a Kerr cavity that would have a positive optical distance but negative Kerr index (the ηa product sign in [24] that defines the type of transverse instabilities remains the same). However the physical mechanisms of negative refraction and negative diffraction are different. Thus, *this intra-cavity geometrical lens arrangement allows for achieving an equivalent left-handed Kerr material in the visible range*. It also allows to continuously tune the diffraction from positive to negative.

The laser source used is a single mode frequency doubled Nd^{3+}:YVO4 laser ($\lambda_0 = 532$ nm) at 5 W. We have split the laser beam into two parts, a pump beam containing 95% of the initial power and a probe beam with 5% of the initial power. The pump beam power is controlled by an electro-optical modulator associated with a Glan polarizer. Then it is shaped by means of two cylindrical telescopes. The resulting beam size ($\sim$200 μm × 2800 μm) gives a "cigar" transverse shape such that only one spot can develop in one of the two directions. This beam is injected inside the cavity may then be considered as one-dimensional. The probe beam polarization is rotated to 90° to prevent interference phenomenon inside the cavity between the two

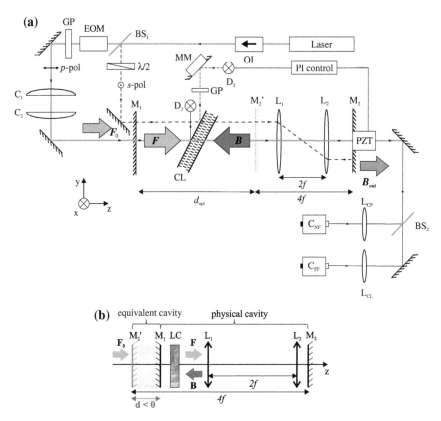

Fig. 1 Inhomogeneous Kerr cavity with negative diffraction: **a** Experimental setup. **b** Schematic representation of the physical cavity and the equivalent cavity, with negative cavity length. *OI* optical isolator; *BS* beam splitter; *EOM* electro-optical modulator; *GP* Glan polarizer; C_1 and C_2 cylindrical lenses; *LC* liquid crystal slice; D_1 and D_2 photodetectors; *MM* motorized mirror; L_1 and L_2 lenses of focal length f: p and s are the polarized components of the pump (*solid line*) and probe (*dashed line*) beams respectively; C_{NF} and C_{FF} near field and far field cameras; M_1 and M_2 are the real cavity mirrors but the optical Perot-Fabry cavity is delimited by M_1 and M_2' mirrors and its length is d

beams (see Fig. 1) The pump beam propagating in the forward direction is monitored at the output of M_2 mirror. We record the near field on C_{NF} and the far field on C_{FF}. The cavity detuning is an important parameter, because it determines the solution type (mono/bistable) of the cavity and also the energy quantity inside the cavity, as shown on the cavity transfer function on Fig. 2. For these reasons, we perform an active stabilization of the cavity phase shift, based on Coen's work and coworkers [25]. We stabilize the cavity length around a reference phase on the beam probe (φ_{ref} on Fig. 2), then we measure the phase shift between the probe and the pump beams to have the real cavity detuning φ, as presented on Fig. 2. Our setup allow to maintain the detuning long enough to realize experiments, it means to have a fix

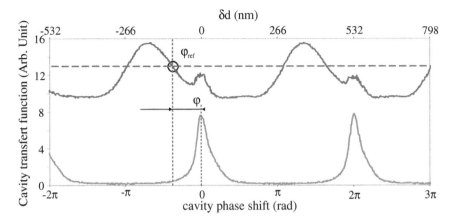

Fig. 2 Transfer function of the optical cavity for: pump beam (*blue line*) and probe beam (*red line*). φ linear phase shift of the cavity. δd cavity length variation

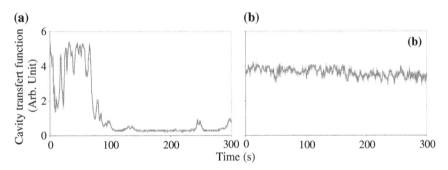

Fig. 3 Temporal evolution of cavity transfer function correlated to cavity phase shift: **a** without stabilization; **b** with stabilization. $d = 5$ mm, $I = 37$ W/cm^2

value $\pm\pi/19$ during many minutes. Without the stabilization, the cavity detuning can evolve quickly and not permit us to do experiments. The typical detuning evolution without and with an active stabilization are presented on Fig. 3.

2.2 Experimental Observations

As the input power is suddenly increased to the upper bistability response branch ($t = 0$ s), the central part of transmitted intensity profile suddenly jumps after some latency time ($t \approx 37$ s on Fig. 4c) to a higher value and invades the surroundings towards the external regions where the field is less intense before stopping its propagation. Finally, the fronts locks to give a localized light state (Fig. 4b). Changing the waist w of the Gaussian forcing or its intensity within the bistability region allows to tune the distance between the bounded fronts and so the localized state extension.

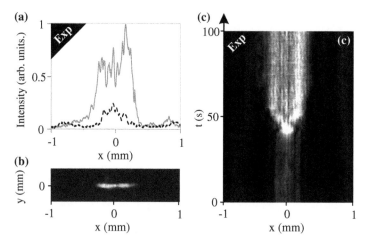

Fig. 4 Experimental front propagation in negative diffractive inhomogeneous Kerr cavity. **a** Transverse cross-section of the initial (final) average localized structure in *dashed black line* (in continue *gray line*). **c** Spatiotemporal response to a step function of the input intensity from the lower to the upper branch of the bistable cycle. **b** Cross section of the localized state. Light region account for high intensity of the light. $I_0 = 433$ W. cm^{-2}, $d = -5$ mm, $\varphi = -0.6$ rad, $w_x = 1400$ μm, $w_y = 100$ μm, $R_1 = 81.8\%$; $R_2 = 81.4\%$

A transverse cut of the transmitted intensity profile is depicted on Fig. 4a obtained in the initial and the final observation period. This figure emphasizes the coexistence of different states in the same region of parameters. In addition, a state emerges from the other because of the inherent fluctuations of the system. Figure 4c shows this phenomenon starting at $t \approx 37$ s. From this instant, the system exhibits two counter-propagating fronts, which will become asymptotically motionless (see Fig. 4c). The motionless front is observed at the location $x_0(expt) \cong \pm 0.17$ w. At this location, the input intensity is only 3 % lower than at the center of the input beam. Thus, we have observed experimentally coexistence between two inhomogeneous states, the noise induces a pair of fronts among these states, which initially are counter-propagate and asymptotically they stop.

3 Analytical Results

3.1 Model Equation

The dynamics of the single longitudinal mode of the bistable system which consist of a Fabry-Perot cavity filled with liquid crystal Kerr like medium and driven by a coherent plane-wave steady can be described by the simple partial differential equation (the LL model [24]) in which we incorporate an an inhomogeneous injected field.

This mean field approach model is valid under the following approximations: the cavity possess a high Fresnel number i.e., large-aspect-ratio system and we assume that the cavity is much shorter than the diffraction and the nonlinearity spatial scales; and for the sake of simplicity, we assume a single longitudinal mode operation. Under these assumptions the space-time evolution of the intracavity field is described by the following partial differential equation

$$\frac{\partial E}{\partial t} = E_{in}(x) - (1 + i\Delta)E + i|E|^2E - i|\alpha|\frac{\partial^2 E}{\partial r^2} + \sqrt{\varepsilon}\xi(x,t) \qquad (1)$$

which includes the effect of diffraction, which is proportional to α; E is the normalized slowly-varying envelope of the electric field, Δ is the detuning parameter, E_{in} is the input field assumed to be real, positive and spatially inhomogeneous. The negative diffraction coefficient is $|\alpha|$. Note that the above model has been derived for a cavity filled with left-handed material operating in negative diffraction regime [21]. ε scales the noise amplitude and $\xi(x,t)$ are Gaussian stochastic processes of zero mean and delta correlation introduced to model thermal noise [26].

3.2 Derivation of F-KPP Equation

The following development is realized for the deterministic case of (1) ($\varepsilon = 0$). The homogeneous steady states of (1) are solutions of $E_i = [1 + i(\Delta - |E_s|^2)]E_s$. The response curve involving the intracavity intensity $|E_s|^2$ as a function of the input intensity $|E_i|^2$ is monostable for $\Delta < \sqrt{3}$ and exhibits a bistable behavior when the detuning $\Delta > \sqrt{3}$. For $\Delta = \sqrt{3}$, we obtain the critical point associated with bistability where the output versus input characteristics have an infinite slope. At the critical point, the coordinate of the intracavity are $E_c = u_c + iv_c$ with $u_c = 3^{1/4}/\sqrt{2}$ and $v_c = -1/3^{1/4}\sqrt{2}$, and the injected field amplitude is $E_{in}^c = 2\sqrt{2}/3^{3/4}$.

The analytical investigation of fronts dynamics connecting two-homogeneous steady states in the framework of the Lugiato-Lefever model (1) is far from the scope of the present chapter. In this section we perform a derivation of a simple bistable model with inhomogeneous injection to study analytically the dynamics of the front connecting the two-homogeneous steady states. To do that, we introduce a small parameter that measures the distance from the critical point $\zeta \ll 1$ and we express the cavity detuning in the form

$$\Delta = \Delta_c(1 + \zeta^2\sigma) \qquad (2)$$

where σ is a quantity of order one. Then, we decompose the envelope of the electric field into its real and imaginary parts: $E = x_1 + ix_2$ and we introduce a new space and time scales as $(x,t) \rightsquigarrow \left(\zeta^2 t/\sigma, 3^{\frac{1}{4}}\zeta x/(\sqrt{\sigma})\right)$. The injected field can be expanded as

$$E_{in} = E_{in}^c \left(1 + \frac{3\zeta^2 I}{4} + \zeta^3 \alpha + \cdots \right) \qquad (3)$$

Let $(u, v) = (x_1, x_2) - (u_c, v_c)$ be the deviations of the real, the imaginary parts of intracavity field with respect to the values of these quantities at the critical point. Inserting these expansion into the LL model and using the above scalings, we obtain

$$\frac{\partial u}{\partial t} = \zeta \left(\frac{u^2}{2} + uv + \frac{v^2}{2}\right) + \zeta^2 \left(\frac{4\alpha}{3} - Iv + \frac{u^2 v}{2} + \frac{v^3}{6}\right) - \frac{1}{\sqrt{3}}\frac{\partial^2 v}{\partial x^2}, \qquad (4)$$

$$\frac{\partial v}{\partial t} = -2(u + v) + \zeta \left(3I - \frac{9u^2}{2} - uv - \frac{v^2}{2}\right) + \frac{\zeta^2}{2}(6Iu - 3u^2 - uv^2) + \sqrt{3}\frac{\partial^2 u}{\partial x^2}. \qquad (5)$$

Our aim is to seek solutions of (4) and (5) in the neighborhood of the critical point associated with the optical bistability. To this end, we expand the cavity field and the injected field as

$$(u, v) = \zeta[(u_0, v_0) + \zeta(u_1, v_1) + \zeta^2(u_2, v_2) + \zeta^3(u_3, v_3) + \cdots] \qquad (6)$$

Inserting these expansions and taking into account of (2) and (3) into the LL model and using the above scalings, we then obtain a hierarchy of linear problems for the unknown functions. At the first order in ζ, we find $u_0 = -v_0$. At the second, we have $v_1 = -u_1 + 3I/2 - u_0^2$. Finally, at the third order, we get a simple bistable model

$$\frac{\partial u}{\partial t} = f(u) + \frac{\partial^2 u}{\partial x^2} \qquad (7)$$

where $u(x, t) = \sqrt{3/(2\sigma)}u_0$ is a scalar field that accounts for the real part of the envelope E,

$$f(u) = \eta + u - u^3 \qquad (8)$$

and $\eta = 4y_2\sigma/3$ controls the relative stability between the equilibria. Note that y_2 is proportional to the pumping E_{in}. Hence, if the pumping is inhomogeneous then the parameter η is also inhomogeneous. For a Gaussian pumping, we consider

$$\eta(x) \equiv \tilde{\eta} + \eta_0 e^{-(x/w)^2}, \qquad (9)$$

where η_0 accounts for the strength of the spatial pumping beam and w is the width of the Gaussian. For $\eta(x) = 0$ both states are symmetric corresponding to the Maxwell point, where a front between these states is motionless.

To perform analytical developments, we use further approximation by taking into account only the first order development close to the center of the optical pumping where the stress is maximum, i.e.,

$$\eta(x) \approx -\tilde{\eta} + \eta_0 \left(1 - (x/w)^2\right) \tag{10}$$

close to the Maxwell point, one can consider the following ansatz for the front solution $u(x, t) = \tanh\left[(x - x_0(t))/\sqrt{2}\right] + H$, where x_0 is the front position and H accounts for small corrections. We substitute the above ansatz for $u(x, t)$ in (7), linearizing in H and imposing the solvability condition, we obtain the kinetic equation for the evolution of the front position

$$\dot{x}_0 = \frac{-3\sqrt{2}}{2}\left[-\tilde{\eta} + \eta_0 \left(1 - \left(\frac{x_0}{w}\right)^2 - \left(\frac{\pi}{\sqrt{6}w}\right)^2\right)\right]. \tag{11}$$

this equation takes into account of corrections imputable to inhomogeneities of the injected signal. The stationary solutions of this equation reads

$$x_0 = \pm w \sqrt{1 - \frac{\tilde{\eta}}{\eta_0} - \frac{\pi^2}{6w^2}}, \tag{12}$$

the term $\pi^2/(6w)$ is negligible for w large. In this case, we have

$$x_0 = \pm w \sqrt{1 - \frac{\tilde{\eta}}{\eta_0}}. \tag{13}$$

The ordinary differential equation (11) admits an exact solution that leads to the following trajectory of the front

$$x_0(t) = \pm a \tanh\left(b(t - t_0)\right), \tag{14}$$

with a, b and t_0 are coefficients depending on η_0, $\tilde{\eta}$ and w. The equilibrium position of the front $x_{0\infty}$ can be inferred from expression (14) for $t \rightarrow \infty$ as $x_{0\infty} = \pm w \sqrt{1 - \tilde{\eta}/\eta_0}$. This equilibrium positions could be obtained directly from (7), assuming $\eta(x_{0\infty}) = 0$ which is the condition for motionless front (Maxwell point). Extending this last property to the initial Gaussian forcing, we get

$$x_{0\infty} = \pm w \sqrt{\ln\left(\frac{\eta_0}{\tilde{\eta}}\right)}. \tag{15}$$

At leading order, (15) recover again the previous expression of $x_{0\infty}$ for the parabolic approximation of the Gaussian profile.

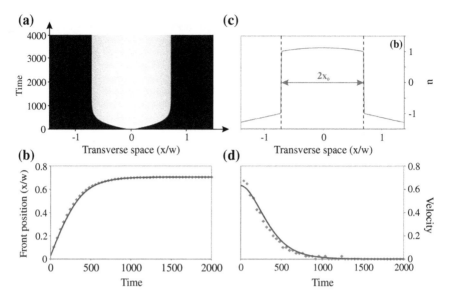

Fig. 5 Numerical study of front pinning with a parabolic forcing. **a** Spatiotemporal diagram; **b** Transverse profile of the front at $t = 4000$; **c** Relative front position $x_0(t)$ (for $x > 0$); **d** Front velocity. *Blue diamonds* are the numerical points and the *red curves* are the analytical predictions. $\eta_0 = 0.6$, $\tilde{\eta} = 0.3$, $w = 350$

3.3 Numerical Results

We conduct numerical simulations of the imperfect pitchfork bifurcation, (7) and (8), with a parabolic spatial injection to compare with the analytical predictions. Initially, we observe the front propagation, as the case with a plane wave. Then, the front speed decreases to become zero, around $t = 1000$, as presented on (Fig. 5a, c, d). At $t = 4000$, the structure is pinned by the pump parabolic profile and presents a bistable structure with a parabolic dependance on the two states (Fig. 5b). We plot on the same graph (Fig. 5c, d), the analytical predictions of the core front position and its speed with the numerical simulations. We have an excellent agreement between both. For the Gaussian case, which is not presented in this document, the hyperbolic tangent allows always to reproduce the front trajectory. We need only to adjust the parameters a and b from (14) to have a good agreement between the predictions and the numerical simulations.

For the LL model, (1), we focus our numerical investigations with a bi-dimensional Gaussian spatial injection and by taking into account the experimental parameters are shown in Fig. 6.

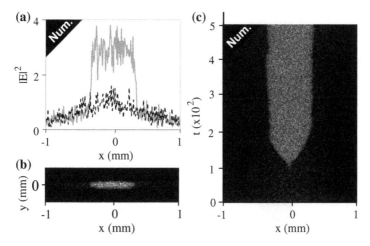

Fig. 6 Front propagation in negative diffractive inhomogeneous Kerr cavity. **a, b** Transverse cross-section of the initial (final) average localized structure in *dashed black line* (in continue *gray line*). **c, d** Spatiotemporal response to a step function of the input intensity from the lower to the upper branch of the bistable cycle. **e, f** Experimental and numerical cross section of the localized state. Light region account for high intensity of the light. **a, c, e** Experiments $I_0 = 433$ W. cm^{-2}, $d = -5$ mm, $\varphi = -0.6$ rad, $w_x = 1400\,\mu$m, $w_y = 100\,\mu$m, $R_1 = 81.8\%$; $R_2 = 81.4\%$. **b, d** Numerical simulation of LL model with $E_0 = 1.9$, $\Delta = 3.0$, $\alpha = 0.001$, $w_x = 1400\,\mu$m, $w_y = 100\,\mu$m, $\varepsilon = 0.4$

We perform numerical simulations with a asymmetric Gaussian forcing with cigar shape $(w_x \gg w_y)$

$$E_{in}(x, y) = E_0 e^{-\left(\left(\frac{x}{w_x}\right)^2 + \left(\frac{y}{w_y}\right)^2\right)}. \tag{16}$$

Furthermore, we take account the inherent fluctuations of the system by the stochastic Lugiato-Lefever model described by (1), $\varepsilon \neq 0$. We use a stochastic Runge-Kutta solver of the order of 2 with additive noise [27]. In this latter case, the temporal step (Δt) is equal to 0.01. Numerical simulations with these ingredients show a quite good agreement with the experimental observations (see Figs. 4 and 6). Hence, the analytical expression (14) can be used to figure out and to characterize the experimental front dynamics. Figure 7 depicts the experimental and numerical temporal evolutions of the front position. It clearly evidences that the expression of (14) reproduces well the front dynamics. Therefore, the effect of a spatial forcing on front propagation is to induce the front moves and stops on an asymptotic position, satisfying a hyperbolic tangent trajectory.

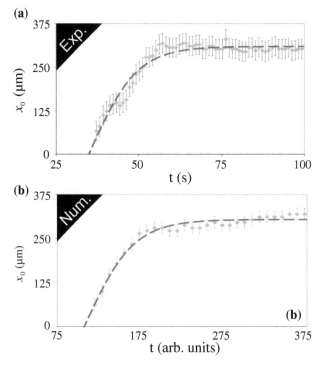

Fig. 7 Temporal evolution of front position $x_0(t)$ corresponding to spatiotemporal diagrams of Figs. 4, 5 and 6. **a** Experiment and **b** numerical simulations of the LL model, (1), used the same parameters considered in Figs. 4, 5 and 6. *Blue diamonds* location values extracted from the smoothed spatiotemporal diagrams. *Dashed red curves* best fit using expression (14). Experimental fit parameters : $a = 308\,\mu\text{m}$, $b = 0.069$, $t_0 = 35.2$ s, numerical fit parameters: $a = 306\,\mu\text{m}$, $b = 0.017$, $t_0 = 107.7$ s

4 Zero Diffraction Cavity

We have explored an interesting limit case, where the diffraction is zero. This limit case delimits the positive diffraction case where exists soliton solution [5, 24] and the negative diffraction, with front propagation [19]. It is natural to wonder which kind of solution exists in this limit. A linear stability study inform us that the two branch of the hysteresis cycle are linearly stable, contrary to the diffraction cases [20, 24]. Consequently, we expect to observe front solution, connecting the two branch of the bistable cycle.

The numerical simulations from (1) have been performed for a zero cavity length, i.e. zero diffraction, reveal the presence of the two states, notwithstanding without front propagation (cf. Fig. 8a). However the experiments realized for a zero diffraction cavity show two fronts, connecting the dark state with the bright state, with slow, asymmetric and irregular speeds (see Fig. 8c). We conclude that the Lugiato-Lefever model is not enough to model the experimental observations for zero diffraction.

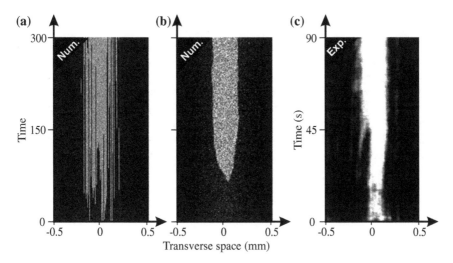

Fig. 8 Spatiotemporal diagrams: **a** 1D LL model, **b** 1D LL model + liquid crystal diffusion, **c** Experiments. Experiments $I_0 = 455$ W.cm^{-2}, $d = 0$ mm, $\varphi = -0.6$ rad, $w_x = 1400$ μm, $w_y = 100$ μm, $R_1 = 81.8$ %; $R_2 = 81.4$ %. Numerical simulation of LL model with $E_0 = 1.95$, $\Delta = 3.0$, $\alpha = 0$, $w_x = 1400$ μm, $\varepsilon = 0.4$

Close to $\alpha = 0$, the diffusion of the liquid crystal molecules cannot be neglected. Consequently, we complete the model by a second equation governing the spatial nonlinear refractive index evolution. the intracavity field E and the nonlinear refractive n are given by

$$\frac{\partial E}{\partial t} = E_{in}(x) - (1 + i\Delta)E + inE - i\,|\alpha|\,\frac{\partial^2 E}{\partial r^2} + \sqrt{\varepsilon}\xi\,(x, t)\,, \qquad (17)$$

$$n - \sigma\frac{\partial^2 n}{\partial r^2} = |E|^2\,, \qquad (18)$$

where σ is the diffusion coefficient. The stochastic numerical simulations realized with this model show a good agreement with the experimental observations, as presented on Fig. 8.

5 Conclusion

We have investigated the formation of localized structures in the Lugiato-Lefever equation in a negative diffraction regime with an inhomogeneous injected beam. We first show how to generate an equivalent left-handed Kerr material in the visible range. Experimentally, we show that the nonlinear dynamical states appearing in a focusing Kerr Fabry-Perot cavity submitted to negative optical feedback are propagating fronts in inhomogeneous medium.

We have performed a reduction of the Lugiato-Lefever equation to a simple bistable model with inhomogeneous injected beam. From this simple model we have derived a simple expression for the speed front. We have used the inhomogeneous spatial pumping in the form of Gaussian beam. The front moves and stops on a asymptotic position. The experimental trajectory of the front position under that forcing follows an hyperbolic tangent law that fully agrees with the prediction from a generic bistable imperfect pitchfork bifurcation model.

Our analysis should be applicable to all fiber resonator [25] with an inhomogeneous injected beam. In this case the coupling is provided by dispersion. When dispersion and diffraction have a comparable influence, three dimensional localized structures can be generated [28–34]. These structures consist of regular 3D lattices or localized bright light bullet traveling at the group velocity of light in the material. We plan to extend our analysis to three dimensional cavities with an inhomogeneous injected beam. In this case, localized light bullet in the negative diffraction and in anomalous dispersion may be stable.

Acknowledgments The authors acknowledge financial support by the ANR International program, project no. ANR-2010-INTB-402-02 (ANRCONICYT39), 'COLORS'. M.G. thanks for the financial support of FONDECYT project 1150507. This research was also supported in part by the Centre National de la Recherche Scientifique (CNRS), the 'Fonds Européen de Développement Economique de Régions' and by the Interuniversity Attraction 463 Poles program of the Belgian Science Policy Office, under 464 Grant No. IAP P7-35 Photonics@be.

References

1. H. Leblond, D. Mihalache, Phys. Rep. **523**, 61 (2013)
2. M. Tlidi, K. Staliunas, K. Panajotov, A.G. Vladimiorv, M. Clerc, Localized structures in dissipative media: from optics to plant ecology. Phil. Trans. R. Soc. A **372**, 20140101 (2014)
3. O. Descalzi, M. Clerc, S. Residori, G. Assanto (eds.), *Localized States in Physics: Solitons and Patterns* (Springer Science & Business Media, 2011)
4. A.J. Scroggie et al., Chaos Solitons Fractals **4**, 1323 (1994)
5. V. Odent, M. Taki, E. Louvergneaux, New J. Phys. **13**, 113026 (2011)
6. Y. Pomeau, Physica D **23**, 3 (1986); M. Tlidi, P. Mandel, R. Lefever, Phys. Rev. Lett. **73**, 640 (1994); P. Coullet, C. Riera, C. Tresser, Prog. Theor. Phys. Supp. **139**, (2000); M.G. Clerc, C. Falcon, Physica A **356**, 48 (2005); U. Bortolozzo, M.G. Clerc, C. Falcon, S. Residori, R. Rojas, Phys. Rev. Lett. **96**, 214501 (2006); M.G. Clerc, E. Tirapegui, M. Trejo, Phys. Rev. Lett. **97**, 176102 (2006); A.G. Vladimirov et al., Opt. express **14**, 1 (2006); M. Tlidi, L. Gelens, Opt. Lett. **35**, 306 (2010); A.G. Vladimirov et al., Phys. Rev. A **84**, 043848 (2011); E. Averlant, M. Tlidi, H. Thienpont, T. Ackemann, K. Panajotov, Opt. Express, **22**, 483 (2014); V. Skarka, N.B. Aleksic, M. Lekic, B.N. Aleksic, B.A. Malomed, D. Mihalache, H. Leblond. Phys. Rev. A **90**, 023845 (2014)
7. K. Staliunas, V.J. Sanchez-Morcillo, Phys. Lett. A **241**, 28 (1998); M. Tlidi, P. Mandel, R. Lefever, Phys. Rev. Lett. **81**, 979 (1998); H. Calisto, M. Clerc, R. Rojas, E. Tirapegui, Phys. Rev. Lett. **85**, 3805 (2000); M. Tlidi, P. Mandel, M. Le Berre, E. Ressayre, A. Tallet, L. Di Menza, Opt. Lett. **25**, 487 (2000); D. Gomila, P. Colet, G. L. Oppo, M. San Miguel, Phys. Rev. Lett. **87**, 194101 (2001); C. Chevallard, M. Clerc, P. Coullet, J.-M. Gilli. Europhys. Lett. **58**, 686 (2002)

8. L. Pismen, *Vortices in Nonlinear Fields: from Liquid Crystals to Superfluids, from Non-Equilibrium Patterns to Cosmic Strings* (Clarendon Press, New York, 1999)
9. J.S. Langer, Rev. Mod. Phys. **52**, 1 (1980)
10. P. Collet, J.P. Eckmann, *Instabilities and Fronts in Extended Systems* (Princeton University Press, Princeton, 1990)
11. W. van Saarloos, P.C. Hohenberg, Phys. D **56**, 303 (1992)
12. P. Coullet, Int. J. Bifurcat. Chaos **12**, 2445 (2002)
13. J.D. Murray, *Mathematical Biology*, 3de edn. (Springer, Berlin, 2003)
14. R.A. Fisher, Ann. Eugen. **7**, 355 (1937)
15. A. Kolmogorov, I. Petrovsky, N. Piskunov, Bull. Univ. Moskou Ser. Int. Se. A **1**, 1 (1937)
16. W. Van Saarloos, Phys. Rep. **386**, 29 (2003)
17. M.G. Clerc, T. Nagaya, A. Petrossian, S. Residori, C. Riera, Eur. Phys. J. D **28**, 435–445 (2004)
18. R.E. Goldstein, G.H. Gunaratne, L. Gil, P. Coullet, Phys. Rev. A **43**, 6700 (1991)
19. V. Odent, M. Tlidi, M.G. Clerc, P. Glorieux, E. Louvergneaux, Phys. Rev. A **90**, 011806(R) (2014)
20. P. Kockaert, P. Tassin, G. Van der Sande, I. Veretennicoff, M. Tlidi, Phys. Rev. A **74**, 033822 (2006)
21. P. Tassin, G.V. der Sande, N. Veretenov, P. Kockaert, I. Veretennico, M. Tlidi, Opt. Express **14**, 9338 (2006)
22. P. Tassin, L. Gelens, J. Danckaert, I. Veretennicoff, G. Van der Sande, P. Kockaert, M. Tlidi, Chaos **17**, 037116 (2007)
23. M. Tlidi, P. Kockaert, L. Gelens, Phys. Rev. A **84**, 013807 (2011)
24. L.A. Lugiato, R. Lefever, Phys. Rev. Lett. **58**, 2209 (1987)
25. S. Coen, M. Tlidi, P. Emplit, M. Haelterman, Phys. Rev. Lett. **83**, 2328 (1999)
26. G. Agez, P. Glorieux, C. Szwaj, E. Louvergneaux, Opt. Commun. **245**, 243 (2005)
27. R.L. Honeycutt, Phys. Rev. A **45**, 600 (1992)
28. M. Tlidi, M. Haelterman, P. Mandel, Quantum and semiclassical optics. J. Eur. Opt. Soc. Part B **10**, 869 (1998)
29. K. Staliunas, Phys. Rev. Lett. **81**, 81 (1998)
30. M. Tlidi, J. Opt. B: Quant. Semiclassical Opt. **2**, 438 (2000)
31. P. Tassin, G.V. der Sande, N. Veretenov, P. Kockaert, I. Veretennicoff, M. Tlidi, Opt. Express **14**, 9338 (2006)
32. N. Veretenov, M. Tlidi, Phys. Rev. A **80**, 023822 (2009)
33. M. Brambilla, T. Maggipinto, G. Patera, L. Columbo, Phys. Rev. Lett. **93**, 203901 (2004)
34. C.-Q. Dai, X.-G. Wang, G.-Q. Zhou, Phys. Rev. A **89**, 013834 (2014)

Splitting, Hatching and Transformation of the Repetition Rate in a Mode-Locked Laser

A.V. Kovalev and E.A. Viktorov

Abstract When the cavity length is varied in a Nd:YVO$_4$ laser with intracavity frequency doubling, the usual mode-locked state can split yielding two mode-locked states with different repetition rates. The two states may coexist, interact and transform from one to the other as the cavity length is varied. The transformation is accompanied by hatching of the beat note signal. Both states are well-balanced: the changes in one state manifest with equivalent amplitudes in the other state. We propose that this effect is an evidence of a bifurcation sequence phenomenon.

1 Introduction

There is a growing interest in the development of compact sources of stable radiofrequency (RF) signals, notably for global navigation satellite systems and telecommunications. Mode-locked lasers are promising devices for the synthesis of high purity microwaves in the sub-GHz and GHz range [1]. Solid-state lasers based on a Nd:YVO$_4$ gain medium with intracavity frequency doubling have previously demonstrated a reliable self-starting regime of mode-locked operation, resulting from the Kerr nonlinearity in the active medium [2, 3], a cascaded $\chi^{(2)}$ lens process and a nonlinear mirror formed by the doubling crystal and an output coupler [4, 5]. Here we report on a transition between two mode-locked states with different repetition rates and we discuss the various bifurcation scenarios which may explain this repetition rate splitting.

A.V. Kovalev · E.A. Viktorov (✉)
ITMO University, Birzhevaya Liniya 14, 199034 St Petersburg, Russia
e-mail: evviktor@ulb.ac.be

A.V. Kovalev
e-mail: avkovalev@niuitmo.ru

© Springer International Publishing Switzerland 2016   87
M. Tlidi and M.G. Clerc (eds.), *Nonlinear Dynamics: Materials,*
Theory and Experiments, Springer Proceedings in Physics 173,
DOI 10.1007/978-3-319-24871-4_6

2 Experimental Setup

The schematic layout of the experimental setup is shown in Fig. 1. The laser active medium was Nd:YVO$_4$ based with intracavity frequency doubling in a KTP crystal and a linear cavity configuration. The cavity length was 109 mm, corresponding to a 1.38 GHz free spectral range. The active element was a right-angle a-cut $1 \times 3 \times 3$ mm^3 1 % at. doped Nd:YVO$_4$ crystal. The KTP crystal was cut for type II second harmonic generation (SHG) in the x-y plane with cut angles $\theta = 90°$ and $\phi = 23.5°$ with dimensions $5 \times 3 \times 3$ mm^3. In order to ensure type II SHG, the active element was rotated so the angle between its c-axis and the KTP crystal z-axis was 45°.

High reflection coatings for 1064 and 532 nm were applied to the end cavity mirror and the rear facet of the active element. Antireflection coatings for 1064 and 532 nm were applied to the front facet of the active element and both facets of the KTP crystal. A flat-spherical output coupler was high reflection coated for 1064 nm radiation and designed for 50 % transmission at 532 nm. It had a 150 mm radius of curvature and was mounted on a ring piezoelectric transducer (PZT) made from modified zirconate-lead titanate. The PZT's susceptibility was 0.8 μm/V. In order to change the cavity length a controlled voltage source was applied to the PZT's contacts.

A commercial 808 nm laser diode was used as a pump source. Its output was directed to the AE rear facet through a forming optical system consisting of two cylindrical fiber lenses. By overlapping the pump waist and the resonator eigenmode, single fundamental transverse mode operation was established. The pump diode power supply was a current source with a variable output current up to 800 mA with an accuracy of 0.5 mA.

Two temperature servo loops were implemented in order to maintain the temperatures of the pump diode and the KTP by means of thermoelectric coolers. The servos enabled a 0.1 °C temperature precision. The input to the servo loops was provided by compact analog temperature sensors with continuous output. Both the KTP and LD

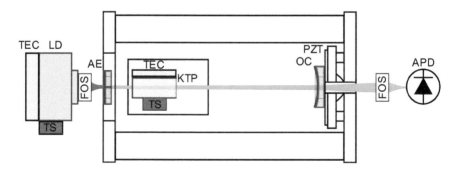

Fig. 1 Experimental setup: *TEC* thermoelectric coolers; *LD* laser diode; *TS* temperature sensor; *FOS* forming optical systems; *AE* Nd:YVO$_4$ active element; *KTP* KTP crystal; *OC* output coupler; *PZT* piezoelectric transducer; *APD* avalanche photodiode

temperatures were set to 25 °C. A low thermal expansion material (invar) was used for the active medium plate, the output coupler holders and the cavity frame. The cavity was covered by an aluminum case and mounted on a 20 mm thick aluminum plate, which was placed on an optical table. The cavity length stability corresponded to an Allan deviation of 4.6×10^{-6} over an observation time of 100 s.

The diverging output laser radiation passed through a telescope in order to shrink the beam and irradiate an avalanche photodiode with bandwidth 1.5 GHz. The photodiode was connected to an electronic spectrum analyzer used to detect the beat note signal. The radiofrequency signal spectra were captured from the ESA by means of a digital capturing system for further analysis.

3 Experimental Results

The slope efficiencies for the fundamental and second harmonic outputs were 1.43 and 9.89 % respectively (left panel in Fig. 2). The lasing threshold was 90 mW of pump power. A stable mode-locked operation involving about 10 cavity modes at 1064 nm was achieved by pumping the Nd:YVO$_4$ with 380 mW, resulting in 28.6 mW average output power at 532 nm.

To examine the laser performance we recorded the radio frequency (RF) spectrum. The mode-locked regime at both harmonics manifested itself as an ultra-narrow beat signal at 1.379 GHz with an RF spectrum linewidth less than 300 Hz FWHM at a 300 Hz resolution bandwidth and 1.3 s sweep time, for both fundamental and second harmonic radiation. The TEM$_{00}$ output was linearly polarized (right panel in Fig. 2).

The repetition rate splitting and transformation occurred at 400 mW pump power. The evolution of the RF signal with cavity length change via the PZT is shown in Fig. 3a. The spectral envelope approximation was implemented in order to track the evolution of the RF parameters (maximum spectral position, maximum spectral intensity and full width at half maximum (FWHM)). The results of the approximation

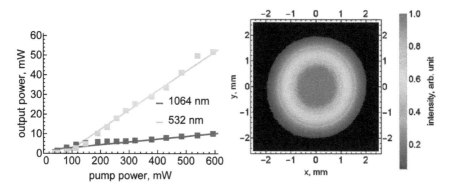

Fig. 2 Laser output characteristics: (*left*) average output power versus pump power for fundamental (*red*) and second (*green*) harmonics; (*right*) beam intensity distribution profile in the far field

(a)   **(b)**

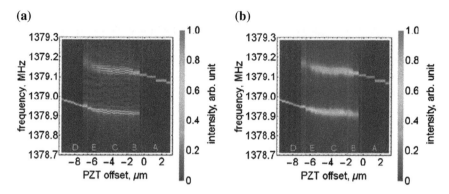

Fig. 3 Cavity length resolved RF spectrum (**a**) and spectral envelope approximation (**b**). An offset is added to the cavity length by the PZT voltage. Resolution bandwidth 3 kHz, sweep time 82.5 ms. The letters denote five stages of the transformation process

are shown in Fig. 3b and are in excellent qualitative agreement with the experimental measurements in Fig. 3a.

Let us follow the transformation of the RF signal with decreasing cavity length. We distinguish five different stages of the transformation process as indicated by the letters in the cavity length resolved RF spectrum (Fig. 3). At the initial cavity length (stage A—no PZT offset), the RF signal at 1379.1 MHz had a narrow ∼300 Hz linewidth (Fig. 4, in blue, Fig. 5A) and corresponded to a "higher" repetition rate mode-locked state (HML).

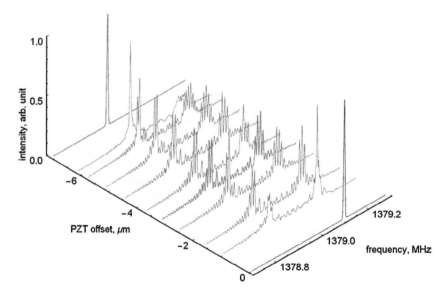

Fig. 4 Snapshots of the RF beat note signal for different PZT offsets, including: 0 μm (*blue*); −3.17 μm (*magenta*); −7.08 μm (*red*) (color online)

As the cavity length was decreased (PZT offset $\sim 1\,\mu$m), the position of the HML line slightly blue-shifted as expected (see Fig. 3), but a much broader, small amplitude RF line appeared at ~ 1378.9 MHz. The appearance of the RF signal at lower frequency cannot be explained by the cavity length decrease and instead represents a repetition rate splitting. As we further decreased the cavity length, the lower repetition rate line became narrower, and grew in amplitude. The initial HML RF line became broader, hatched and its amplitude decreased (Fig. 5B).

At a PZT offset of $-3.17\,\mu$m (as shown by the magenta line in Fig. 4), the two RF lines were equally broad with the same amplitude, and featured a slow hatching modulation (Fig. 5C). This indicates the coexistence of two mode-locked states at the higher (HML) and lower (LML) repetition rates. The slow modulation hatching of the RF lines is not regular and depends on the transformation stage.

With further cavity length decrease (stage D), the LML RF line became narrow, dominant, and finally, solitary at a PZT offset of $-7.08\,\mu$m (stage E). This completed the transformation of the HML state into the LML state.

While both RF lines demonstrated the conventional blue shifts with cavity length decrease, the difference between the two beat signals remained largely the same at 166.5 kHz for the full range of the cavity length change. The maximum dispersion-related frequency shift is ~ 10 kHz and cannot, therefore, explain the effect of the splitting.

Let us now analyze the transformation process using the parameters extracted from the RF spectra. The evolution of the parameters versus PZT offset is shown in Fig. 6 and confirms a five-stage character of the transformation.

The evolution of the maximum spectral position is remarkably similar for both the HML state and the LML state (Fig. 6 (top)); in fact they are virtually identical.

Both states show gradual blue shifts with cavity length decrease in stages A, B, D and E, but the maximum spectral position remains unchanged in stage C.

The evolution of the maximum spectral intensity in Fig. 6 (middle) shows a similar invariability in stage C, but is very different in stages B and D. There, the maximum spectral intensity of the HML state sharply decreases while the LML state shows a sharp increase of the amplitude at these stages. The process is therefore reciprocal and antisymmetric as shown in Fig. 6 (middle).

The evolution of the FWHM as shown in Fig. 6 (bottom) further confirms the interchange of the transformation process. Both RF lines are identically narrow at the beginning (HML state, stage A) and at the end (LML state, stage E) of the transformation process. In stages B and D the linewidths are either sharply increasing (HML state) or sharply decreasing (LML state). Both linewidths are nearly identical in stage C.

Remarkably, the slow modulation frequency of the RF lines hatching demonstrates an evolution that can also be related to the same stages of the transformation process. It is shown in Fig. 7. There is no hatching in the initial (HML) and final (LML) states. Similar to the other spectral parameters in Fig. 6, the hatching frequency sharply changes in stages B and D, but remains relatively unchanged in stage C. We will discuss the transformation process in the next section.

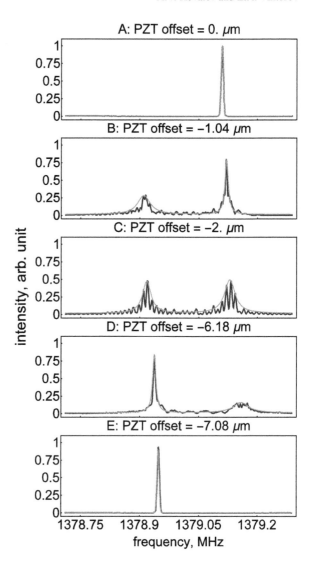

Fig. 5 RF spectra and spectral envelopes from the five stages of the transformation process

4 Discussion

This splitting of the repetition rate and the transformation of the HML state into the LML state with cavity length decrease has not been reported before to the best of our knowledge. The frequency splitting effect in actively mode-locked fiber laser reported in [6] originates from an alternation of the pulse timing in the repetition-rate doubled regime and is therefore very different.

Conventional analysis of the passively mode-locked operation considers the appearance of the pulse train to be the result of a Hopf bifurcation. The pulse train

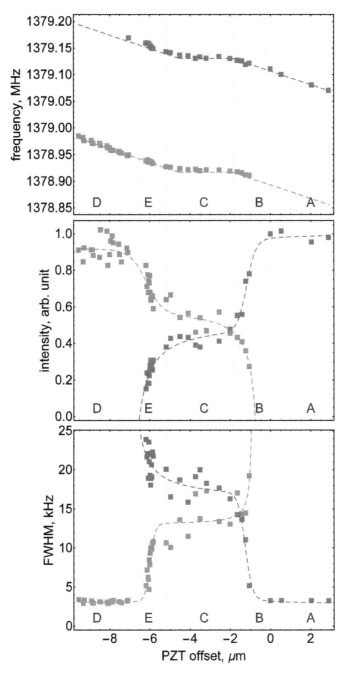

Fig. 6 The parameters extracted from RF spectra (3 kHz resolution bandwidth) at the five stages of the transformation process: maximum spectral position (*top*); maximum spectral intensity (*middle*); FWHM (*bottom*). Magenta (*orange*) corresponds to the mode-locked state at higher (*lower*) repetition rate (color online)

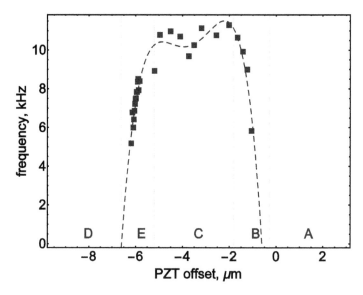

Fig. 7 The hatching frequency extracted from RF spectra in the five stages of the transformation process

normally corresponds to a single branch in the bifurcation analysis. It has been very recently reported that a bistability consisting of two branches of mode-locked operation with slightly different repetition rates can be formed. The two branches coexist and collide with a resultant important impact on the timing jitter [7]. This bistability has not been observed in our experiments. We propose that the observed splitting of the repetition rate results from a secondary Hopf bifurcation with the two Hopf bifurcations being close (or possibly degenerate).

A more complicated scenario [8] includes a sequence of bifurcations in which the limit cycle bifurcates to a 2-torus and then undergoes an inverse bifurcation to a limit cycle, transverse to the original limit cycle. The fine details supporting this scenario cannot be revealed in our experiment, but the five-stage transformation process is suggestive of a sequence of bifurcations. Both stages of the fast transformations (B and D) are relatively narrow in terms of the parameter range, but are unlikely to indicate a subcritical character of the bifurcation as we did not observe any bistability between the mode-locked states. The slow transformation stage C is slightly asymmetric. It starts at the point where the maximum spectral intensity and the FWHM are equal for both mode-locked states (Fig. 6), and the hatching frequency is the highest (Fig. 7). The transformation in this stage relates to a slow change of the RF spectra parameters, but ends with no visible critical point before the fast transformation stage D.

5 Conclusions

In conclusion, we find that in a passively mode-locked Nd:YVO$_4$ laser with intra-cavity frequency doubling, operation may consist of two mode-locked states with different repetition rates. The two states may coexist, interact, and transform from one to the other. An offset was added to the cavity length by a PZT voltage in order to distinguish between the two states and revealed a five-stage transformation process. We propose that the effect is evidence of a bifurcation sequence phenomenon. The effect may prove to be important in the design of mode-locked lasers in several applications.

References

1. J.J. McFerran, E.N. Ivanov, A. Bartels, G. Wilpers, C.W. Oates, S.A. Diddams, L. Hollberg, Low-noise synthesis of microwave signals from an optical source. Electron. Lett. **41**, 650–651 (2005)
2. H.C. Liang, R.C.C. Chen, Y.J. Huang, K.W. Su, Y.F. Chen, Compact efficient multi-GHz Kerr-lens mode-locked diode-pumped Nd:YVO$_4$ laser. Opt. Exp. **16**, 21149–21154 (2008)
3. H.C. Liang, Y.J. Huang, W.C. Huang, K.W. Su, Y.F. Chen, High-power, diode-end-pumped, multigigahertz self-mode-locked Nd:YVO$_4$ laser at 1342 nm. Opt. Lett. **35**, 4–6 (2010)
4. H. Iliev, D. Chuchumishev, I. Buchvarov, V. Petrov, Passive mode-locking of a diode-pumped Nd:YVO$_4$ laser by intracavity SHG in PPKTP. Opt. Express **18**, 5754–5762 (2010)
5. V. Aleksandrov, T. Grigorova, H. Iliev, A. Trifonov, I.C. Buchvarov, $\chi^{(2)}$-lens mode-locking of a high average power Nd:YVO$_4$ Laser. Conference on Lasers and Electro-Optics, SM4F.3 (2014)
6. R. Kiyan, O. Deparis, O. Pottiez, P. Megret, M. Blondel, Frequency splitting in repetition-rate doubled rational harmonic actively mode-locked pulse train. Conference on Lasers and Electro-Optics, 247–248 (2001)
7. A. Pimenov, T. Habruseva, D. Rachinskii, S.P. Hegarty, G. Huyet, A.G. Vladimirov, Effect of dynamical instability on timing jitter in passively mode-locked quantum-dot lasers. Opt. Lett. **39**, 6815–6818 (2014)
8. J. Simonet, E. Brun, R. Badii, Transition to chaos in a laser system with delayed feedback. Phys. Rev. E **52**, 2294–2301 (1995)

Spontaneous Symmetry Breaking in Nonlinear Systems: An Overview and a Simple Model

Boris A. Malomed

Abstract The paper combines two topics belonging to the general theme of the spontaneous symmetry breaking (SSB) in systems including two basic competing ingredients: the self-focusing cubic nonlinearity and a double-well-potential (DWP) structure. Such systems find diverse physical realizations, chiefly in optical waveguides, made of a nonlinear material and featuring a transverse DWP structure, and in models of atomic BEC with attractive inter-atomic interactions, loaded into a pair of symmetric potential wells coupled by tunneling across the separating barrier. With the increase of the nonlinearity strength, the SSB occurs at a critical value of the strength. The first part of the paper offers a brief overview of the topic. The second part presents a model which is designed as the simplest one capable to produce the SSB phenomenology in the one-dimensional geometry. The model is based on the DWP built as an infinitely deep potential box, which is split into two wells by a delta-functional barrier at the central point. Approximate analytical predictions for the SSB are produced for two limit cases: strong (deep) or weak (shallow) splitting of the potential box by the central barrier. Critical values of the strength of the nonlinearity at the SSB point, represented by the norm of the stationary wave field, are found in both cases (the critical strength is small in the former case, and large in the latter one). For the intermediate case, a less accurate variational approximation (VA) is developed.

1 Introduction and an Overview of the Topic

Properties of collective excitations in physical systems are determined by the interplay of several fundamental ingredients, viz., spatial dimension, external potential acting on the corresponding physical fields (or wave functions), the number of independent components of the fields, the underlying dispersion relation for linear

B.A. Malomed (✉)
Department of Physical Electronics, School of Electrical Engineering,
Faculty of Engineering, Tel Aviv University, 69978 Tel Aviv, Israel
e-mail: malomed@post.tau.ac.il

© Springer International Publishing Switzerland 2016
M. Tlidi and M.G. Clerc (eds.), *Nonlinear Dynamics: Materials,
Theory and Experiments*, Springer Proceedings in Physics 173,
DOI 10.1007/978-3-319-24871-4_7

excitations, and, finally, the character of the nonlinear interactions of the fields. In particular, the shape of the external potentials determines the system's symmetry, two most common types of which correspond to periodic (alias *lattice*) potentials and double-well potentials (DWPs), the latter featuring the symmetry between two wells separated by a potential barrier. The well are coupled by the tunneling of fields across the barrier, which is an essentially linear effect.

One of fundamental principles of quantum mechanics (that, by itself, is a strictly linear theory) is that the ground state (GS) of the quantum system exactly follows the symmetry of the potential applied to the system. On the other hand, excited states may realize other representations of the same symmetry [1]. In particular, the GS wave function for a quantum particle trapped in the one-dimensional DWP is symmetric, i.e., even, with respect to the double-well structure, while the first excited state always features the opposite parity, being represented by an antisymmetric (spatially odd) function. A similarly feature of Bloch wave functions supported by periodic potentials is that the state at the bottom of the corresponding lowest Bloch band features the same periodicity, while generic Bloch functions are quasi-periodic ones, with the quasi-periodicity determined by the quasi-momentum of the excited states.

While the quantum-mechanical Schrödinger equation is linear for the single particle, the description of ultracold rarefied gases formed by bosonic particles (i.e., the Bose-Einstein condensate, BECs) is provided by the Gross-Pitaevskii equation (GPE), which, in the framework of the mean-field approximation, takes into regard effects of collisions between the particles, by means of a cubic term added, to the Schrödinger equation for the single-particle wave function [2]. The repulsive or attractive forces between the colliding particles are accounted for by, respectively, the self-defocusing, alias self-repulsive, or self-focusing, i.e., self-attractive, cubic term in the GPE. Similarly, the nonlinear Schrödinger equation (NLSE) with the self-focusing or defocusing cubic term (alias the Kerr or anti-Kerr one, respectively) models the transmission of electromagnetic waves in nonlinear optical media [3].

As well as their linear counterparts, the GPE and NLSE include external potentials, which often feature the DWP symmetry. However, the symmetry of the GS in models with the self-focusing nonlinearity (i.e., the state minimizing the energy at a fixed number of particles in the bosonic gas, or fixed total power of the optical beam in the photonic medium—in both cases, these are represented by a fixed norm of the respective wave function) follows the symmetry of the underlying potential structure only as long as the nonlinearity remains weak enough. A generic effect, which sets in with the increase of the strength of the nonlinearity, i.e., effectively, with the increase of the norm, is *spontaneous symmetry breaking* (SSB). In its simplest form, the SSB in terms of the BEC implies that the probability to find the boson in one well of the trapping DWP structure is larger than in the other. This, incidentally, implies that another basic principle of quantum mechanics, according to which the GS cannot be degenerate, is no longer valid either in nonlinear models of the quantum origin, such as the GPE: obviously, the SSB which takes place in the presence of the DWP gives rise to a degenerate pair of two mutually symmetric ground states, with the maximum of the wave function observed in either potential well. In terms of optics, the SSB makes the light power trapped in either core of the DWP-shaped dual-core waveguide larger than in the mate core. Thus, the SSB is a fundamental effect common to diverse models of the quantum and classical origin alike, which combine the wave propagation, nonlinear self-focusing, and symmetric trapping potentials.

It should be stressed that the same nonlinear system with the DWP potential always admits a symmetric state coexisting with the asymmetric ones; however, past the onset of the SSB, the symmetric state no longer represents the GS, being unstable against small symmetry-breaking fluctuations. Accordingly, in the course of the spontaneous transition from the unstable symmetric state to a stable asymmetric one, the choice between the two mutually degenerate asymmetric states is governed by small perturbations, which "push" the self-attractive system to place, at random, the maximum of the wave function in the left or right potential well.

In systems with the self-defocusing nonlinearity, the ground state is always symmetric and stable. In this case, the SSB manifests itself in the form of the spontaneous breaking of the *antisymmetry* of the first excited state (the spatially odd one, which has exactly one zero of the wave function, at the central point, in the one-dimensional geometry). The state with the spontaneously broken antisymmetry also features a zero, which is shifted from the central position to the left or right, the sign of the shift being randomly selected by initial perturbations.

Historically speaking, the SSB concept for nonlinear systems of the NLSE type was, probably, first proposed by Davies in 1979 [4], although in a rather abstract mathematical form. In that work, a nonlinear extension of the Schrödinger equation for a pair of quantum particles, interacting via a three-dimensional isotropic potential, was addressed, and the SSB was predicted in the form of the spontaneous breaking of the GS rotational symmetry. Another early prediction of the SSB was reported in the *self-trapping model*, which is based on a system of linearly coupled ordinary differential equations with self-attractive cubic terms [5] . The latter publication had brought the concept of the SSB to the attention of the broad research community.

An important contribution to theoretical studies of the SSB was made by work [6], which addressed this effect in the model for the propagation of CW (continuous-wave) optical beams in dual-core nonlinear optical fibers (alias nonlinear directional couplers), with the underlying symmetry between the linearly coupled cores . In the scaled form, the corresponding system of propagation equations for CW amplitudes u_1 and u_2 in the two cores is

$$i\frac{du_1}{dz} + f\left(|u_1|^2\right)u_1 + \kappa u_2 = 0,$$

$$i\frac{du_2}{dz} + f\left(|u_2|^2\right)u_2 + \kappa u_1 = 0,$$

(1)

(in this case, "CW" implies that the amplitudes do not depend on the temporal variable), where z is the propagation distance, κ the coefficient accounting for the inter-core linear coupling through the mutual penetration of evanescent fields from each core into the mate one, and $f\left(|u_{1,2}|^2\right)$ is a function of the intensity of the light

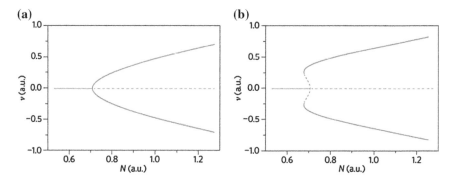

Fig. 1 Diagrams for standard supercritical (**a**) and subcritical (**b**) spontaneous-symmetry-breaking bifurcations, as per [28]. *Continuous* and *dashed lines* depict, respectively, stable and unstable solution branches. Total norm N and asymmetry parameter v (see (3)) are shown in arbitrary units (a.u.)

in each core which represents its intrinsic nonlinearity. In the simplest case of the Kerr (cubic) self-focusing nonlinearity, which corresponds to

$$f\left(u^2\right) = |u|^2, \tag{2}$$

this system gives rise to the symmetry-breaking *bifurcation* of the *supercritical* type [7]. This bifurcation destabilizes the symmetric state and, simultaneously, gives rise to a pair of stable asymmetric ones, which are mirror images of each other, corresponding to interchange $u_1 \leftrightarrow u_2$, as shown in Fig. 1a. In the figure, the asymmetry and the total norm, which characterizes the strength of the nonlinearity, are defined as

$$v \equiv \left(|u_1|^2 - |u_2|^2\right) / \left(|u_1|^2 + |u_2|^2\right), \quad N \equiv \left(|u_1|^2 + |u_2|^2\right). \tag{3}$$

On the other hand, the *saturable* nonlinearity, in the form of $f\left(|u|^2\right) = |u|^2 / (I_0 + |u|^2)$, where $I_0 > 0$ is a constant which determines the intensity-saturation level, gives rise to a *subcritical* symmetry-breaking bifurcation. In the latter case, the branches of asymmetric states, which originate at the point of the stability loss of the symmetric mode, originally evolve *backward* (in terms of the total power, $|u_1|^2 + |u_2|^2$), being unstable, and then turn forward, getting stable at the turning point, see Fig. 1b. This SSB scenario implies that the pair of stable asymmetric states emerge *subcritically*, at a value of the total power smaller than the one at which the symmetric mode becomes unstable. In terms of statistical physics, the super- and subcritical bifurcations may be classified as phase transitions of the second and first kinds, respectively.

The next step in the studies of the SSB phenomenology in models of dual-core nonlinear optical fibers and similar systems was the consideration of the fields depending on the temporal variable, τ. In that case, assuming the anomalous sign of

the group-velocity dispersion in each core of the fiber, equations (1) are replaced by
a system of NLSEs with the linear coupling:

$$i\frac{\partial u_1}{\partial z} + \frac{1}{2}\frac{\partial^2 u_1}{\partial \tau^2} + f\left(|u_1|^2\right)u_1 + \kappa u_2 = 0,$$

(4)

$$i\frac{\partial u_2}{\partial z} + \frac{1}{2}\frac{\partial^2 u_2}{\partial \tau^2} + f\left(|u_2|^2\right)u_2 + \kappa u_1 = 0.$$

The same system, with variable τ replaced by transverse coordinate x, models the
spatial-domain evolution of electromagnetic fields in dual-core planar waveguides,
in which case the second derivatives represent the paraxial diffraction, instead of the
group-velocity dispersion.

The uncoupled NLSEs with the Kerr self-focusing nonlinearity (2) give rise to
commonly known solitons [3]. The corresponding SSB bifurcation may destabilize
obvious symmetric soliton solutions of system (4),

$$u_1 = u_2 = \eta \operatorname{sech}(\eta\tau)\exp\left(\left(\frac{1}{2}\eta^2 + \kappa\right)z\right),$$

(5)

where η is an arbitrary real amplitude of the soliton. The bifurcation replaces the
symmetric soliton mode (5) by asymmetric two-component modes. The critical value
of the soliton's peak power, η^2, at which the SSB instability of the symmetric solitons
sets in under the action of the Kerr nonlinearity, was found in an exact form, $\eta^2_{\text{crit}} =
4/3$, in [8]. The transition to asymmetric solitons, following the instability onset, was
first predicted, by means of the variational approximation, in [9, 10]. Then, it was
found that, on the contrary to the supercritical bifurcation of the CW states in system
(1) with the Kerr self-focusing nonlinearity, the SSB bifurcation of the symmetric
soliton in system (4) is subcritical [11, 12].

An independent line of the analysis of the SSB had originated from the studies of
GPE-based models for atomic BECs trapped in DWP structures. The scaled form of
the corresponding GPE for the mean-field wave function, $\psi(x,t)$, is

$$i\frac{\partial \psi}{\partial t} = -\frac{1}{2}\frac{\partial^2 \psi_1}{\partial x^2} - g\,|\psi|^2\,\psi + U(x)\psi,$$

(6)

where $g < 0$ and $g > 0$ correspond to the repulsive and attractive collision-induced
nonlinearity, respectively. The DWP can be taken, for instance, as

$$U(x) = U_0\left(x^2 - a^2\right)^2,$$

(7)

with positive constants U_0 and a^2.

The GPE (6) can be reduced to the two-mode system, similar to the system of
(1) (with z replaced by t), by means of the tight-binding approximation [13], which

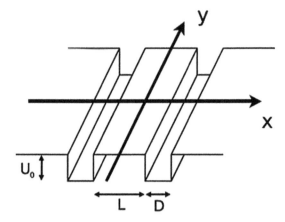

Fig. 2 An example of the quasi-one-dimensional double-trough potential, built of two parallel potential toughs with the rectangular profile, as per [17]

adopts $\psi(x, t)$ in the form of a superposition of two stationary wave functions, ϕ, corresponding to the states trapped separately in the two potential wells, centered at $x = \pm a$:

$$\psi(x, t) = u_1(t)\phi(x - a) + u_2\phi(x + a). \tag{8}$$

In particular, this approximation implies that the nonlinearity acts on each amplitude u_1 and u_2 also separately, while the coupling between them is linear.

The analysis of the SSB in the models based on the GPE (6) was initiated in works [14, 15]. Most often, the BEC nonlinearity (on the contrary to the self-focusing Kerr terms in optics) is self-repulsive, which, as mentioned above, gives rise to the spontaneous breaking of the antisymmetry of the odd states, with $\psi(-x) = -\psi(x)$, while the GS remains symmetric. Further, the GPE may be extended by adding an extra spatial coordinate, on which the DWP does not depend, thus giving rise to a two-dimensional GPE with a quasi-one-dimensional *double-trough potential*, which is displayed in Fig. 2. In the latter case, the self-attractive nonlinearity (which, although being less typical in BEC, is possible too) creates bright matter-wave solitons, which self-trap in the free direction [16]. Accordingly, bright symmetric solitons are possible in the double-trough potential, and they are replaced, via a subcritical SSB bifurcation, by stable asymmetric ones at a critical value of the total norm of the mean-field wave function (which determines the effective strength of the self-attractive nonlinearity) [17].

The above discussion was focused on static symmetric and asymmetric modes in nonlinear systems featuring the DWP structure. The analysis of dynamical regimes, usually in the form of oscillations of the norm of the wave function between two wells of the DWP, i.e., roughly speaking, between the two equivalent asymmetric states existing above the SSB point, has been developed too. Following the straightforward analogy with Josephson oscillations of the electron wave function in tunnel-coupled superconductors [18, 19], the possibility of matter-wave oscillations in *bosonic Josephson junctions* was predicted [20].

Similar to the situation in many other areas of nonlinear science, the variety of theoretically predicted results concerning the SSB phenomenology by far exceeds the number of experimental findings. Nevertheless, some manifestations of the SSB have been reported in experiments. In particular, the self-trapping of a macroscopically asymmetric state of the atomic condensate of ^{87}Rb atoms with repulsive interactions between them, loaded into the DWP (which may be considered, as mentioned above, as a spontaneous breaking of the antisymmetry of the lowest excited state, above the symmetric GS) and Josephson oscillations in the same system, were reported in work [21]. On the other hand, the SSB of laser beams coupled into an effective transverse DWP created in a self-focusing photorefractive medium (where the nonlinearity is saturable, rather than strictly cubic) has been demonstrated in work [22]. Still another experimental observation of the SSB effect in nonlinear optics was the spontaneously established asymmetric regime of operation of a symmetric pair of coupled lasers [23]. More recently, symmetry breaking was experimentally demonstrated in a symmetric pair of nanolaser cavities embedded into a photonic crystal [24] (although the latter system is a dissipative one, hence its model is essentially different from those outlined above, cf. work [25], where the SSB effect for dissipative solitons was formulated in terms of linearly-coupled complex Ginzburg-Landau equations with the cubic-quintic nonlinearity). An observation of a related effect of the spontaneous breaking of the chiral symmetry in metamaterials was reported too [26].

Many results for the SSB phenomenology and related Josephson oscillations, chiefly theoretical ones, but also experimental, obtained in various areas of physics (nonlinear optics and plasmonics, BECs, superconductivity, and others) are represented by a collection of articles published in topical volume [27].

2 A Simple Model for the Spontaneous Symmetry Breaking (SSB) in a Double-Well Potential (DWP)

2.1 Formulation of the Model

The objective of this section is to introduce what may be the simplest model which admits the SSB in a system combining the self-attractive nonlinearity and a DWP structure. In a sketchy form, the model was mentioned in work [28], but it was not elaborated there. The account given here is not complete either, as only approximate analytical results are included. A full presentation, completed by relevant numerical results, will be given elsewhere.

The model is schematically shown in Fig. 3. It is built as an infinitely deep potential box, with the DWP structure created by means of the delta-functional barrier created in the center, cf. the cross section of the double-trough potential displayed in Fig. 2. The respective scaled form of the GPE is given by equation (6) with $g > 0$ and $U(x) = \varepsilon\delta(x)$, where the delta-functional potential corresponds to $U_b \to \infty$, $a \to 0$, while the strength of the barrier, $\varepsilon \equiv U_b a$, is kept fixed. The edges of the potential

Fig. 3 A sketch of the double-well-potential (DWP) structure under the consideration (as per paper [28]): an infinitely deep potential box ($U_0 \to \infty$), of width $L \equiv 1$, is split in the middle by a narrow tall barrier, $\varepsilon\delta(x)$, see equation (10). Even and odd wave functions of the ground and first excited states, in the absence of the spontaneous symmetry breaking, are shown by the *continuous* and *dashed curves*, respectively

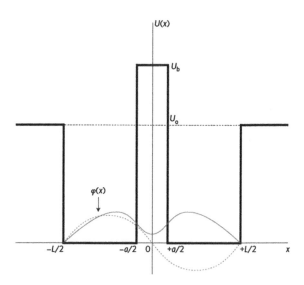

box at points $x = \pm 1/2$ are represented by the boundary conditions (b.c.)

$$\psi\left(x = \pm\frac{1}{2}\right) = 0. \qquad (9)$$

Stationary states with chemical potential μ are looked for as $\psi(x, t) = e^{-i\mu t}\phi(x)$, with real function $\phi(x)$ obeying the following stationary equation with the respective b.c.:

$$\mu\phi = -\frac{1}{2}\frac{d^2\phi}{dx^2} - g\phi^3 + \varepsilon\delta(x)\phi, \quad \phi\left(x = \pm\frac{1}{2}\right) = 0. \qquad (10)$$

The delta-functional barrier at $x = 0$ implies that $\phi(x)$ is continuous at this point, while its derivative features a jump:

$$\frac{d\phi}{dx}\Big|_{x=+0} - \frac{d\phi}{dx}\Big|_{x=-0} = 2\varepsilon\phi(x = 0). \qquad (11)$$

It is possible to fix $g \equiv 1$ in (6) and (10) by means of scaling, but it is more convenient, for the sake of the subsequent analysis, to keep $g > 0$ as a free parameter. The strength of the nonlinearity is determined by product gN, where the total norm of the wave function is defined as the sum of the norms trapped in the left and right potential wells (cf. (3)):

$$N = \left(\int_{-1/2}^{0} + \int_{0}^{+1/2}\right)\phi^2(x)dx \equiv N_- + N_+, \qquad (12)$$

Before proceeding to the analysis of the SSB in the nonlinear model, it is relevant to briefly discuss its linear counterpart, with $g = 0$ in (10). Spatially symmetric (even) solutions of this equation are looked for as

$$\phi_{\text{even}}^{(\text{lin})}(x) = A \sin \left(\sqrt{2\mu} \left(\frac{1}{2} - |x| \right) \right),$$ (13)

where A is an arbitrary amplitude, and eigenvalue μ is determined by the equation following from the jump condition (11):

$$\tan \left(\sqrt{\mu/2} \right) = -\sqrt{2\mu}/\varepsilon.$$ (14)

It is easy to see that, with the increase of ε from 0 to ∞, the lowest eigenvalue μ_0, corresponding to the GS of the linear model, monotonously grows from $\mu_0 \, (\varepsilon = 0) = \pi^2/2$ to

$$\mu_0 \, (\varepsilon = \infty) = 2\pi^2.$$ (15)

Similarly, the eigenvalue of the first excited symmetric state, $\mu_2 \, (\varepsilon)$, monotonously grows from to $\mu_2 \, (\varepsilon = 0) = 9\pi^2/2$ to $\mu_2 \, (\varepsilon = \infty) = 8\pi^2$. Located between eigenvalues μ_0 and μ_2, is $\mu_1 = 2\pi^2$, which corresponds to the lowest excited state, i.e., the first antisymmetric (spatially odd) eigenfunction, $\phi_{\text{odd}}^{(\text{lin})}(x) = A \sin \left(\sqrt{2\mu_1} x \right)$. Naturally, μ_1 coincides with the limit value (15) of μ_0, and it does not depend on ε, as the odd eigenfunction vanishes at $x = 0$, where the δ-function is placed.

2.2 An Analytical Solution for the SSB Point in the Strongly-Split DWP (Large ε)

The main objective of the analysis is to predict the critical norm which gives rise to the SSB, through the competition between the self-focusing, which favors the spontaneous accumulation of the wave function in one well, and the linear coupling between the wells, which tends to distribute the wave function evenly between them. An approximate analytical solution to this problem can be obtained in the case of weakly coupled potential wells, which corresponds to large ε (a very tall central barrier). In this case, weak nonlinearity, i.e., a small amplitude of the wave function, is sufficient to induce the SSB. In turn, the small amplitude implies that solutions to (10) vanishing at $x = \pm 1/2$ are close to eigenmodes (13) of the linearized version of the same equation, i.e., the approximate solutions may be sought for as

$$\phi(x) = A_\pm \sin \left(k_\pm \left(\frac{1}{2} - |x| \right) \right),$$ (16)

where signs $\pm$ pertain to the regions of $x < 0$ and $x > 0$, respectively, $k_\pm$ being appropriate wavenumbers. The substitution of *ansatz* (16) into the condition of the continuity of the wave function at $x = 0$, and the jump condition (11) for the first derivative, yields the following relations between amplitudes $A_\pm$ and the wavenumbers:

$$A_+ \sin\left(\frac{1}{2}k_+\right) = A_- \sin\left(\frac{1}{2}k_-\right),\tag{17}$$

$$A_- k_- \cos\left(\frac{1}{2}k_-\right) - A_+ k_+ \cos\left(\frac{1}{2}k_+\right) = 4\varepsilon A_\pm \sin\left(\frac{1}{2}k_\pm\right).\tag{18}$$

Further, in the same small-amplitude limit, the cubic term in (10) may be approximated as follows, neglecting the third harmonic contained in it:

$$\left[A_\pm \sin\left(k_\pm\left(\frac{1}{2} - |x|\right)\right)\right]^3 \approx \frac{3}{4} A_\pm^3 \sin\left(k_\pm\left(\frac{1}{2} - |x|\right)\right),\tag{19}$$

which, in turn, implies an effective shift of the chemical potential in equation (10) and determines the corresponding wavenumbers in equation (16):

$$k_\pm = \sqrt{2\left(\mu + \frac{3}{4}g A_\pm^2\right)}.\tag{20}$$

In the limit of $\varepsilon \to \infty$, wave functions (16) must vanish at $x = 0$, hence the respective GS corresponds to $k_\pm = 2\pi$, i.e., to the above-mentioned value (15) of the chemical potential. In the same limit, the norm (12) of the GS is

$$N = \left(A_-^2 + A_+^2\right)/4.\tag{21}$$

At large but finite ε, the GS chemical potential is sought for as

$$\mu = 2\pi^2 - \delta\mu, \text{ with } \delta\mu \ll 2\pi^2.\tag{22}$$

Next, the substitution of this expression into equation (20), expanding it for small $\delta\mu$ and $A_\pm^2$, and inserting the result into equations (17) and (18) leads to equations which take a relatively simple form at the point of the onset of the SSB bifurcation, i.e., in the limit of $A_+ - A_- \to 0$ (the vanishingly small factor $(A_+ - A_-)$ then factorizes out and cancels in the expanded version of (17)):

$$N_{cr} = \frac{8\pi^2}{3g}\varepsilon^{-1}, \ \left(A_\pm^2\right)_{cr} = 2N_{cr}, \ \delta\mu = 12\pi^2\varepsilon^{-1}.\tag{23}$$

This result was mentioned, without the derivation, in [28]. Thus, as expected, the critical value of the norm at the SSB point decays ($\sim \varepsilon^{-1}$) with the increase of ε. The substitution of $\delta\mu$ from (23) into (22) suggests that this asymptotic solution is actually valid for $\varepsilon \gg 6$.

2.3 An Analytical Solution for the SSB in the Weakly-Split DWP (Small ε): The Soliton Approximation

The case of small ε, opposite to that considered above, implies that the central barrier splitting the confined box into the two potential wells is weak, hence the effective coupling between the wells is strong. According to the general principles of the SSB theory [27], strong nonlinearity, i.e., large norm N, is necessary to compete with the strong coupling. Large N, in turn, implies that the wave field self-traps into a narrow NLSE soliton [3],

$$\phi_{\text{sol}}(x - \xi) = \frac{1}{2}\sqrt{g}N\,\text{sech}\left(\frac{g}{2}N\,(x - \xi)\right), \tag{24}$$

where ξ is the coordinate of the soliton's center, the respective chemical potential is

$$\mu_{\text{sol}} = -(gN)^2/8. \tag{25}$$

and it is assumed that N is large enough to make the soliton's width much smaller than the size of the box ($L \equiv 1$ in Fig. 3), i.e.,

$$gN \gg 1. \tag{26}$$

The soliton is repelled from the edges of the potential box. In compliance with the b.c. in (10), this may be interpreted as the repulsive interaction with two *ghost solitons* generated as mirror images (with opposite signs) of the real one (24) with respect to the edges of the box:

$$\phi_{\text{ghost}}(x) = -\sqrt{g}\,(N/2)\left[\text{sech}\left(\frac{g}{2}N\left(x - \frac{1}{2} + \xi\right)\right) + \text{sech}\left(\frac{g}{2}N\,(x + 1 + \xi)\right)\right]. \tag{27}$$

The potential of the interaction between two far separated NLSE solitons is well known [30–34]. In the present case, the sum of the two interaction potentials, corresponding to the pair of the ghosts, amounts to the following effective potential accounting for the repulsion of the real soliton from edges of the box in which it is confined:

$$U_{\text{box}}(\xi) = g^2 N^3 \exp\left(-\frac{g}{2}N\right)\cosh\,(Ng\xi). \tag{28}$$

On the other hand, the soliton is repelled by the central barrier, the respective potential being [33]

$$U_{\text{barrier}}(\xi) = \varepsilon \phi_{\text{sol}}^2 (\xi = 0) = \frac{\varepsilon g}{4} N^2 \text{sech}^2 \left(\frac{g}{2} N \xi \right),
\qquad (29)$$

where the latter expression was obtained neglecting the deformation of the soliton's shape. A straightforward analysis of the total effective potential, $U(\xi) = U_{\text{box}}(\xi) + U_{\text{barrier}}(\xi)$, demonstrates that the position of the soliton placed at $\xi = 0$, which represents the symmetric state in the present case, is stable, i.e., it corresponds to a minimum of the net potential, at

$$8gN \exp \left(-\frac{g}{2} N \right) > \varepsilon,
\qquad (30)$$

or, in other words, at

$$N > N_{\text{cr}} \approx (2/g) \ln (16/\varepsilon)
\qquad (31)$$

(the underlying assumption that ε is small was used to derive equation (31) from (30)). With the increase of N, the SSB bifurcation takes place at $N = N_{\text{cr}}$, when the potential minimum at $\xi = 0$ turns into a local maximum. At $0 < (N - N_{\text{cr}})/N_{\text{cr}} \ll 1$, the center of the soliton spontaneously shifts to either of two asymmetric positions, which correspond to a pair of emerging potential minima at $\xi \neq 0$:

$$\xi = \pm \sqrt{(N - N_{\text{cr}})/(gN_{\text{cr}}^2)}.
\qquad (32)$$

The latter result explicitly describes the SSB bifurcation of the supercritical type, which occurs in the present setting.

2.4 The Variational Approximation for the SSB in the Generic DWP

A possibility to develop a more comprehensive, albeit coarser, analytical approximation for solutions of (10) in the generic case (when the strength of the splitting barrier, ε, is neither specifically large nor small) is suggested by the variational approximation (VA) [29]. To this end, note that (10) can be derived from the Lagrangian,

$$L = \int_{-1/2}^{+1/2} \mathscr{L}(x) dx, \quad \mathscr{L} = \frac{1}{2} \left(\frac{d\phi}{dx} \right)^2 - \mu \phi^2 - \frac{g}{2} \phi^4 + \varepsilon \delta(x) \phi^2.
\qquad (33)$$

Aiming to apply the VA for detecting the SSB onset point, one can adopt the following *ansatz* for the GS wave function:

$$\phi(x) = a\cos(\pi x) + b\sin(2\pi x) + c\cos(3\pi x), \tag{34}$$

with each term satisfying b.c. in (10). Real constants a, c, and b must be predicted by the VA. The SSB is accounted for by terms $\sim b$ in ansatz (34), hence the onset of the SSB is heralded by the emergence of a solution with infinitesimal b, branching off from from the symmetric solution with $b = 0$, similar to how the onset of the SSB bifurcation in terms of ansatz (16) is signaled by the emergence of infinitesimal $(A_+ - A_-)$ in the above analysis.

The total norm (12) of ansatz (34) is $N = (1/2)\left(a^2 + c^2 + b^2\right)$, while its asymmetry at $b \neq 0$ may be quantified by

$$\Theta \equiv \frac{N_+ - N_-}{N} = \frac{16}{15\pi}\frac{b\,(5a - 3c)}{a^2 + c^2 + b^2}. \tag{35}$$

A straightforward consideration demonstrates that, for all values of a, b, and c, expression (35) is subject to constraint $|\Theta| < 1$, as it must be. When $b = 0$, the theorem that the spatially symmetric GS cannot have nodes, i.e., $\phi(x) \neq 0$ at $|x| < 1/2$, if applied to ansatz (34), easily amounts to the following constraint:

$$-1 < c/a < 1/3. \tag{36}$$

Further, the substitution of ansatz (34) into Lagrangian (33) yields

$$L = \left(\frac{1}{4}\pi^2 - \frac{1}{2}\mu + \varepsilon\right)a^2 + \left(\pi^2 - \frac{1}{2}\mu\right)b^2 + \left(\frac{9}{4}\pi^2 - \frac{1}{2}\mu + \varepsilon\right)c^2 + 2\varepsilon ac$$
$$-\frac{g}{4}\left(\frac{3}{4}a^4 + a^3c + 3a^2b^2 + 3a^2c^2 - 3ab^2c + \frac{3}{4}b^4 + 3b^2c^2 + \frac{3}{4}c^4\right), \tag{37}$$

from which three variational equations follow:

$$\partial L/\partial(b^2) = 0, \tag{38}$$

$$\partial L/\partial a = \partial L/\partial c = 0. \tag{39}$$

In the general form, these equations are rather cumbersome. However, being interested in the threshold at which the SSB sets in, one may set $b = 0$ (after performing the differentiation with respect to b^2 in (38)), which lead to the following system of three equations for three unknowns a, b, and μ:

$$2\pi^2 - \mu = \frac{3g}{2}\left(a^2 - ac + c^2\right), \tag{40}$$

$$\left(\frac{1}{2}\pi^2 - \mu + 2\varepsilon\right)a + 2\varepsilon c - \frac{g}{4}\left(3a^3 + 3a^2c + 6ac^2\right) = 0, \tag{41}$$

$$\left(\frac{9}{2}\pi^2 - \mu + 2\varepsilon\right)c + 2\varepsilon a - \frac{g}{4}\left(a^3 + 6a^2c + 3c^3\right) = 0. \tag{42}$$

In particular, equations (41) and (42) with $g = 0$, while (40) is dropped, offer an additional application: they predict the chemical potential of the GS in the linear system ($g = 0$), as the value at which the determinant of the linearized version of (41) and (42) for a and c vanishes:

$$\mu_0 = \frac{5}{2}\pi^2 + 2\varepsilon - 2\sqrt{\pi^4 + \varepsilon^2} \tag{43}$$

(recall that $\mu_0(\varepsilon)$ could not be found above in an exact form). The latter approximation is meaningful once it yields the chemical potential of the GS smaller than the above-mentioned exact value $2\pi^2$ corresponding to the lowest excited state. This condition holds at

$$\varepsilon < (15/8)\pi^2 \approx 18.5. \tag{44}$$

Further, it is easy to check that equations (40)–(42) yield no physical solutions at $\varepsilon = 0$, in agreement with the obvious fact that the SSB does not occur when the central barrier is absent, i.e., the potential is not split into two wells.

Finally, a particular exact solution of (40)–(42) (which includes a particular value of ε) can be found by setting $c = 0$, i.e., assuming that the third harmonic vanishes in ansatz (34):

$$a^2 = 3\pi^2/(2g),\ \mu = -\pi^2/4,\ \varepsilon = 3\pi^2/16 \approx 1.85. \tag{45}$$

A noteworthy feature of this particular solution is that it has $\mu < 0$. Indeed, (10) suggests that a sufficiently strong nonlinear term should make the chemical potential negative, as corroborated by equation (25).

A consistent analysis of the VA for the present model, and its comparison with numerical results will be reported elsewhere.

3 Conclusion

The objective of this paper was two-fold. First, a short overview was given of the general topic of the SSB (spontaneous symmetry breaking) in nonlinear one-dimensional models featuring the competition of the self-focusing cubic nonlinearity and DWP (double-well-potential) structure. Physically relevant examples of such systems are offered by nonlinear optical waveguides with the transverse DWP structure, and by BEC trapped in two symmetric potential wells coupled by tunneling of atoms. The SSB occurs at a critical value of the nonlinearity strength, i.e., of the field's norm (which is tantamount to the total power, in the case of the trapped optical beam).

The second part of the paper reported a particular model, which is the simplest one capable to grasp the SSB phenomenology: an infinitely deep potential box, split into two wells by a delta-functional barrier set at the center. Approximate analytical results predicting the SSB point have been presented for two limit cases, viz., the strong or weak splitting of the potential box by the central barrier. In both cases, critical values of the norm at the SSB point have been found, being, respectively, small and large. For the generic (intermediate) case, a coarser approach based on the VA (variational approximation) has been developed. The detailed analysis of the VA and comparison of the predictions with numerical results will be reported separately.

References

1. D. Landau, E.M. Lifshitz, *Quantum Mechanics* (Nauka Publishers, Moscow, 1974)
2. S. Giorgini, L.P. Pitaevskii, S. Stringari, Rev. Mod. Phys. **80**, 1215 (2008); H.T.C. Stoof, K.B. Gubbels, D.B.M. Dickrsheid, *Ultracold Quantum Fields* (Springer, Dordrecht, 2009)
3. Y.S. Kivshar, G.P. Agrawal, *Optical Solitons: From Fibers to Photonic Crystals* (Academic Press, San Diego, 2003)
4. E.B. Davies, Symmetry breaking in a non-linear Schrödinger equation. Commun. Math. Phys. **64**, 191–210 (1979)
5. J.C. Eilbeck, P.S. Lomdahl, A.C. Scott, The discrete self-trapping equation. Physica D **16**, 318–338 (1985)
6. A.W. Snyder, D.J. Mitchell, L. Poladian, D.R. Rowland, Y. Chen, Physics of nonlinear fiber couplers. J. Opt. Soc. Am. B **8**, 2101–2118 (1991)
7. G. Iooss, D.D. Joseph, *Elementary Stability Bifurcation Theory* (Springer, New York, 1980)
8. E.M. Wright, G.I. Stegeman, S. Wabnitz, Solitary-wave decay and symmetry-breaking instabilities in two-mode fibers. Phys. Rev. A **40**, 4455–4466 (1989)
9. C. Paré, M. Fłorjańczyk, Approximate model of soliton dynamics in all-optical couplers. Phys. Rev. A **41**, 6287–6295 (1990)
10. A.I. Maimistov, Propagation of a light pulse in nonlinear tunnel-coupled optical waveguides. Kvant. Elektron. **18**, 758–761 [Sov. J. Quantum Electron. **21**, 687–690 (1991)]
11. N. Akhmediev, A. Ankiewicz, Novel soliton states and bifurcation phenomena in nonlinear fiber couplers. Phys. Rev. Lett. **70**, 2395–2398 (1993)
12. B.A. Malomed, I. Skinner, P.L. Chu, G.D. Peng, Symmetric and asymmetric solitons in twincore nonlinear optical fibers. Phys. Rev. E **53**, 4084 (1996)
13. G.L. Alfimov, P.G. Kevrekidis, V.V. Konotop, M. Salerno, Wannier functions analysis of the nonlinear Schrödinger equation with a periodic potential. Phys. Rev. E **66**, 046608 (2002)
14. G.J. Milburn, J. Corney, E.M. Wright, D.F. Walls, Quantum dynamics of an atomic Bose-Einstein condensate in a double-well potential. Phys. Rev. A **55**, 4318–4324 (1997)
15. A. Smerzi, S. Fantoni, S. Giovanazzi, S.R. Shenoy, Quantum coherent atomic tunneling between two trapped Bose-Einstein condensates. Phys. Rev. Lett. **79**, 4950–4953 (1997)
16. V.M. Pérez-García, H. Michinel, H. Herrero, Bose-Einstein solitons in highly asymmetric traps. Phys. Rev. A **57**, 3837–3842 (1998)
17. M. Matuszewski, B.A. Malomed, M. Trippenbach, Spontaneous symmetry breaking of solitons trapped in a double-channel potential. Phys. Rev. A **75**, 063621 (2007)

18. G. Schön, A.D. Zaikin, Quantum coherent effects, phase transitions, and the dissipative dynamics of ultra small tunnel junctions. Phys. Rep. **198**, 237–412 (1990)
19. A.V. Ustinov, Solitons in Josephson junctions. Physica D **123**, 315–329 (1998)
20. S. Raghavan, A. Smerzi, S. Fantoni, S.R. Shenoy, Coherent oscillations between two weakly coupled Bose-Einstein condensates: Josephson effects, π oscillations, and macroscopic quantum self-trapping. Phys. Rev. A **59**, 620–633 (1999)
21. M. Albiez, R. Gati, J. Fölling, S. Hunsmann, M. Cristiani, M.K. Oberthaler, Direct observation of tunneling and nonlinear self-trapping in a single bosonic Josephson junction. Phys. Rev. Lett. **95**, 010402 (2005)
22. P.G. Kevrekidis, Z. Chen, B.A. Malomed, D.J. Frantzeskakis, M.I. Weinstein, Spontaneous symmetry breaking in photonic lattices: theory and experiment. Phys. Lett. A **340**, 275–280 (2005)
23. T. Heil, I. Fischer, W. Elsässer, J. Mulet, C.R. Mirasso, Chaos synchronization and spontaneous symmetry-breaking in symmetrically delay-coupled semiconductor lasers. Phys. Rev. Lett. **86**, 795–798 (2000)
24. P. Hamel, S. Haddadi, F. Raineri, P. Monnier, G. Beaudoin, I. Sagnes, A. Levenson, A.M. Yacomotti, Spontaneous mirror-symmetry breaking in coupled photonic-crystal nanolasers. Nat. Photonics **9**, 311–315 (2015)
25. A. Sigler, B.A. Malomed, Solitary pulses in linearly coupled cubic-quintic Ginzburg-Landau equations. Physica D **212**, 305–316 (2005)
26. M. Liu, D.A. Powell, I.V. Shadrivov, M. Lapine, Y.S. Kivshar, Spontaneous chiral symmetry breaking in metamaterials. Nat. Commun. **5**, 4441 (2014)
27. B.A. Malomed (ed.), *Spontaneous Symmetry Breaking, Self-Trapping, and Josephson Oscillations* (Springer, Berlin, 2013)
28. B.A. Malomed, Symmetry breaking in laser cavities. Nat. Photonics **9**, 287–289 (2015)
29. B.A. Malomed, Variational methods in nonlinear fiber optics and related fields. Prog. Opt. **43**, 71–193 (2002)
30. V.I. Karpman, V.V. Solov'ev, A perturbation approach to the 2-soliton systems. Physica D **3**, 487–502 (1981)
31. J.P. Gordon, Interaction forces among solitons in optical fibers. Opt. Lett. **8**, 596–598 (1983)
32. F.M. Mitschke, L.F. Mollenauer, Experimental observation of interaction forces between solitons in optical fibers. Opt. Lett. **12**, 355–357 (1987)
33. Y.S. Kivshar, B.A. Malomed, Dynamics of solitons in nearly integrable systems. Rev. Mod. Phys. **61**, 763–915 (1989)
34. B.A. Malomed, Potential of interaction between two- and three-dimensional solitons. Phys. Rev. E **58**, 7928–7933 (1998)

Experimental Spatiotemporal Chaotic Textures in a Liquid Crystal Light Valve with Optical Feedback

Marcel G. Clerc, Gregorio González-Cortés and Mario Wilson

Abstract Macroscopic systems subjected to external forcing exhibit complex spatiotemporal behaviors as result of dissipative self-organization. Based on a nematic liquid crystal layer with spatially modulated input beam and optical feedback, we set up a two-dimensional pattern forming system which exhibits a transition from stationary to spatiotemporal chaotic patterns. Using an adequate projection of spatiotemporal diagrams, we determine the largest Lyapunov exponent. This exponent allows us to characterize the transition presented by this system. This exponent and Fourier transforms lead to a reconciliation of experimental observations of spatiotemporal complexity and theoretical developments.

1 Introduction

Macroscopic systems maintained far from equilibrium through the injection and dissipation of energy and matter exhibit spatially coherent structures, *patterns*, which are ubiquitous in Nature [1–6], these structures appear as a way to optimize energy transport and momenta [2–5]. Patterns are the result of the interplay between the linear gain and the nonlinear saturation mechanisms. In many physical systems, these structures are stationary and emerge as a spatial instability of a uniform state when a control parameter is changed and surpasses a critical value, which usually corresponds to an imbalance of forces. Thus, these bifurcations correspond to spontaneous

M.G. Clerc · M. Wilson (✉)
DFI, Facultad de Ciencias Físicas y Matemáticas, Universidad de Chile,
Blanco Encalada 2008, Santiago, Chile
e-mail: mario.wilson@ing.uchile.cl

M.G. Clerc
e-mail: marcelclerc@gmail.com

G. González-Cortés
Departamento de Física, Facultad de Ciencias, Universidad de Chile,
Las Palmeras 3425, Ñuñoa, Santiago, Chile
e-mail: gregorio.gonzalez@ug.uchile.cl

© Springer International Publishing Switzerland 2016   113
M. Tlidi and M.G. Clerc (eds.), *Nonlinear Dynamics: Materials,*
Theory and Experiments, Springer Proceedings in Physics 173,
DOI 10.1007/978-3-319-24871-4_8

symmetry breaking. As the parameters of the system are changed, stationary patterns can become unstable and bifurcate to more complex patterns, even into aperiodic dynamics states [4, 7–9]. This behavior is characterized by complex spatiotemporal dynamics exhibited by the pattern and a continuous coupling between spatial modes in time. Complex spatiotemporal dynamics of patterns have been observed, for example, in fluids [10–12], chemical reaction-diffusion systems [13], cardiac fibrillation [14], electroconvection [15], fluidized granular matter [16], nonlinear optical cavity [17] and in a liquid crystal light valve [18], to mention a few. In most of these studies, complex behavior is characterized by spatial and temporal Fourier transforms, wave vector distribution, filtering spatiotemporal diagrams, power spectrum of spatial mode, length distributions, Poincaré maps and number of defects as a function of the parameters. However, in these experimental studies spatiotemporal complexity is not characterized by rigorous tools of dynamical system theory as Lyapunov exponents [19]. These exponents characterize the exponential sensitivity of the dynamical behaviors under study and in turn gives a characteristic time scale on which one has the ability to predict the time evolution of the system. Hence, when the largest Lyapunov exponent is positive (negative) the system under study is chaotic (stationary). Indeed, from experimental data in spatially extended systems, it is a thorny task to infer the value of Lyapunov exponents.

In this chapter we report the experimental study of the spatiotemporal chaotic dynamics of two-dimensional patterns. Based on a nematic liquid crystal layer with spatially modulated input beam and optical feedback, we set up a two-dimensional pattern forming system which exhibit a transition from stationary to spatiotemporal chaotic pattern. By means of using an adequate projection of spatiotemporal diagrams, we calculate the largest Lyapunov exponent. This exponent allows us to characterize the bifurcation diagram displayed by this system.

2 Experimental Setup: Liquid Crystal Light Valve with Optical Feedback

A flexible experimental setup that exhibits pattern formation in nonlinear optics is the liquid crystal light valve with optical feedback [20]. Particularly, in this setup has been reported regular patterns as hexagonal, stripe and zig-zag structures, super lattices, localized structures, dislocations, disclinations, spiral states, domain walls between patterns, front propagation, and quasi-crystals (see review [20] and references therein). This setup contains a liquid crystal light valve (LCLV) inserted in an optical feedback system (see Fig. 1). The LCLV is composed of a nematic liquid crystal film sandwiched in between a glass and a photoconductive plate over which a dielectric mirror is deposed. The liquid crystal film (LC) is planar aligned (nematic director **n** parallel to the walls), with a thickness $d = 15\,\mu$m. The liquid crystal filling our LCLV is a nematic LC-654, produced by NIOPIK (Moscow) [21]. It is a mixture of cyano-biphenyls, with a positive dielectric anisotropy $\Delta\varepsilon = \varepsilon_\parallel - \varepsilon_\perp = 10.7$ and

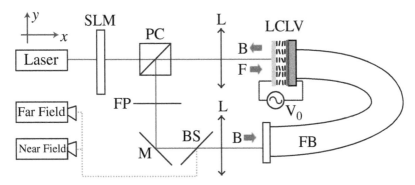

Fig. 1 Schematic representation of the experimental setup. LCLV stands for the liquid crystal light valve, L stands for two lenses with a focal distance $f = 25$ cm, M is a mirror, FB is an optical fiber bundle, PC is a polarizing cube, BS represents a beam splitter, FP indicates the Fourier plane, and SLM is a spatial light modulator driven by a computer. F and B represent the forward and the backward beam, respectively

large optical birefringence, $\Delta n = n_{\parallel} - n_{\perp} = 0.2$, where $\varepsilon_{\parallel}$ and $\varepsilon_{\perp}$ are the dielectric permittivities $\parallel$ and $\perp$ to $\mathbf{n}$, respectively, and $n_{\parallel}$ and are $n_{\perp}$ are the extraordinary ($\parallel$ to $\mathbf{n}$) and ordinary ($\perp$ to $\mathbf{n}$) refractive index [22]. Transparent electrodes over the glass plates permit the application of an electrical voltage across the liquid crystal layer. The photoconductor behaves like a variable resistance, which decreases for increasing illumination. The feedback is obtained by sending back onto the photoconductor, through a fiber bundle (FB), the light which has passed through the liquid crystal layer and has been reflected by the dielectric mirror. This light beam experiences a phase shift which depends on the liquid crystal orientation and, on its turn, modulates the effective voltage that locally applies to the liquid crystal sample.

Transparent electrodes deposited over the cell walls allow an external voltage application V_0 across the LC layer. Molecules tend to orient along the direction of applied electric field, which—on its turn—changes local and dynamically by following the illumination spatial distribution present in the photoconductor wall of the cell. When LC molecules reorient, due to their birefringent nature, they induce a refractive index change. Thus, the LCLV acts as a manageable Kerr medium, causing a phase variation $\phi = 2kd\,\Delta n \cos^2 \theta$ in the reflected beam proportional to the intensity of the beam I_w incoming on the photoconductive side, where θ is the average of molecular reorientation. Here, $k = 2\pi/\lambda$ is the optical wave number. A schematic picture of the performed experiment is depicted in Fig. 1.

The LCLV is illuminated by an expanded He-Ne laser beam, $\lambda = 633$ nm, with 3 cm transverse diameter and power $I_{in} = 6.5$ mW/cm^2, linearly polarized along the vertical axis. Once shone into the LCLV, the beam is reflected by the dielectric mirror deposed on the rear part of the cell and, thus, sent to the polarizing cube (PC), which due to the phase-change the light has suffered in the reflection, will send the light into the feedback loop. In order to close the feedback loop, a mirror (M) and an optical fiber bundle (FB) are used, these elements assure the light to reach the back (photoconductor) part of the valve. The PC introduces polarization interference between the ordinary and extraordinary waves propagating in the LC layer, ensuring bistability between differently orientated states of the LC molecules. In the feedback loop, a 4-f array is placed in order to obtain a self-imaging configuration and access to the Fourier plane (FP), this array is constructed with 2 identical lenses (L) with focal length $f = 25$ cm placed in such a way that both sides of the LCLV are conjugated planes. Thanks to this configuration the free propagation length in the feedback loop can be easily adjusted. For the performed experiments a length of $d = -4$ cm was used. A spatial light modulator (SLM) was placed in the input beam optical path with a 1 : 1 imaging between the SLM and the frontal part of the LCLV. The SLM is a LC display with 1 inch diagonal size and 1024×768 pixels resolution, each pixel is coded in 8 bits of intensity level and the whole system is controlled through an external computer. With the aid of a specialized software a square mask was produced and sent to the SLM, which acts as a programmable filter able to impose arbitrary border conditions to the input beam. For a uniform mask of 160 gray-value, the typical input intensity would be $I_w = 0.83$ mW/cm^2. To obtain the shape used in the experiments, a two-dimensional mask, $I(x, y)$, was produced with rectangular or square shape. The system dynamics is controlled by adjusting the external voltage V_0 applied to the LCLV.

In brief, the dynamics exhibited by LCLV with optical feedback, is that the liquid crystal molecular orientation change the phase of light emerging from the LCLV, which in turn—optical feedback—induces a voltage that reorients the liquid crystal molecules. Therefore, thanks to the optical circuit, the liquid crystal molecular orientation self-induces a non local spatiotemporal dynamics.

3 Experimental Observation: From Stationary to Disordered Patterns

The presented dynamics in the LCLV has been explored using an intensity mask of zero-level intensity everywhere except for a central square part with length 2.5 mm. The intensity coming out of the SLM is spatially modulated as $I_{in} = I_0(x, y) + b_0$, where both I_0 and b_0 can be controlled by changing the mask created in the SLM, where $\{x, y\}$ are the transverse coordinates of the sample, I_0 is measured when imposing a square uniform 2-D mask with side a_0 a given gray-value, that is,

$$I_0(x, y) = \begin{cases} I_0 & |x| < a_0, |y| < a_0 \\ b_0 & else \end{cases}$$

and b_0 is constant throughout the sample. In the presented configuration $I_0 = 0.9$ mW/cm^2 and $b_0 = 0.1$ mW/cm^2. This average intensity was constant along all the conducted experiments. As was mentioned in the former section, the response of the LC molecules depends on I_0 and V_0, which are the natural control

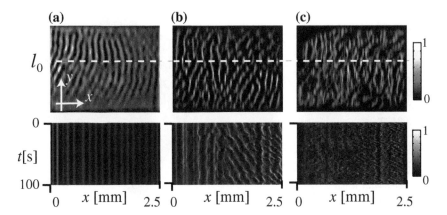

Fig. 2 Spatial texture in liquid crystal light valve with optical feedback with different tensions. *Top* panels correspond to snapshots of the patterns obtained with the CCD camera. *Bottom* panels account for the projected spatiotemporal diagrams of observed dynamics. This projection process corresponds to consider spatiotemporal evolution of an arbitrary line of the pattern. **a** periodic regime at $V_0 = 3.2V_{rms}$, **b** quasi-periodic dynamics at $V_0 = 4.7V_{rms}$, and **c** chaotic behavior at $V_0 = 3.8V_{rms}$

parameters of LCLV with optical feedback. Due that I_0 is kept constant, the main control parameter is V_0; this parameter varies between 3 and 5 V_{rms}.

The minimum range of V_0 has been fixed, starting with the appearance of stationary patterns. For different V_0 values, the dynamical behavior obtained in the LCLV was recorded with a charge-coupled device (CCD) camera. Figure 2a shows a typical snapshot of observed pattern and its respective spatiotemporal evolution. This evolution is characterized by projected spatiotemporal diagram, which is constructed by picking arbitrarily one line—transversal to the rolls direction—in the illuminated zone and superposing it as time evolves. From this figure, we can infer that the system exhibits stationary stripe patterns. These patterns are a consequence that in the Fourier plane (cf. FP in Fig. 1), through a slit, we can filter modes and break the rotational symmetry. In this way, we forced the system just to present stripe patterns in a given direction.

Increasing the voltage V_0, the pattern begins to oscillate in a complex manner. Figure 2b shows a representative snapshot of observed pattern and its respective spatiotemporal evolution. We observe in the projected spatiotemporal diagram locally waves, oscillations and spatiotemporal dislocations. A similar dynamic has been reported in one-dimensional inhomogeneous spatial non-equilibrium systems [23–25]. Where waves are understood as a result of the gradients generated by inhomogeneities. The spatiotemporal dislocations as consequence of a combination of a phase instability and an advection process caused by an inhomogeneous drift force. In our experimental system these inhomogeneities and drift force can be caused by the inherent imperfections of the experimental system and inhomogeneities induced by the filter in the Fourier plane. Therefore this kind of dynamical behavior is expected.

Fig. 3 Experimental
Spatiotemporal diagram of
liquid crystal light valve with
optical feedback. **a** 3-D
spatiotemporal diagram
constructed with collected
data taken at $V_0 = 4.3V_{rms}$.
The spatial and temporal
dimensions are
2.5×2.5 mm^2 and 70 s,
respectively. **b** and **c**
Projected spatiotemporal
diagrams extracted from (**a**)
reconstructed with y and x
spatial coordinates,
respectively

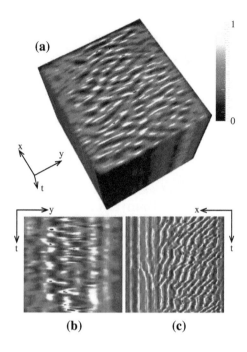

Notwithstanding, in two spatial dimensions, there is no theory that explains the type of observed dynamics. From a dynamic point of view, the complex dynamics observed by this pattern is constantly repeated over time. Which leads us to infer that this kind of behavior could be of quasi-periodic type [4].

Further increasing the voltage V_0, the dynamics exhibited by the pattern becomes more and more complex. Figure 2c shows a representative snapshot of observed spatiotemporal pattern and its respective spatiotemporal evolution. Clearly in the projected spatiotemporal diagram, we detected that there is an intermittent behavior. That is, the pattern exhibits aperiodic oscillations invaded by large fluctuations, generating several spatial and temporal dislocations. Likewise, the system exhibits a spatiotemporal highly complex disorder. Which one usually associates spatiotemporal chaos of patterns [26–28]. To illustrate the complex dynamics displayed by the system, in Fig. 3 is shown the spatiotemporal diagram observed in the chaotic region. For each V_0 value the dynamics was recorded, after a 5 min transient, during 50 min at 5 frames per second, obtaining 15,000 frames per value. In order to have a better behavior appreciation different projection spatiotemporal diagrams were constructed. Figure 3 shows an example taken at $V_0 = 4.3V_{rms}$; a 3-D spatiotemporal diagram is shown, with two projected spatiotemporal diagram for each spatial coordinate. This is a characteristic property of phase turbulence of modes.

As an additional characterization tools support—the Fourier spectra—were calculated showing that, the dynamics change between stripe patterns, quasi periodicity and spatiotemporal chaotic textures. Figure 4 shows the Fourier spectra of 3 different regimes. The stationary pattern is characterized by a dominant frequency f. The

Fig. 4 Fourier spectra for 3
different voltage values. In
light gray (*red*) the Fourier
transform of a stationary
pattern, in gray (*blue*) a quasi
periodic one, it can be
observed that new
incommensurable
frequencies appear (f', f'',
f''' and f''''). In *black*, the
Fourier spectrum of a chaotic
texture, it can be observed
that the spectrum broadens

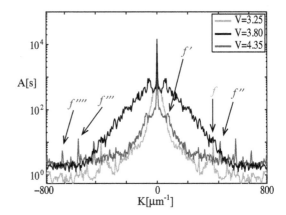

width of this peak is due to temperature fluctuations and dynamics of defects such as
dislocations and boundary grains. The quasi periodic texture is characterized by the
appearance of inconmensurable frequencies, f', with respect to the main frequency
and its harmonics. The spatiotemporal chaotic texture is characterized by presenting
spectrum enlargement as a result of the interaction between the main incommensurable modes [29].

In the next section, based on tools of dynamical system theory, we will characterize rigorously the experimental transition from stationary to spatiotemporal chaotic
textures.

4 Characterization of Observed Dynamics

4.1 Lyapunov Exponents

A characterization of complex dynamics like chaos and spatiotemporal chaos can be
done by means of Lyapunov exponents [19]. These exponents measure the growth
rate of generic small perturbations around a given trajectory in finite-dimensional
dynamical systems. Hence, the Lyapunov exponents characterize the linear response
around a given trajectory. There are as many exponents as the dimension of the
system under study. Additional information about the complexity of the system could
be obtained from the exponents, for instance, the dimension of the strange attractor
(spectral dimensionality) or dynamic disorder measures (entropy) [30].

In the theoretical framework, the analytical study of Lyapunov exponents is a
titanic endeavor and in practice inaccessible, then the logical strategy is a numerical
derivation of the exponents. To study numerically the Lyapunov exponents is necessary to discretize the differential equation. From experimental data, in the case of
low-dimensional dynamical systems, by means of recognition of initial conditions

one can determine the largest Lyapunov exponent (LLE). This exponent accounts for the greatest exponential growth and it defined by

$$\lambda_0 = \lim_{t \to \infty} \lim_{\Delta_0 \to 0} \frac{1}{t} \ln \left[\frac{||u(x,t) - u'(x,t)||}{||u(x,t_o) - u'(x,t_o)||} \right], \tag{1}$$

where $u(x,t)$ and $u'(x,t)$ are given fields, $\Delta_o \equiv ||u(x,t_0) - u'(x,t_0)||$ and $||f(x,t)||^2 \equiv \int |f(x,t)|^2 dx$ is a norm. $\Delta(t) \equiv ||u(x,t) - u'(x,t)||$ stands for the global evolution of the difference between the fields.

When λ_0 is positive or negative, the perturbation of a given trajectory is characterized by an exponential separation or approach, respectively. Hence, attractors such as stationary patterns or uniform equilibria are characterized by negative λ_0. Conversely, complex behaviors such as chaos and spatiotemporal chaos will exhibit positive λ_0. Dynamical behaviors with zero largest Lyapunov exponent correspond to equilibria with invariant directions, such as periodic or quasi-periodic solutions and non-chaotic attractors characterized with polynomial growth rate [30]. Therefore, the largest Lyapunov exponent is an exceptional order parameter for characterizing transitions between stationary and complex spatiotemporal dynamics.

4.2 Experimental Measurements of Largest Lyapunov Exponent

To show if the system under study presents chaos the largest Lyapunov exponent must be estimated. To seed light into this problem the projected spatiotemporal diagrams are specially helpful. Experimentally, to estimate the LLE it is mandatory to have two close initial conditions (two fields initially very close) and observe if their evolution diverge at large times. The implemented method needs, as a first step, to find two close fields (see lines 1 and 2 in Fig. 5a, b) along the projected spatiotemporal diagrams and compute their difference Δ_0. The temporal evolution of the difference should be given by $\Delta(t) \approx \Delta_0 e^{\lambda t}$ for large t (cf Fig. 5c). Where λ_0 stands for the largest Lyapunov exponent. Due to the complexity of evolution of the difference between fields—clearly the number of positive Lyapunov exponents are numerous—we will consider at least two unstable growth directions, that is $\Delta(t) \approx ae^{bt} + ce^{dt}$. Figure 5 shows the implemented method used to estimate largest Lyapunov exponent.

4.3 Transition Between Stationary to Spatiotemporal Textures

To study the route to spatiotemporal complexity exhibited by the spatial texture in the liquid crystal light valve with optical feedback, a bifurcation diagram was constructed with the obtained largest Lyapunov exponents as can be seen in Fig. 6.

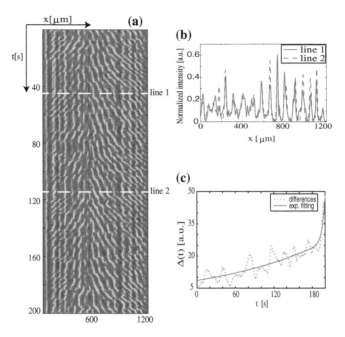

Fig. 5 Largest Lyapunov exponent estimation for liquid crystal light valve with optical feedback with $V = 3.75V_{rms}$. **a** Projected spatiotemporal diagram with two close initial conditions marked by line 1 and line 2. **b** Instantaneous profile of the intensity field in line 1 and 2, respectively. **c** Temporal evolution of global difference $\Delta(t)$, *dots* stand for the experimental data and the *continuous curve* is the exponential fitting $\Delta(t) \approx ae^{bt} + ce^{dt}$ with $a = 0.1163$, $b = 0.6258$, $c = 7.598$ and $d = 0.01465$

The system starts with stationary stripe patterns at $V_0 = 3.0V_{rms}$ and the dynamics remains unchanged until the applied voltage reached $V_0 = 3.5V_{rms}$. In this voltage range striped patterns are stationary (cf. Fig. 6). At $V_0 = 3.5V_{rms}$, the largest Lyapunov exponent goes to zero, meaning that the system exhibits a bifurcation. Experimentally, we observed that the steady pattern changes to a periodic regime, i.e. the envelope of the patterns exhibits a limit cycle, jumping into a chaotic behavior when the voltage is increased. The chaotic regime is highlighted by the colored area in Fig. 6. The chaotic behavior remains until the mean intensity in the LCLV destroys the chaotic attractor due to destructive interference at $V_0 = 3.9V_{rms}$ (see Fig. 6), causing a crisis. Once the light is recovered in the LCLV the system enter in an intermittence regime between chaotic and quasi periodic behavior.

Another tool of dynamical system theory utilized to characterize geometrically the different attractors is the phase space reconstruction [31]. This method is based on that given a temporary signal, one builds an attractor of the system by taking the signal at different periodic times (arbitrary periodic separation τ) and build the vector $(I(x,t), I(x,t+\tau,x), I(x,t+2\tau), \dots)$ with x a fixed position. Figure 7 shows three different attractors projected in three dimensions using this embedding

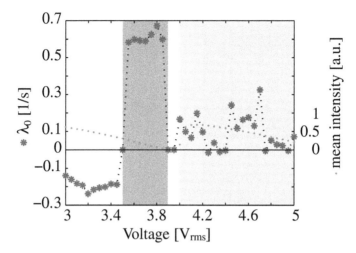

Fig. 6 Bifurcation diagram of the liquid crystal light valve with optical feedback constructed with the estimated largest Lyapunov exponent as functions of the voltage V_0. The diagram is clearly separated in three dynamical regimes, stationary patterns shadowed in *white*, chaos shadowed in *gray* (*blue*) and intermittence between chaos and quasi periodicity shadowed in *light gray* (*yellow*) area. The *stars* correspond to the calculated largest Lyapunov exponent while the *light gray* (*green*) points show the normalized mean intensity in the LCLV

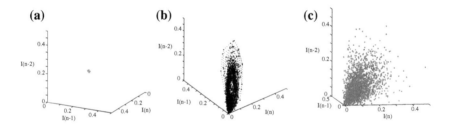

Fig. 7 Phase space reconstruction of attractor of liquid crystal light valve with optical feedback. **a** Fixed point dynamics observed at $V_0 = 3.4V_{rms}$, **b** Quasiperiodic orbit obtained at $V_0 = 4.65V_{rms}$ and **c** Projection in 3 dimensions of a strange attractor achieved at $V_0 = 3.8V_{rms}$

method. For low voltage a fixed point can be observed (see Fig. 7a), this behavior is observed when the stationary pattern is displayed in the LCLV. Points dispersion accounts for the inherent fluctuations of the system. Increasing the tension V_0, the phase space reconstruction exhibits a torus (cf. Fig. 7b), which accounts for the quasi periodic behavior observed by $V_0 = 4.65V_{rms}$. For larger voltage, the phase space reconstruction exhibits a strange attractor as it is illustrated in Fig. 7c.

5 Conclusions

Far from equilibrium system exhibit complex spatiotemporal self-organization. Based on a nematic liquid crystal layer with spatially modulated input beam and optical feedback, we set up a two-dimensional pattern forming system which exhibit a transition from stationary to spatiotemporal chaotic pattern. We have characterized experimentally in a rigorous manner this transition, using an adequate projection of spatiotemporal diagrams, we determine the largest Lyapunov exponent. This exponent allows us to characterize the route to spatiotemporal chaos of patterns.

To our knowledge most experimental studies have been focused on the qualitative description of complex behavior of patterns kept out of equilibrium. Exhibiting a clear distancing of the theoretical efforts, which have developed tools to characterize complexity such as Lyapunov exponents, Kolmogorov-Sinai entropy, Poincaré maps, Lyapunov spectrum, strange attractors, geometry of manifolds, fractal dimensions and distributions. Using the concept of largest Lyapunov exponent, we have reconciled the theory with experimental observations. Certainly new concepts of the theory of dynamical systems must be developed to more efficiently characterize experimentally the spatiotemporal complex behaviors. Notwithstanding, largest Lyapunov exponent and power spectrum can allow distinguished well established dynamical behaviors as turbulence and spatiotemporal chaos, which are often confused.

Acknowledgments M.G.C. acknowledges the support of FONDECYT N°1150507. M.W. acknowledges the support of FONDECYT N°3140387.

References

1. P. Ball, *The Self-Made Tapestry: Pattern Formation in Nature* (Oxford University Press, New York, 1999)
2. G. Nicolis, I. Prigogine, *Self-Organization in Non Equilibrium Systems* (Wiley, New York, 1977)
3. L.M. Pismen, *Patterns and Interfaces in Dissipative Dynamics* (Springer Series in Synergetics, Berlin, 2006)
4. M.C. Cross, P.C. Hohenberg, Pattern formation outside of equilibrium. Rev. Mod. Phys. **65**, 851–1112 (1993)
5. M. Cross, H. Greenside, *Pattern Formation and Dynamics in Nonequilibrium Systems* (Cambridge University Press, New York, 2009)
6. M. Tlidi, K. Staliunas, K. Panajotov, A.G. Vladimirov, M.G. Clerc (2014) Localized structures in dissipative media: from optics to plant ecology. Philos. Trans. R. Soc. Lond. A: Math. Phys. Eng. Sci. **372**, 20140101 (2027)
7. G. Nicolis, Introduction to Nonlinear Science (Cambridge University press, Cambridge, 1995)
8. P. Coullet, J. Lega, Defect-mediated turbulence in wave patterns, Europhys. Lett. **7**, 511–516 (1988). P. Coullet, L. Gil, J. Lega, Defect-mediated turbulence, Phys. Rev. Lett. **62**, 1619–1622 (1989). A form of turbulence associated with defects, Physica **37 D**, 91–103(1989)
9. G. Goren, J.P. Eckmann, I. Procaccia, Scenario for the onset of space-time chaos. Phys. Rev. E **57**, 4106–4134 (1998)

10. W. Decker, W. Pesch, A. Weber, Spiral defect chaos in Rayleigh-Benard convection. Phys. Rev. Lett. **73**, 648–651 (1994); B. Echebarria, H. Riecke, Defect chaos of oscillating hexagons in rotating convection, Phys. Rev. Lett. **84**, 4838–4841 (2000); K.E. Daniels, E. Bodenschatz, Defect turbulence in inclined layer convection, Phys. Rev. Lett. **88**, 034501(2002)
11. M. Miranda, J. Burguete, Experimentally observed route to spatiotemporal chaos in an extended one-dimensional array of convective oscillators. Phys. Rev. E **79**, 046201 (2009)
12. P. Brunet, l. Limat, Defects and spatiotemporal disorder in a pattern of falling liquid columns. Phys. Rev. E **70**, 046207 (2004)
13. Q. Ouyang, J.M. Flesselles, Transition from spirals to defect turbulence driven by a convective instability. Nature (London) **379**, 143–146 (1996)
14. A. Garfinkel, M.L. Spano, W.L. Ditto, J.N. Weiss, Controlling cardiac chaos. Science **257**, 1230–1235 (1992)
15. S.Q. Zhou, G. Ahlers, Spatiotemporal chaos in electroconvection of a homeotropically aligned nematic liquid crystal. Phys. Rev. E **74**, 046212 (2006)
16. S.J. Moon, M.D. Shattuck, C. Bizon, D.I. Goldman, J.B. Swift, H.L. Swinney, Phase bubbles and spatiotemporal chaos in granular patterns. Phys. Rev. E **65**, 011301 (2001)
17. G. Huyet, J.R. Tredicce, Spatio-temporal chaos in the transverse section of lasers. Physica. D **96**, 209–214 (1996)
18. N. Verschueren, U. Bortolozzo, M.G. Clerc, S. Residori, Spatiotemporal chaotic localized state in liquid crystal light valve experiments with optical feedback. Phys. Rev. Lett. **110**, 104101 (2013); Phil. Trans. R. Soc. A. **372**, 20140011 (2014)
19. P. Manneville, *Dissipative Structures and Weak Turbulence* (Academic Press, San Diego, 1990)
20. S. Residori, Patterns, fronts and structures in a liquid-crystal-light-valve with optical feedback. Phys. Rep. **416**, 201–272 (2005)
21. Further details can be found on the web site of the manufacturer : http://www.niopik.ru
22. P.G. de Gennes, J. Prost, *The Physics of Liquid Crystals*, 2nd edn. (Clarendon Press, Oxford, 1993)
23. M.G. Clerc, C. Falcon, M.A. Garcia-Nustes, V. Odent, I. Ortega, Emergence of spatiotemporal dislocation chains in drifting patterns. CHAOS **24**, 023133 (2014)
24. E. Louvergneaux, Pattern-dislocation-type dynamical instability in 1D optical feedback Kerr Media with Gaussian Transverse pumping. Phys. Rev. Lett. **87**, 244501 (2001)
25. S. Bielawski, C. Szwaj, C. Bruni, D. Garzella, G.L. Orlandi, M.E. Couprie, Advection-induced spectrotemporal defects in a free-electron laser. Phys. Rev. Lett. **95**, 034801 (2005)
26. G. Nicolis, *Introduction to Nonlinear Science* (Cambridge University Press, Cambridge, 1995)
27. M.G. Clerc, N. Verschueren, Quasiperiodicity route to spatiotemporal chaos in one-dimensional pattern-forming systems. Phys. Rev. E. **88**, 052916 (2013)
28. K.E. Daniels, E. Bodenschatz, Defect turbulence in inclined layer convection. Phys. Rev. Lett. **88**, 034501 (2002)
29. Y. Kuramoto, *Chemical Oscillations, Waves, and Turbulence* (Springer, New York, 1984)
30. E. Ott, *Chaos in Dynamical Systems*, 2nd edn. (Cambridge University Press, Cambridge, 2002)
31. H. Abarbanel, *Analysis of Observed Chaotic Data* (Springer, New York, 1996)

Chiral Modes in 2D PT-Symmetric Nanostructures

M. Botey, R. Herrero, M. Turduev, I. Giden, H. Kurt and K. Staliunas

Abstract We propose a 2-dimensonal PT-symmetric photonic honeycomb structure that supports chiral Block-like modes being its excitation input and frequency dependent. We show that while the fundamental resonant frequency excites the counter-clockwise mode, the clockwise mode is excited at the corresponding second harmonic frequency from the same input channel. The geometry is derived as the simplest nontrivial extension from 1-dimensional PT-symmetric systems to provide asymmetric coupling between harmonic wave components of the electromagnetic field. The PT-symmetric honeycomb is generated by a closed set of three lattice vectors that enable the simultaneous resonance of two disjoint triads of wavevectors in a circular coupling. As a basic effect, we numerically show the measurable asymmetric transmission of Gaussian light beams incident on such a finite-sized structure with a hexagonal shape, at the fundamental and second harmonic resonant frequencies.

1 Introduction

Optical systems with periodic gain and loss modulations provide a new platform to develop synthetic materials with novel properties [1, 2]. In given conditions, combining index modulations and balanced loss and gain profiles with specific symmetries, such systems become classical analogues of quantum systems described by non Her-

M. Botey (✉) · R. Herrero · K. Staliunas
DONLL, Departament de Física i Enginyeria Nuclear,
Universitat Politècnica de Catalunya, Barcelona, Spain
e-mail: muriel.botey@upc.edu

M. Turduev
Department of Electrical and Electronics Engineering, TED University, Ankara, Turkey

I. Giden · H. Kurt
Department of Electrical and Electronics Engineering, TOBB University of Economics and Technology, Ankara, Turkey

K. Staliunas
Institució Catalana de Recerca i Estudis Avançats (ICREA), Barcelona, Spain

© Springer International Publishing Switzerland 2016   125
M. Tlidi and M.G. Clerc (eds.), *Nonlinear Dynamics: Materials,*
Theory and Experiments, Springer Proceedings in Physics 173,
DOI 10.1007/978-3-319-24871-4_9

mitian Parity-Time- (PT-) symmetric Hamiltonians. PT symmetry requires that the complex potential, $U(\vec{r}) = U^{Re}(\vec{r}) + i U^{Im}(\vec{r})$, obeys the symmetry requirement $U(\vec{r}) = U^*(-\vec{r})$, which means that the real part of the potential is an even function, $U^{Re}(\vec{r}) = U^{Re}(-\vec{r})$, whereas the imaginary part is odd, $U^{Im}(\vec{r}) = -U^{Im}(-\vec{r})$. Although the imaginary part of the potential is generally difficult to obtain in nature, this is not the case in optics. Therefore, PT-symmetric optical systems rapidly demonstrated intriguing new features in simple 1-dimensional (1D) systems such as PT phase transitions [4], realizations of unidirectional invisible mediums [5, 6] or unidirectional waveguide transmitters [7, 8].

Complex crystals can also be realized in 2D, with simultaneous modulations of gain-loss and index. Indeed, recent works on systems with gain-loss modulations in two dimensions [1, 2], and also on complex 2D crystals [9–13] where the gain-loss and index are simultaneously modulated, have shown micro- and nanophotonics to be a platform for developing synthetic materials with novel beam propagation effects.

In the present paper, we propose a 2D PT-symmetric structure which is derived as the simplest nontrivial extension of 1D PT-symmetric systems, holding novel properties inherent to its 2D character. The resulting 2D PT-symmetric honeycomb geometry is generated by a closed set of three lattice vectors enabling the simultaneous resonance of two disjoint triads of wavevectors in a circular coupling, leading to asymmetric coupling between harmonic wave components of the electromagnetic field. The system is described by a coupled-mode approach, which provides the Bloch-like modes of the complex modulation of the potential, and supported by numerical calculations on light propagation within it. We show that the system holds two Bloch-like modes with opposite chirality. At resonance, the circular coupling constantly amplifies one of the two chiral eigenmodes being its excitation input dependent. As a basic effect, we numerically show the measurable asymmetric clockwise and counterclockwise transmission of a Gaussian light beam incident on such a finite-sized structure with a hexagonal shape. However, at the fundamental and second harmonic resonant frequencies, modes with opposite chirality are excited from the same input channel. This last result hints that the proposed structure could have applications in nonlinear optics. Therefore, the proposed geometry could be further extended to systems with higher order rotational symmetry, yielding to nontrivial 2D PT-symmetric quasi-crystals either for lineal or nonlinear optical applications.

2 Derivation of the 2D PT-Symmetric Honeycomb Structure

For the design of a 2D structure enabling PT-symmetric coupling, we shall start by simply remaining that at the basis of PT-symmetry in optics there is a complex refractive index, $n(\vec{r}) = n^{Re}(\vec{r}) + i n^{Im}(\vec{r})$, fulfilling $n(\vec{r}) = n^*(\vec{r})$. In the simplest case, we can consider a 1D optical harmonic potential in the form: $n(x) = n_0 + \Delta n(x) = n_0 + \Delta n[\cos(qx) + i \sin(qx)]$, which may be more conveniently expressed as:

$$n(x) = n_0 + \Delta n \, exp\,(iqx), \qquad\qquad (1)$$

where n_0 is the refractive index of background dielectric embedding medium, q is the reciprocal lattice vector of the modulation, and Δn is the real amplitude of the complex index modulation. The real part of the refractive index may be regarded as a 1D Photonic Crystal (PhC), a Bragg mirror characterized by its reciprocal lattice vectors $\pm q$, see Fig. 1a. As it is well known, at resonance, for $k = \pm q/2$, a Bragg mirror behaves as a perfect bidirectional reflector. The lattice periodicity of the material provides the corresponding wavenumber, reflecting $+k = +q/2$ to $-k = q/2 - q = -q/2$ and $-k = -q/2$ to $+k = -q/2 + q = +q/2$, see the schematic representation on the right column of Fig. 1a. On the contrary, when combined with a balanced gain-loss modulation with the same periodicity but spatially displaced by a quarter-period, as shown in Fig. 1b, the structure unidirectionally couples a wave with wavevector $-k = -q/2$ to $+k = -q/2 + q = q/2$. In this 1D PT-symmetric case the lattice only holds one lattice wave-vector $+q$. Therefore, at resonance, the structure holds asymmetric light transmission, i.e., light incident from the right is reflected, while light incident from the left with the same wave-vector is not reflected, breaking the left-right symmetry of wave propagation [4–8]. It is worth to note that the realization of a non-harmonic, stepwise, 1D PT-symmetric structure, as shown on the second column of Fig. 1b, requires alternated layers of four different materials featuring: high index with loss, low index with loss, low index with gain and high index with gain.

Keeping this basic principle in mind, the simplest choice to generate a PT-symmetric complex crystal in two-dimensions would be the direct extension of the 1D PT-symmetry as expressed in (1) to 2D in the form:

$$n\,(\vec{r}) = n_0 + \Delta n_x \exp\,(iq_x x) + \Delta n_y \exp\,(iq_y y) \qquad\qquad (2)$$

which simply factorizes the PT symmetries in both quadratures but does not lead to intrinsic 2D properties. Therefore, we intend to build a non factorizable 2D PT-symmetric structure, leading indeed to new properties. As the simplest nontrivial case, we chose a triangular lattice,

$$n\,(\vec{r}) = n_0 + \Delta n \sum_{j=AB,BC,CA} \exp\,(i\vec{q}_j \cdot \vec{r}), \qquad\qquad (3)$$

generated by three vectors symmetrically rotated by angles of $2\pi/3$ with respect to one another, namely, $q_{AB,CA} = (q/2, \pm q\sqrt{3}/2)$ and $q_{BC} = (-q, 0)$, being Δn the real amplitude of the complex modulation. Note that considering only the real part of (3) leads to the corresponding dielectric PhC with 6-fold symmetry, as represented by Fig. 1c. At resonance, $|k_{A,B,C}| = q\sqrt{3}/3$, such a PhC structure symmetrically couples all plane wave components directed along the symmetry axes, as schematically shown on the right column of Fig. 1c. However, for the complex lattice described by (3), the coupling is analogous to the one given by (1), being PT-symmetric in any direction. Such a complex lattice exhibits a 3-fold symmetry, i.e. the lattice vectors are q_{AB}, q_{BC} and q_{CA}, but not $-q_{AB}$, $-q_{BC}$, $-q_{CA}$. This can be expected to produce new peculiarities.

In order to design a realistic 2D honeycomb structure, one may replace the lower refractive index areas by finite-sized cylinders and embed them in a high refractive index background, see the second column of Fig. 1c. If the cylinders alternatively exhibit gain and loss we end up with a PT-symmetric honeycomb structure by combining only three materials: low refractive index gain cylinders, low refractive index lossy cylinders and a high refractive index background, see the second column of Fig. 1d. To illustrate the PT-symmetric character of such a structure, we shall compare its Fourier transform with the one of the analogous real structure. The third column of Fig. 1 depicts the reciprocal space, Fourier transform, of the step-wise structures of the second column. We find that indeed Fig. 1c supports six lattice vectors, $\pm q_{AB}$, $\pm q_{BC}$, $\pm q_{CA}$, in a 6-fold symmetry, which, at resonance, enable the symmetric coupling between the wavevectors k_A, k_B, k_C, see Fig. 1c. Nevertheless, the reciprocal space of the arrangement of cylinders, shown in Fig. 1d, reproduces the

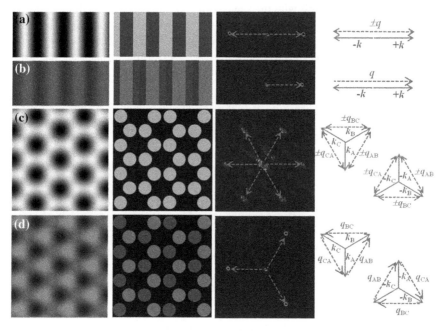

Fig. 1 Comparison of four different structures: **a** 1D PhC, real index modulation where *black* stands for high index and *white* for low index, **b** 1D PT-symmetric structure, where *red* areas denote gain and *blue* loss, **c** 2D PhC honeycomb structure **d** 2D PT-symmetric complex honeycomb structure. *First row* schematic representation of the harmonic optical potentials. *Second row* corresponding implementations with structures finite-sized domains of different materials. *Third row* reciprocal space showing the lattice vectors. *Forth row* coupling of wave-vectors at resonance, in figures (**a**) and (**c**) the presence of $\pm q$ wavevectors guarantees the symmetric coupling; in (**d**) the closed set of wavevectors, $q_{AB} + q_{BC} + q_{CA} = 0$, couples the two disjoint triads $\{k_A, k_B, k_C\}$ and $\{-k_A, -k_B, -k_C\}$

three points in the configuration proposed in (3). Such 3-fold symmetry is generated by only three reciprocal lattice vectors, namely q_{AB}, q_{BC}, q_{CA}, which form a closed set, $q_{AB} + q_{BC} + q_{CA} = 0$. Therefore, at resonance, this closed set of wavevectors enables the asymmetric coupling of the two disjoint triads: $\{k_A, k_B, k_C\}$ and $\{-k_A, -k_B, -k_C\}$, see the last column of Fig. 1d. Therefore, we expect that a wave incident on the structure with one of those wavevectors will generate either a clockwise or a counter-clockwise chiral mode which will be constantly amplified through the coupling. As a basic consequence, such a wave incident on the honeycomb PT-symmetric structure is expected to exhibit asymmetric left-right transmission. Apart from the three points indicating the lattice vectors, q_{AB}, q_{BC}, q_{CA}, other higher-order harmonics of the complex distribution appear if we zoom out the Fourier transform of our non-harmonic PT-symmetric modulation of the optical potential. We will show later that such harmonics are the responsible for the second harmonic asymmetric transmission.

The triangular lattice is seemingly the simplest nontrivial case of a nonfactorizable 2D PT-symmetric complex crystal. Further nontrivial cases could be realized for higher odd-fold rotational symmetry, which would also yield nontrivial 2D PT-symmetric quasi-crystals. Here we consider only this triangular case. Finally, note that the structure has no net gain since half of the cylinders are lossy, and the other half provide gain.

3 Chiral Excitation and Asymmetric Transmission

Next, we numerically check whether the proposed system displays the predicted properties of the proposed 2D complex PT-symmetric system, in particular the asymmetric flow of light and input dependent chiral excitation. Differently from 1D, we expect the asymmetric coupling between wavevectors to rotate the input by $2\pi/3$, depending on the input channel. We consider two finite-sized structures with hexagonal shape, see Fig. 2a, c, with the distributions represented on the second column of Fig. 1c, d in order to explore the optical properties by realistic numerical calculations. Numerical tests are performed numerically using the well-established 2D and 3D FDTD technique [15].

First, we analyze the propagation of a short broadband pulse incident on the structure from the top, in the vertical direction, and calculate the transmitted intensity on two detectors, symmetrically located on both sides of the structure, namely T1 and T2 in Fig. 2a, c. The resulting transmittance scanned for low frequencies for the corresponding dielectric and complex structures are depicted in Fig. 2b, d, respectively. For the PhC structure of Fig. 2a an incident wave k_A couples symmetrically to k_B and k_C, as schematically shown on Fig. 1c. We can observe that the field distribution depicted in Fig. 2b obtained by numerical FDTD simulation, is perfectly symmetric. On the other side, Fig. 2d demonstrates at a glance the expected asymmetric left-right transmission for the PT-symmetric structure. While in Fig. 2b both curves are perfectly superimposed, in Fig. 2d we clearly appreciate differences in

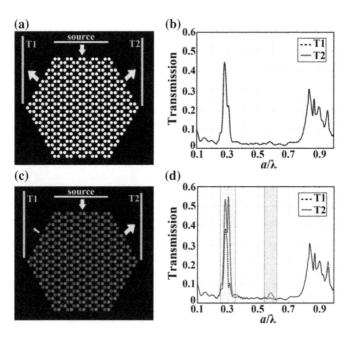

Fig. 2 a, b Schematic representations of the hexagons, indicating the positions of both detectors and the position of the source. For the numerical calculations we take $n_0 = 1.3$, $n = 1.1 \pm i\,0.1$, and cylinders of radius $R = 0.45\,a$, being 'a' the center-to-center distance between cylinders. The side of the hexagon is $13a$. Transmission to the right and left detectors, in arbitrary normalized units, as a function of frequency, in reduced frequency units $a/\lambda \approx$, from: **c** the dielectric structure with no gain/loss modulation and **d** the 2D PT-symmetric honeycomb structure. The *shaded areas* indicate the fundamental and second harmonic resonant frequencies, where symmetry is broken

transmission arising at the frequencies: $a/\lambda \approx 0.3$ and 0.6, where "a" is the center-to-center distance between cylinders. The fundamental resonant frequency of the structure corresponds precisely to $a/\lambda \approx 0.3$, note that $q = 4\pi/3a$. At $a/\lambda \approx 0.3$ plane waves are asymmetrically coupled by the lattice periodicity and, as expected, transmitted to the right detector. On the other hand, at this frequency, transmission to the left detector is much smaller for the PT-symmetric distribution due to the symmetry breaking. However, at the second harmonic resonant frequency $a/\lambda \approx 0.6$ the situation is completely different. In this case, transmission to the left detector is enhanced when compared to the purely PhC case on Fig. 2b. This later case can be attributed to higher order harmonics and will be further studied in detail. Finally, for higher frequencies we also find that the situation depicted in Fig. 2d is very similar to the same transmittance in Fig. 2b, since far from resonance no symmetry breaking is predicted.

As a following step, we analyze the transmission of a Gaussian beam, with central frequency being at the fundamental resonance. The numerical results on the electric field distribution, when the Gaussian beam is incident from above the hexagonal shaped structure are shown in Fig. 3. For comparison we consider both the PhC

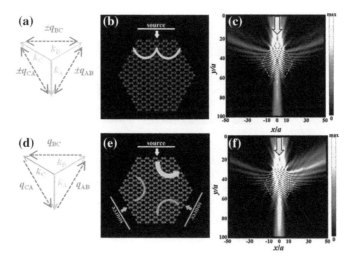

Fig. 3 Comparison of the transmission of a Gaussian pulse (width $14a$), incident from above, through an *hexagonal* shaped structure containing: a PhC (real) hexagonal lattice (*first row*) and the 2D PT-symmetric (complex) structure of Fig. 2c. **a, d** Schematic representation of the coupling at resonance, provided by the structures when *light* is incident from above. **b, e** Structures and flow of light within them. The *grey arrows* in (**e**) indicate equivalent input channels. **c, f** Normalized intensity distributions for an incident Gaussian beam on (**b, e**)

honeycomb structure, and the corresponding 2D PT-symmetric structure. Figure 3a schematically represents how the incident wave k_A is both coupled to k_B and k_C, being such coupling symmetric within a PhC honeycomb. Therefore, under the incidence of a Gaussian beam from above the PhC hexagonally shaped structure, the flow of light to the right and to the left is expected to be perfectly symmetric, see Fig. 3b. Figure 3c depicts the FDTD numeric simulation of the beam propagation which indeed shows how the beam is transmitted forward, as well as to the right and to the left. The situation comes to be completely different for a PT-symmetric honeycomb structure. In this case, the asymmetric coupling provided by the lattice, represented in Fig. 3d, determines the asymmetric beam transmission of Fig. 3f. The incident wave k_A is coupled to k_B while k_A is not coupled to k_C (or equivalently k_B is not coupled to k_A). The wavevector k_A excites a counterclockwise mode. Therefore, when light is incident from above, at resonance, transmission to the left channel is enhanced, while transmission to the right channel is reduced. The beam is also partially transmitted forward. Finally, we also find that the field distribution upon the incidence of a Gaussian pulse is very similar in both structures far from resonance, where no symmetry braking occurs, as predicted by Fig. 2.

We want to note that the 2D PT-symmetric finite-sized hexagon has three equivalent input channels for which the counterclockwise mode is excited and light is mostly transmitted to the left, as indicated in Fig. 3e. On the contrary, light flows to the right for the other three input channels as a clockwise mode is excited. Figure 4 shows such clockwise mode, compared to the light flow in a reference real PhC

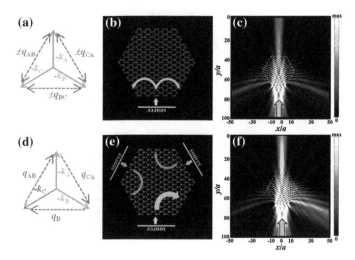

Fig. 4 Comparison of the transmission of a Gaussian pulse (width 14a, incident from below, trough an *hexagonal shaped* structure containing the PhC (real) and 2D PT-symmetric (complex) structures of Fig. 3. **a**, **d** Schematic representation of the coupling at resonance, provided by the structures when light is incident from *below*. **b**, **e** Structures and flow of *light* within them. The *thinner arrows* in **e** indicate equivalent input channels. **c**, **f** Normalized intensity distributions for an incident Gaussian beam on **b**, **e**

hexagonal structure. When light is incident from below with a resonant wavevector $-k_A$, transmission is symmetric for the case of the real PhC (with no gain/loss) as schematically represented in Fig. 4a. However for the PT-symmetric structure $-k_A$ is only coupled to $-k_B$, exciting the clockwise mode, see Fig. 4d. Hence, in Fig. 4f transmission to the right is enhanced and consequently transmission to the left is suppressed in this case. The comparison of the numeric results for the PT-symmetric and real PhC honeycomb structure are shown in Fig. 4c, f, respectively.

Besides being input dependent, the chiral excitation turns out to be frequency dependent in the PT-symmetric honeycomb structure. As already pointed out, the symmetry is broken not only at the first harmonic frequency but also at the second harmonic frequency but with opposite chirality. We now compare the light propagation at the first and second harmonic frequencies, $a/\lambda \approx 0.3$ and 0.6. We expect that a wavevector $2k_A$ incident from above will excite a clockwise mode as shown in Fig. 5a (contrary to what occurs for k_A, see Fig. 5e). Indeed, if we examine the spectral transmission plot of Fig. 2d, we observe how at $a/\lambda \approx 0.6$ clockwise transmission is enhanced while at $a/\lambda \approx 0.3$ counterclockwise transmission is higher. In order to increase this asymmetry at $a/\lambda \approx 0.6$, we have here considered a bigger hexagon, with a twice side length 26a, as shown in Fig. 5a. Transmissions from the 13a and 26a hexagons are very similar, except for the increase of the asymmetry at the second harmonic frequency. Also, both hexagons show the same points on the reciprocal space, as containing the same structure, but the higher order points are more clearly appreciated on Fig. 5c, see Fig. 5b, f. A deeper inspection of Fig. 5c shows there are

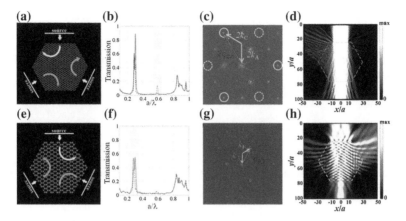

Fig. 5 Light propagating for the first resonant and second harmonic frequencies. *Top row*, light flow at $a/\lambda \approx 0.6$: **a** PT-symmetric honeycomb structure and input channels for the excitation of the clockwise mode **b** Counter-clockwise/clockwise asymmetric transmission, to detectors located as in Fig. 2c. **c** Second order wavevectors coupling. The *solid circles* indicate the pints corresponding to $-2q_{CA}, -2q_{AB}$ and $-2q_{BC}$; while the *dotted circles* are empty. **d** FDTD numerical transmission for a Gaussian beam with central frequency at $a/\lambda \approx 0.6$. *Bottom row*, light flow at $a/\lambda \approx 0.3$: **e** PT-symmetric *honeycomb structure* and input channels for the excitation of the counterclockwise mode. **f** *Left-right* asymmetric transmission, to detectors located as in Fig. 2c. **g** Wavevectors coupling at $a/\lambda \approx 0.3$ **h** FDTD numerical simulation transmission for a Gaussian beam with central frequency at $a/\lambda \sim 0.3$

points corresponding to the lattice vectors $-2q_{CA}, -2q_{AB}$ and $-2q_{BC}$; while the points corresponding to $+2q_{CA}, +2q_{AB}$ and $+2q_{BC}$ are missing. Therefore, in this case the wavenumber $-2q_{CA}$ couples $2k_A$ *to* $2k_C$ exciting a clockwise mode, see Fig. 5c. Indeed, a Gaussian pulse with central carrier frequency $2k_A$ is transmitted to the right as shown in a numeric FDTD simulation on Fig. 5d. This situation may be compared to the previous analysis at $a/\lambda \approx 0.3$, shown in Fig. 5e–h.

4 Chiral Block-Like Modes

For a PhC, the Bloch modes are defined as localized electromagnetic states of the periodic media that are invariant in propagation. However, in a complex system described by a non-Hermitian Hamiltonian, complex Bloch-like modes may either amplify or decay in time. In what follows, we calculate such Bloch-like modes analytically considering the proposed simple case of a harmonic PT-symmetric complex crystal of triangular symmetry. We consider an incident plane wave with a polarization perpendicular to the plane of the crystal and a wavevector directed vertically, $\vec{k} = (0, -k)$, near resonance: $\vec{k} = \vec{k}_A + \Delta \vec{k}$. The small variations are considered to be in the same incident direction: $\Delta \vec{k} = (0, -\Delta k_y)$. Disregarding the second time derivatives, the wave equation can be written as:

$$- 2i\omega \partial_t \vec{E} = \frac{c^2}{n(\vec{r})^2} \nabla^2 \vec{E} + \omega^2 \vec{E}. \tag{4}$$

We expand the electric field into the first three harmonics of the field, which are resonant in the lattice, namely: $\vec{k}_A = (0, -k_0)$, $\vec{k}_B = \vec{k}_A + \vec{q}_{AB}$, and $\vec{k}_C = \vec{k}_A - \vec{q}_{CA}$, and obtain, for the TM polarization:

$$E = \sum_{j=A,B,C} a_j \exp \left[i \left(\vec{k}_j + \Delta \vec{k} \right) \cdot \vec{r} \right]. \tag{5}$$

Introducing the expansion (5) into (4) yields coupled equations between the field expansion amplitudes, a_A, a_B, a_C:

$$-i\frac{n_0}{k_0 c} \partial_t \begin{pmatrix} a_A \\ a_B \\ a_C \end{pmatrix} = \begin{pmatrix} \vec{k}_A \cdot \Delta \vec{k} & \Delta n/n_0 & 0 \\ 0 & \vec{k}_B \cdot \Delta \vec{k} & \Delta n/n_0 \\ \Delta n/n_0 & 0 & \vec{k}_C \cdot \Delta \vec{k} \end{pmatrix} \begin{pmatrix} a_A \\ a_B \\ a_C \end{pmatrix}. \tag{6}$$

The dispersion diagrams, i.e., the temporal eigenvalues and the associated Bloch-like modes, are obtained by diagonalization of the matrix in (6). Figure 6a, b display the real and imaginary parts, respectively, of the matrix eigenvalues for the three Bloch-like modes close to the edge of the Brillouin zone, i.e., at resonance between lattice vectors. The temporal evolution of the Bloch mode is defined by the complex matrix eigenvalues with a factor $in_0/k_0 c$. As expected, sufficiently far from resonance, all the eigenvalues are real-valued (where the asymmetry of the coupling is not pronounced).

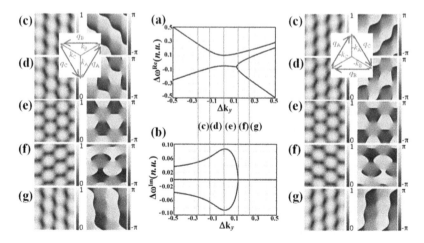

Fig. 6 Calculated dispersion diagrams and Bloch-like modes. **a** Real and **b** imaginary parts of the matrix eigenvalues $\Delta\omega$, where Δk_y, on the horizontal axis, is the distance from resonance in k_0 units. **c–g** Amplitude, and phase, of the amplified chiral Bloch mode for $\Delta k_y = -0.25, -0.125$, 0, 0.125 and 0.25 respectively, when illuminated from above (*left*) and from below (*right*)

Close to resonance, the PT-phase transition occurs, and we obtain Bloch-like modes with complex eigenvalues, one with a negative imaginary part and hence amplified in time. Therefore, in an extended structure, after a finite propagation time, the field distribution is expected to exhibit the amplitude and phase corresponding to this amplified mode. The amplitudes and phases of the most amplified Bloch modes, as calculated analytically from (6), are depicted in Fig. 6c–g, at different positions close to the PT-transition point, $\Delta k_y = 0$, and for the cases where light is either incident from above or below the structure.

To check the analytic predictions, we analyze the field evolution after excitation by a relatively long Gaussian pulse with central frequency at resonance, and spectrum narrower than the width of the transmission resonance peak in Fig. 2d. After a sufficiently long excitation (of the large hexagon honeycomb configuration), the incident radiation is redistributed among all the coupled harmonics approaching a stationary distribution. In such final distribution we expect to observe the growing Bloch-like modes. The analytically calculated amplitude and phase of the amplified chiral Bloch-like mode is shown in Fig. 7a, b, when the structure is illuminated from above. The numeric filed distribution presented in Fig. 7c is used to extract the amplitude and phase of the Bloch mode shown in Fig. 7d, e, respectively. The results show a good agreement with the analytically calculated amplified Bloch-like mode. The small differences may be attributed to the simplified model used (not accounting for the real shape of the scatterer) and the interplay between higher-order harmonics, as well as mainly to the finite size of the structure.

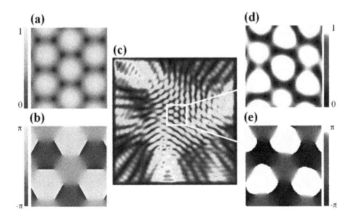

Fig. 7 Comparison of the electric field distribution of the analytic Block-like mode and the numeric simulation of the propagation of a Gaussian beam with carrier frequency $a/\lambda = 0.303$, and 10a-beam width, incident from above. **a, b** Amplitude/phase of the Block-like mode for $\Delta k_y = 0$ of Fig. 6e, when light is incident from above. **c** Field distribution obtained directly from the FDTD numeric simulations. **d, e** Amplitude/phase corresponding to the stationary distribution within the hexagon corresponding to the magnified $6a \times 6a$ region area of figure **c**

5 Implementation Proposal

We finally propose a possible realization of the investigated 2D PT-symmetric complex structure, where the predicted effects could be measured. Figure 8a illustrates a silicon slab with a honeycomb lattice of alternating p–n and n–p semiconductor junctions. We numerically analyze such device with full 3D FDTD simulations, performed using the LUMERICAL software package [16]. We assume that the device is illuminated by a broadband pulse with a Gaussian profile, with a source 7 μm width and 0.5 μm height and analyze light transmission on two detectors symmetrically placed on either side of the structure. The calculated normalized transmission spectra are depicted in Fig. 8b. A measurable clockwise/counter-clockwise asymmetry is observed in the transmission near resonance at the wavelength $\lambda = 1.501$ μm (wavelength in vacuum). The steady-state electric field distributions at the cross-sectional xy plane ($z = 0$) and yz plane ($x = 0$) are shown in Fig. 8c, d respectively. The electric field snapshot in Fig. 8d shows the asymmetric clockwise and counter-clockwise light transmission at resonance. Furthermore, the cross-sectional field distribution

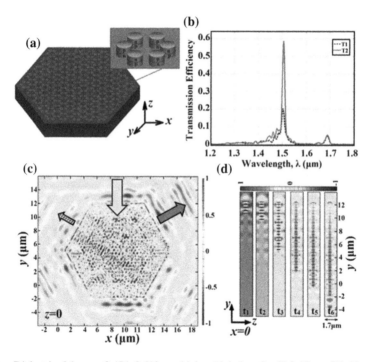

Fig. 8 **a** Dielectric slab, $n_0 = 3.474$, 0.612 μm high, with holes of radii 0.45 μm filled by p-n/n-p semiconductor junctions, $n = 3.460 \pm 0.007\text{i}$; a = 1.0 μm, where *circles* indicate alternated p/n n/p junctions. **b** Clockwise–counterclockwise normalized transmission when light is incident from above. **c** Electric field distribution at $z = 0$, the *big arrow* indicates the input channel. **d** Snapshots at the cross-sectional plane $x = 0$, at increasing times, showing confinement of the field

depicted in Fig. 8c provides evidence of the vertical confinement and guiding of the beam while propagation within the slab. As a result, the out-of-plane losses would be almost negligible for this specific design.

6 Conclusions

To conclude, we have designed a 2D honeycomb complex structure being PT-symmetric as the simplest nontrivial extension from a 1D PT-symmetric optical potential, and analyzed the propagation of light within the designed configuration. We predict that while 1D PT-symmetric structures hold asymmetric reflection at resonance, an analogous 2D PT-symmetric holds an asymmetric chirality, in the sense that, close to resonance, the clockwise courter-clockwise symmetry is broken. As being generated by a closed set of lattice vectors, the proposed 2D PT-symmetric structure holds chiral Block-like modes which, when excited, are constantly amplified within the complex structure. We also show that the fundamental and second harmonic frequency excite modes with opposite chirality. As a consequence, we numerically demonstrate that a hexagonally shaped PT-symmetric honeycomb structure transmits asymmetrically light beams incident on it at the fundamental or second order resonance. The Block-like modes are derived from a coupled mode analysis, and the predicted field and phase distributions at resonance are in good agreement with FDTD simulations. Finally, following the proposed scheme, we design and analyze a feasible configuration where the predicted effects could be measured.

It may be expected that a new generation of synthetic optical components enabling an on-chip control on light propagation could rely on such non-reciprocal optical systems. Indeed, the results hint the way to the realization of passive 2D optical insulators of even transistors, which could even be implemented as active devices.

Acknowledgments We acknowledge financial support by Spanish Ministerio de Ciencia e Innovación, and European Union FEDER through project FIS2011-29731-C02-01 and -02 and and Generalitat de Catalunya (2009 SGR 1168). H.K. acknowledges partial support of the Turkish Academy of Science.

References

1. K. Staliunas, R. Herrero, R. Vilaseca, Subdiffraction and spatial filtering due to periodic spatial modulation of the gain-loss profile. Phys. Rev. A **80**, 013821 (2009)
2. M. Botey, R. Herrero, K. Staliunas, Light in materials with periodic gain-loss modulation on a wavelength scale. Phys. Rev. A **82**, 013828 (2010)
3. C.M. Bender, S. Boettcher, Real spectra in non-Hermitian Hamiltonian having PT symmetry. Phys. Rev. Lett. **80**, 5243–5246 (1998)
4. A. Guo et al., Observation of PT-symmetry breaking in complex optical potentials. Phys. Rev. Lett. **103**, 93902 (2009)

5. S. Longhi, Invisibility in PT-symmetric complex crystals. J. Phys. A, Math. Theor. **44**, 485302 (2011)
6. Z. Lin, H. Ramezani, T. Eichelkraut, T. Kottos, H. Cao, D.N. Christodoulides, Unidirectional Invisibility Induced by PT-Symmetric Periodic Structures. Phys. Rev. Lett **106**, 213901 (2011)
7. C.E. Ruter, K.G. Makris, R. El-Ganainy, D.N. Christodoulides, M. Segev, D. Kip, Observation of parity-time symmetry in optics. Nat. Phys. **6**, 192–195 (2010)
8. L. Feng, Y.-L. Xu, W.S. Fegadolli, M.-H. Lu, J.E. Oliveira, V.R. Almeida, Y.-F. Chen, A. Scherer, Experimental demonstration of a unidirectional reflectionless parity-time metamaterial at optical frequencies. Nat. Mater. **12**, 108–113 (2012)
9. R. Herrero, M. Botey, M. Radziunas, K. Staliunas, Beam shaping in spatially modulated broad-area semiconductor amplifiers. Opt. Lett. **37**, 5253–5255 (2012)
10. M. Radziunas, M. Botey, R. Herrero, K. Staliunas, Intrinsic beam shaping mechanism in spatially modulated broad area semiconductor amplifiers. Appl. Phys. Lett. **103**, 132101 (2013)
11. N. Kumar, L. Maigyte, M. Botey, R. Herrero, K. Staliunas, Beam shaping in two-dimensional metallic photonic crystals. JOSA B **31**, 686–690 (2014)
12. A. Cebrecos, R. Picó, V. Romero-García, A.M. Yasser, L. Maigyte, R. Herrero, M. Botey, V.J. Sánchez-Morcillo, K. Staliunas, Enhanced transmission band in periodic media with loss modulation. Appl. Phys. Lett. **105**, 204104 (2014)
13. R. Herrero, M. Botey, K. Staliunas, Nondiffractive-nondiffusive beams in complex crystals. Phys. Rev. A **89**, 063811 (2014)
14. M. Turduev, M. Botey, I. Giden, R. Herrero, H. Kurt, E. Ozbay, K. Staliunas, Two-dimensional complex parity-time-symmetric photonic structures. Phys. Rev. A **91**, 023825 (2015)
15. Simulations are performed using the open source software MEEP, initially developed at MIT (http://ab-initio.mit.edu/) for electromagnetic field calculations
16. Lumerical FDTD Solutions, Inc. http://www.lumerical.com

Stabilization of Broad Area Semiconductor Amplifiers by Spatially Modulated Potentials

S. Kumar, W.W. Ahmed, R. Herrero, M. Botey,
M. Radziunas and K. Staliunas

Abstract We propose the stabilization of the output beam of Broad Area Semiconductor (BAS) amplifiers through the introduction of a spatially periodic modulated potential. We show that a periodic modulation of the pump profile in transverse and longitudinal directions, under certain 'resonance' condition, can solve two serious problems of BAS amplifiers (and possibly lasers), which are (i) the lack of an intrinsic spatial mode selection mechanism in linear amplification regimes and (ii) the modulation instability (also called Bespalov-Talanov instability) in nonlinear regimes. The elimination of these two drawbacks can significantly improve the spatial quality of the emitted beam in BAS amplifiers.

1 Introduction

Broad Area Semiconductor (BAS) amplifiers and lasers, are relevant light sources used in many applications. Their main advantage is their planar geometry that enables an efficient pump access to the entire volume of the active amplifying medium, allowing a high conversion efficiency. BAS devices, however, suffer from a serious disadvantage of poor spatial and temporal quality of the emitted beam [1]. If no special mechanisms are incorporated in the design, such as different schemes of optical injection [2, 3] or optical feedback [4, 5] amongst others, the emission exhibits spatiotemporal fluctuations which leads to a broad and noisy optical and angular

S. Kumar · W.W. Ahmed · R. Herrero (✉) · M. Botey · K. Staliunas
Departament de Fisica i Enginyeria Nuclear, Universitat Politècnica de Catalunya
(UPC), Colom 11, E-08222 Terrassa, Barcelona, Spain
e-mail: ramon.herrero@upc.edu

M. Radziunas
Weierstrass Institute for Applied Analysis and Stochastics, Leibniz Institute in
Forschungsverbund Berlin e.V., Mohrenstrasse 39, 10117 Berlin, Germany

K. Staliunas
Institució Catalana de Recerca i Estudis Avançats (ICREA), Passeig Lluis
Companys 23, E-08010 Barcelona, Spain

© Springer International Publishing Switzerland 2016   139
M. Tlidi and M.G. Clerc (eds.), *Nonlinear Dynamics: Materials,*
Theory and Experiments, Springer Proceedings in Physics 173,
DOI 10.1007/978-3-319-24871-4_10

spectrum [6, 7]. The deterioration of the spatial structure primarily affects BAS lasers, however is also problematic for BAS amplifier in strong amplification regimes: if the spatial modulations do not evolve along the single propagation along relatively short amplifying media for lower pump powers, for higher pump powers the modulation can strongly reduce the beam spatial quality.

The poor spatial quality of the emitted beam in BAS amplifiers and lasers is primarily due to two different physical phenomena. Firstly, it is the lack of an intrinsic spatial mode selection mechanism in linear stage of amplification, i.e. the absence of a natural angular selection due to the large aspect-ratio of such devices. Secondly, the strongly pumped semiconductor material typically displays self-focusing nonlinearity due to the refractive index dependence on the population inversion through linewidth enhancement factor α_H (also called Henry factor). This effect introduces inhomogeneity in the system and gives birth to Modulation Instability (MI) observed in nonlinear regimes [8]. The MI in technical terms results in multi-transverse mode operation, which eventually decreases the beam quality.

In this work, we demonstrate theoretically that a suppression of the instabilities can be achieved by introducing a two-dimensional (2D) modulation of the gain function on a scale of several wavelengths. This technically can be realized using a periodic grid of electrodes for electrically pumped semiconductors (Fig. 1). The two main results are the substantial improvement of the spatial quality of the amplified beam in linear regimes (the beam quality factor, M^2, can be reduced down to unity corresponding to a Gaussian beam), and the suppression of the MI in nonlinear regimes.

Previous studies show that periodic modulations of the complex potential on the wavelength scale, corresponding to Gain and refractive Index Modulations (GIM) in the optical case, can lead to particular beam propagation effects, such as self-collimation, spatial (angular) filtering, beam focalization and a nondiffractive-nondiffusive amplification of very narrow beams [9–11]. In the modulated semiconductor media, the simultaneous in-phase GIM structure is imposed due to linewidth enhancement factor, α_H, which related the modulation of gain with the modulation of refractive index (index is proportional to the carrier inversion).

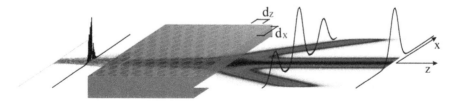

Fig. 1 Semiconductor amplifier with modulated pump by periodic electrodes with longitudinal and transverse periods $d_\perp$ and $d_{||}$. The incident low quality beam is filtered while propagating through the structure. Sideband components are present beyond the device output due to the spatial modulation

2 Mathematical Model

We consider a commonly used stationary model for semiconductor amplifiers consisting of two nonlinearly coupled equations for the carrier density N and the complex amplitude $A(x, z)$ of the TE-polarized electric field. The field with wavenumber k_0 propagates under paraxial approximation through the material interacting with the carriers via the carrier dependent gain and the refractive index [12, 13]:

$$
\begin{aligned}
\frac{\partial A}{\partial z} &= \frac{i}{2k_0} \frac{\partial^2 A}{\partial x^2} + \left[N(1 - i\alpha_H) - (1 + \gamma) \right] A \\
\frac{D}{k_0^2} \frac{\partial^2 N}{\partial x^2} &+ P(x, z) = N + BN^2 + CN^3 + (N - 1)|A|^2
\end{aligned}
\tag{1}
$$

where D is the carrier diffusion, B and C are the spontaneous and Auger recombination coefficients in the carrier rate equation. The paraxial propagation of the field includes the internal loss coefficient γ and the Henry factor α_H. P is the pump that in our case will be periodically modulated in the transverse and longitudinal space directions as $P(x, z) = p_0 + 4m' \cos(q_x x) \cos(q_z z)$. The geometry is defined by the normalized longitudinal, $q_z = 2\pi/d_z$, and transverse, $q_x = 2\pi/d_x$, components of the lattice wave-vectors (the reciprocal lattice vectors). The lattice vectors define the adimensional geometry factor $Q = 2k_0 q_z/q_x^2 = 2d_x^2/\lambda d_z$ (λ: the optical field wavelength in semiconductor) that localizes the resonance between the longitudinal and transverse field harmonics for $Q = 1$. The physical meaning of geometry factor Q can be related with Talbot effect: at $Q = 1$ the period of longitudinal Talbot oscillations coincide with the longitudinal modulation period.

Neglecting the spontaneous and Auger coefficients for small enough N values, and considering the same modulation frequencies for the carrier density, we can write N as:

$$
N = \frac{P(x, z) + |A|^2 + D'N_0}{1 + |A|^2 + D'}
\tag{2}
$$

where $N_0 = (p_0 + |A|^2)/(1 + |A|^2)$ is the carrier density for a non-modulated pump and $D' = Dq_x^2/k_0^2$ is the normalized carrier diffusion. After adiabatic elimination of N the field evolution along propagation direction can be written as:

$$
\frac{\partial A}{\partial z} = \frac{i}{2k_0^2} \frac{\partial^2 A}{\partial x^2} + \left[\frac{P(x, z) - 1 + D'(N_0 - 1)}{1 + |A|^2 + D'} (1 - i\alpha_H) - i\alpha_H - \gamma \right] A
\tag{3}
$$

The complex amplitude of the electric field, $A(x, z)$, evolves in paraxial approximation experiencing diffraction, nonlinearities due to the gain and refractive index dependence on the carrier density (which itself depends on the field intensity), and linear losses.

3 Angular Filtering of the Beam in the Linear Regime

We study the beam propagation in the linear regime, suitable for relatively weak
fields. Considering small field amplitudes, the field propagation can be written as

$$\frac{\partial A}{\partial z} = \frac{i}{2k_0} \frac{\partial^2 A}{\partial x^2} + \left[g + 4m \left(1 - i\alpha_H \right) \cos(q_\perp x) \cos(q_\parallel z) \right] A \qquad (4)$$

where $m = m'/(1 + D')$ and $g = \left(p_0 + D'N_0 \right) \left(1 - i\alpha_H \right) /(1 + D') - (1 + \gamma)$ is a
constant part associated to an exponential growth/decay of the homogeneous part of
the field [14].

For analytical estimations, we apply a harmonic expansion of the field in terms
of the periods of pump modulation:

$$A(x, z) = e^{ik_x x} \left[a_0(z) + a_{-1}(z) e^{-iq_x x - iq_z z} + a_1(z) e^{+iq_x x - iq_z z} + \cdots \right] \qquad (5)$$

inserting it in (1), we obtain the evolution of mode amplitudes along propagation
direction:

$$\begin{aligned} da_0/dz &= \left(-i\frac{k_x^2}{k_0} + g \right) a_0 + m \left(1 - i\alpha_H \right) \left(a_1 + a_{-1} \right) \\ da_{\pm 1}/dz &= \left(-i\frac{(k_x \pm q_x)^2}{k_0} + g \right) a_0 + m \left(1 - i\alpha_H \right) a_0 \end{aligned} \qquad (6)$$

The eigenmodes of (6) correspond to the Bloch modes propagating in z direction
according to the factor $\exp\left(ik_z z \right) = \exp\left(\left(ik_{z,\mathrm{Re}} - k_{z,\mathrm{Im}} \right) z \right)$ where the complex propagation wavenumbers k_z are the system eigenvalues and the exponential grow/decay
of modes is associated to its negative/positive imaginary part $k_{z,\mathrm{Im}}$. The analysis is
analogous to that performed for a pure GL modulation in [9] and for photonic crystals
in [15, 16].

The quantitative characteristics of the linear filtering are summarized in Fig. 2
considering a typical wavelength of $1\,\mu\mathrm{m}$ in vacuum, with effective refractive index
$n = 3$ of the semiconductor material, gain coefficient of $10^4\,\mathrm{m}^{-1}$ that corresponds
to $m = 10^{-4}$, and a transverse period $d_x = 4\,\mu\mathrm{m}$. For these particular parameters,
the longitudinal period is $d_z = 96\,\mu\mathrm{m}$ at $Q = 1$.

The imaginary part of the eigenvectors, $k_{z,\mathrm{Im}}$, is negative for one of the Bloch
modes, the one with amplitude maxima located at gain areas of the GIM media.
The most efficient spatial filtering regime is obtained for $Q \approx 1$, when only one
Bloch mode is amplified, and the narrow single-peak gain profile of this mode in
the dispersion relation implies beam amplification only in a finite angular range. In
this way, the radiation is spatially filtered by the amplification of transverse modes
propagating at small angles to the optical axis while modes at larger angles are less
amplified or damped [17].

The filtering effect depends critically on the geometry parameter Q [18]. At the resonance condition, $Q = 1$, the dispersion curves of the three Bloch modes have a simple analytic form, which allows to estimate the spatial filtering effect. It is determined by the half-width of the gain peak, which for the previously used parameters corresponds to $0.5°$ inside the semiconductor.

The beam intensity increases by a factor of 80 in the specific case shown in Fig. 2 and the intensity losses to side-bands are always compensated. The width of the propagated beam inside the BAS amplifier shows a square root dependence on the propagation distance that evidences the diffusive propagation and associated spatial filtering effect.

We analyze the spatial filtering effect considering amplification of a noisy Gaussian beam (Fig. 2). The quantitative characteristics of the filtering are summarized in Fig. 3. Behind the amplifier, the sidebands generated by the GIM rapidly separate from the central beam allowing us to calculate the width, W, and divergence, θ, of the central part of the beam. We numerically propagate the beam in backward direction to reach the focal plane and calculate there the beam quality factor, $M^2 = W\theta\pi/\lambda$. The obtained values can be compared with the quality factor for a perfect Gaussian beam, $M^2 = 1$.

Beam quality factors about 1.1 can be achieved with devices having about 10-longitudinal modulation periods d_z, what is less than 1 mm for parameters used. The smallest quality factors are obtained near the resonance condition ($Q = 1$) as shown in Fig. 3a for two different amplifiers having 4 and 16 longitudinal periods with an initial beam noise level of $M^2 = 5.1$. The quality factor reaches minimal values of 1.014 at $Q \approx 1.1$ for the long amplifier, and 1.6 at $Q \approx 1.4$ for the less efficient short amplifier. The shift of the efficient filtering range to higher Q values

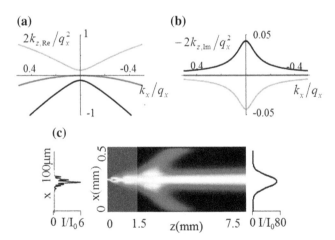

Fig. 2 Real (**a**) and imaginary (**b**) parts of the longitudinal wavenumbers $k_\parallel$ for $Q = 1$ (spatial dispersion curves). **c** Input beam profile with the beam quality factor $M^2 = 5.1$ beam intensity propagating along the BAS amplifier (*yellow part*) and beyond it in free space simulated using (1)

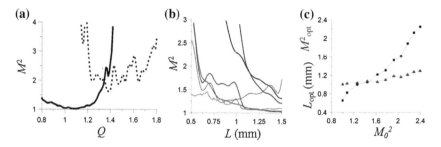

Fig. 3 **a** Beam quality factor dependence on the geometry factor Q for a short amplifier with length
$L \approx 4d_z$ (*dashed*) and a longer one $L \approx 16d_z$ (*solid*) for noisy input beams with $M_0^2 = 5.1$, the
same input beam of Fig. 2. **b** Quality factor of the output beam as a function of the BAS length
for $M^2 = 5.1$ and $Q = 0.8$ (*black*), 0.9 (*blue*), 1.0 (*red*), 1.1 (*purple*), 1.2 (*orange*), 1.3 (*green*). **c**
Optimal BAS length as a function of input beam quality factor M^2 and the corresponding output
beam quality factor M_{opt}^2, for $Q = 1.2$

for shorter amplifiers is related to an interplay between a broad angular gain profile
and a strong beam shape distortion induced at Q values larger than unity. This beam
distortion can be explained by the double-peak shape in the imaginary part of the
dispersion curve that appears for $Q > 1.1$. A first consequence is the distortion of
the amplified beam. The analysis of propagation eigenvalues and series expansions,
in the limit $|Q - 1| \ll 1$, allows to estimate the appearance of this double-peak at
$Q = 1 + |\alpha_H| m / q_z$, which is in agreement with the numerical simulations. The
output beam quality as a function of the device length for different geometry factor
values is shown in Fig. 3b. It can be easily observed that the larger is the BAS length,
the stronger is the filtering effect. For long structures, M^2 rapidly decreases and
approaches unity.

For $Q > 1$, the filtering effect reduces M^2 along the amplifier length while the
beam dispersion induces a distortion to the beam and M^2 increases for long enough
BAS amplifiers. Thus, an optimal length, L_{opt}, corresponding to the maximal output
beam quality M_{opt}^2 can be determined. This optimal length not only depends on the
amplifier parameters but also on the initial beam quality, M_0^2 as shown in Fig. 3c.
The smaller is the initial beam quality, the larger is the optimal length of the BAS
amplifier, while the obtained output quality M_{opt}^2 also increases but always remains
around unity.

4 Supression of Modulation Instability in BAS Amplifiers

In the previous section, we showed that in the linear operating regime of a BAS amplifier, the modification of the dispersion profile through a periodically modulated gain
results in an anisotropic net gain of Bloch-like modes in the semiconductor media
with consequent angular filtering of the spatial modes and beam homogenization.

However, the amplitude of this propagating field grows along the amplifier and non-linearities rapidly take an important role in the system's dynamics. The linear spatial filtering effect cannot guarantee the beam's quality any more due to the emergence of MI seeded by intrinsic noise in the nonlinear system.

MI is at the basis of spontaneous spatial pattern formation in many spatially extended nonlinear systems [19, 20]. Patterns arise when, in the initial stage of propagation the maximal symmetric homogeneous state loses its stability with respect to the modes of spatial modulation. These unstable modes grow and lead to stationary patterns or to the development of spatio-temporal periodic and chaotic dynamics via secondary bifurcations [21, 22]. Indeed, the spatio-temporal behavior of the system is closely related with the field dispersion which, in particular, plays an important role in the stability of the homogeneous solution. It follows clearly from this point that the manipulation of MI for the stability enhancement of the homogenous state, is possible by modifying the dispersion of the system.

In order to analyze the effect of modification of the dispersion on the homogeneous solution of the system, we use the Complex Ginzburg-Landau Equation (CGLE), which universally describes the MI in many different pattern-forming systems [23]. Considering just one spatial dimension, the CGLE can be written in the simple form:

$$\partial_z A = i\partial_{xx} A + (1 - ic)\left(1 - |A|^2\right) A \qquad (7)$$

where the threshold parameter and the dispersion coefficient have been normalized to unity by scaling time and space coordinates. In this form, the CGLE has only one parameter, the nonlinearity coefficient c.

The linear stability analysis of the homogeneous stationary solution ($A = 1$) is performed by introducing a small amplitude perturbation for every transverse wavevector k_x in the form: $A(x, z) = 1 + a(z)\cos(k_x x)$ with $|a| \ll 1$. The linearization of the CGLE for each perturbation mode leads to the spectrum of eigenvalues $\lambda(k_x)$, that can be written as a function of the spatial dispersion $k_z(k_x)$ introduced by the spatial derivatives of the system ($k_z = -k_x^2$ for the Laplacian operator $i\partial_{xx}^2 A$). The analytical expression for the obtained eigenvalues is of the form:

$$\lambda_{1,2}(k_x) = -1 \pm \sqrt{1 - 2ck_z - k_z^2} \qquad (8)$$

showing that for a given nonlinearity, the system has real, positive eigenvalues when $k_z(k_x)$ are in the interval $-2c < k_z(k_x) < 0$. The presence of a real, positive eigenvalues for a particular spatial mode k_x signifies a modulationally unstable mode, i.e.: the development of MI at that spatial frequency. Figure 3a represents the real part of eigenvalues as a function of the longitudinal wavenumber for a given nonlinear coefficient. Thus, for a fixed c, the presence of MI depends on the shape of the spatial dispersion that in the case of (8) is a parabolic function corresponding to the Laplacian operator $i\partial_{xx}^2 A$. This normal dispersion relation, represented in Fig. 3b by the solid red parabola, ensures transverse modes with positive eigenvalues and the presence of MI. Hence, the MI can be suppressed by modifying the spatial dispersion to prevent the presence of k_z within the unstable frequency range.

As discussed in the last section, the spatial dispersion can be tailored by introducing a small-scale periodic modulation of the potential. The new spatial dispersion is obtained disregarding homogeneous terms and system nonlinearities and only considering the normal diffraction of each mode (red parabolas in Fig. 4b) and their coupling through the modulated potential:

$$\partial_z A = i \partial_{xx}^2 A + 4im \cos(q_x x) \cos(q_z z) A \tag{9}$$

This equation is analogous to (4), but, in contrast to previously considered case, has a pure real potential amplitude, m. The choice of appropriate modulation parameters can modify the spatial dispersion allowing to avoid the frequency range associated to the instability. The solid blue lines in Fig. 3b correspond to a distorted dispersion given by a real potential amplitude, m, showing no k_z values in the unstable range.

Recently, it has been shown that the MI suppression by spatial modulations of the system potential is possible in BAS amplifiers [24]. This would allow for the control of the very onset of unstable pattern dynamics by decreasing the range of unstable wavenumbers or totally eliminating the instability.

The linear approximation of the previous section is only applicable to small field amplitudes while a model that takes into account the system nonlinearities is necessary for the characterization and manipulation of the MI. Considering small carrier diffusion and maintaining the system nonlinearity, (3) is simplified to the field equation:

$$\frac{\partial A}{\partial z} = \frac{i}{2k_0^2} \frac{\partial^2 A}{\partial x^2} + s \left[\frac{P(x, z) - 1}{1 + |A|^2} (1 - i\alpha_H) - i\alpha_H - \gamma \right] A \tag{10}$$

where s denotes the inverse absorption length in a cold semiconductor media, that introduces a space renormalization ($x \rightarrow x/\sqrt{s}, z \rightarrow z/s$). The similarity of this

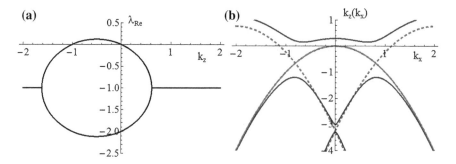

Fig. 4 **a** Real part of eigenvalues as a function of the dispersion relation $k_z(k_x)$ calculated from (8) with c$=0.5$. **b** Dispersion relation for normal parabolic diffraction (*solid red*) and dispersion relations of first harmonics (*dashed red*) for the modulated system of (9) with $q_x = 1.0$, $q_z = 2.0$. Dispersion relation of the system for a modulation amplitude $m = 0.7$ (*blue*)

system to the CGLE (7) is easily found by expanding the nonlinearity and just taking the first terms, obtaining:

$$\frac{\partial A}{\partial z} = \frac{i}{2k_0^2} \frac{\partial^2 A}{\partial x^2} + s \left[(P - 1)(1 - i\alpha_H)(1 - |A|^2) - i\alpha_H - \gamma \right] A \qquad (11)$$

where the linewidth enhancement factor of the semiconductor, α_H, is equivalent to the nonlinear coefficient c and pump P becomes a multiplying factor of the nonlinearity. Linear terms $-i\alpha_H - \gamma$ correspond to a homogeneous phase shift and homogeneous field attenuation, respectively, and do not have any effect on the stability of a pure amplifying system.

The stationary homogeneous solution of unmodulated (10) has a field amplitude $A_0 = \sqrt{(p_0 - 1)/\gamma - 1}$ and represents the plane wave propagating in a saturated semiconductor media. The linear stability analysis is performed in an analogous way of (7). We perturb the homogeneous solution with weak transverse perturbations such that the field function has the form: $A = A_0 (1 + a_+ \exp(ik_x x + \lambda z) + a_- \exp(-ik_x x + \lambda z))$ where $|a_\pm| \ll 1$. This leads to the following expression for the growth exponent, λ, of the transverse perturbation mode,

$$\lambda(k_x) = \text{Re} \left\{ -\frac{c_1}{2} \pm i \frac{k_x}{2k_0} \sqrt{k_x^2 - 2\alpha_H c_1 k_0 - (c_1 k_0/k_x)^2} \right\} \qquad (12)$$

where $c_1 = 2s\gamma (p_0 - \gamma - 1)/(p_0 - 1)$. Equation (12) predicts a long wave instability with the maximum growth exponent $\lambda_{\max} = c_1 \left(\sqrt{1 + \alpha_H^2} - 1 \right)/2$ at modulation wavenumber $k_x = \sqrt{\alpha_H c_1 k_0}$. The instability is closely related with the factor α_H that for semiconductor lasers ranges between 1.5 and 8 [25, 26].

A good indicator of the total instability is the integral of the positive part of the growth spectra: $L = \int_{k_x, \lambda > 0} \lambda(k_x) \, dk_x$, which we refer to as the *global instability* and use to characterize the full or partial stabilization of the system. Note that we only consider spatial instabilities, since, in the case of single pass amplifiers, temporal instabilities can be disregarded.

We introduce a transverse and longitudinal 2D spatial modulation in the system with the aim of suppressing the MI of the system. The gain and refractive index modulation is introduced through the same periodically modulated pump considered for the linear BAS model: $P(x, z) = p_0 + 4m' \cos(q_x x) \cos(q_z z)$. Large modifications of the dispersion curve occurs near the resonance condition, $Q \approx 1$. In this parameter range, only three spatial harmonics are relevant and their spatial dispersion curves cross at small transverse wavenumbers, $k_x \approx 0$, see three red parabolas in Fig. 4b .

The steady state of the modulated system is a periodic pattern (or Bloch mode), following the modulation of the gain profile with a given field amplitude stabilized by the nonlinearity. It can be obtained by the expansion of the field into three main harmonics of modulation:

$$A(x, z) = A_{0,0} + A_{-1,-1} e^{-iq_x x - iq_z z} + A_{1,-1} e^{iq_x x - iq_z z} \qquad (13)$$

The steady solution is a locked state of the harmonics $A_{0,0}$, $A_{-1,-1}$, $A_{1,-1} \neq 0$, which can be determined by numerical integration of (10) using the expansion of (13).

We perform the linear stability analysis numerically, through a modified Floquet procedure for each transverse mode [23, 24]. For the spatial homogenous system, only k_x and $-k_x$ are coupled, but the introduction of the modulated potential causes a linear coupling between the spatial harmonics of the perturbation, k_x, $k_x \pm q_x$, $k_x \pm 2q_x$, ... increasing the dimensionality of the problem and the number of calculated Floquet multipliers. The eigenvalues λ of the stationary homogeneous solution are calculated from the logarithm of the Floquet multipliers and their real part λ_{Re}, the Lyapunov exponent, indicates the growth or decay of harmonics over one period.

Figure 5 summarizes the stability analysis in parameter space (Q, m).

Around the resonance condition, $Q \approx 1$, the instability is significantly reduced although its complete suppression is not possible. The remaining weak instability can be associated to small k_x values generating long waves (LW), Fig. 5b, or to larger k_x values generating short waves (SW), Fig. 5c.

For the considered model, a remaining weak LW instability persists in all cases, for all parameters, even for the cases of strongest suppression (Fig. 5d). This result is different for the considered CGLE where the complete suppression of MI is found [23]. Two main differences are responsible of the different behaviors, the more complex nonlinearity of BAS amplifiers and the complex amplitude of the modulated potential intrinsically introduced by the linewidth enhancement factor, α_H.

Numerical integrations along the BAS amplifier confirm the impossibility to obtain the stable regimes showing the complete suppression of MI. Figure 6b shows the field propagation along the BAS with a strong suppression of the MI but still appearing long waves with a narrow unstable spectrum and field oscillations (same parameters of Fig. 5d).

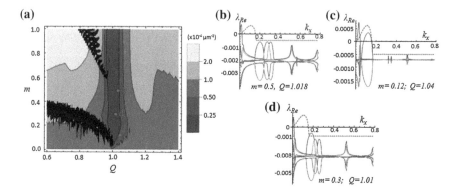

Fig. 5 Linear stability analysis by modified Floquet method: **a** map of the global instability L in the parameter space (Q, m). **b–d** Spectra of the real part of the eigenvalues or growth exponents for different sets of parameters: **b** remaining LW instability, **c** remaining SW and LW instability, **d** case of strongest suppression. In all cases the curves of growth exponents are compared to the unmodulated case (*red-dashed curve*). k_x and λ_{Re} are both expressed in μm^{-1}

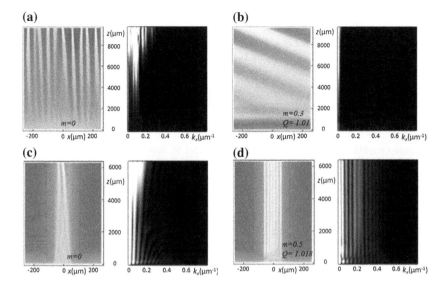

Fig. 6 Propagation along the unmodulated BAS amplifier (**a**, **c**) and for the modulated case (**b**, **d**). **a**, **b** correspond to BAS amplifiers with periodic boundary conditions, infinite width, and **c**, **d** correspond to a finite-width BAS. The plots show the intensity (*left*) and angular spectrum (*right*) of the optical field. For a clearer representation, the small space scale modulations have been filtered out. Parameters: ($p = 1.2$, $\alpha_H = 2.0$, $\gamma = 0.01$, $k_0 = 2\pi$, $s = 0.01$, width $= 128\,\mu$m)

In order to avoid the instability given by the small unstable spectrum range, we can consider BAS amplifiers of small enough width d, where the instability becomes irrelevant because the unstable part of the spectrum $\Delta k_x < 2\pi/d$ cannot develop. Figure 6c, d shows the field evolution through a BAS amplifier of 128 microns width. As shown in the figure, the modulation completely stabilizes the system leading to a stable, non-diverging beam (Fig. 6d).

5 Conclusions

To conclude, we show that a 2D spatial modulation in BAS amplifiers can lead to substantially improvement of the spatial structure of the amplified beam in the linear regime. The modulated pump current introduces index and gain modulations in the semiconductor, and, consequently, implies large variations in the spatial dispersion relation profile of the system and, thus, becomes an efficient mechanism to reduce and even suppress the modulation instability in the nonlinear regime.

In the linear regime, we show that proposed modulation induces a spatial filtering and significant quality improvement of the optical beam during a single-pass through a BAS amplifier with technically realizable modulation periods and a length on the order of one millimeter.

In the nonlinear regime, MI is not completely eliminated for infinitely extended systems but the gain modulation reduces the angular spectrum of the growing modes. For BAS amplifiers with widths of hundred micrometer, the MI is shown to be completely eliminated.

We expect that the proposed beam shaping and MI suppression in BAS amplifiers could be extended to BAS lasers to overcome the generally poor spatial quality of such relevant light sources.

Acknowledgments We acknowledge financial supported by Spanish Ministerio de Educación y Ciencia and European FEDER (project FIS2011-29734-C02-01). The work of M. Radziunas was supported by the Einstein Center for Mathematics Berlin under project D-OT2.

References

1. T. Burkhard, M.O. Ziegler, I. Fischer, W. Elsäßer, Spatio-temporal dynamics of broad area semiconductor lasers and its characterization. Chaos, Solitons Fractals **10**, 845–850 (1999)
2. L. Goldberg, M.K. Chun, Injection locking characteristics of a 1 W broad stripe laser diode. Appl. Phys. Lett. **53**, 1900–1902 (1988)
3. M. Radziunas, K. Staliunas, Spatial "rocking" in broad-area semiconductor lasers. Europhys. Lett. **95**, 14002 (2011)
4. V. Raab, R. Menzel, External resonator design for high-power laser diodes that yields 400mW of TEM00 power. Opt. Lett. **27**, 167–169 (2002)
5. S.K. Mandre, I. Fischer, W. Elsässer, Control of the spatiotemporal emission of a broad-area semiconductor laser by spatially filtered feedback. Opt. Lett. **28**, 1135–1137 (2003)
6. J. Marciante, G. Agrawal, Nonlinear mechanisms of filamentation in broad-area semiconductor lasers. IEEE J. Quantum Electron. **32**, 590 (1996)
7. H. Adachihara, O. Hess, E. Abraham, P. Ru, J.V. Moloney, Spatiotemporal chaos in broad-area semiconductor lasers. JOSA B **10**, 658 (1993)
8. F. Prati, L. Columbo, Long-wavelength instability in broad-area semiconductor lasers. Phys. Rev. A **75**, 053811 (2007)
9. K. Staliunas, R. Herrero, R. Vilaseca, Subdiffraction and spatial filtering due to periodic spatial modulation of the gain-loss profile. Phys. Rev. A **80**, 013821 (2009)
10. M. Botey, R. Herrero, K. Staliunas, Light in materials with periodic gain-loss modulation on a wavelength scale. Phys. Rev. A **82**, 013828 (2010)
11. R. Herrero, M. Botey, K. Staliunas, Nondiffractive-nondiffusive beams in complex crystals. Phys. Rev. A **89**, 063811 (2014)
12. G.P. Agrawal, N.A. Olsson, Self-phase modulation and spectral broadening of optical pulses in semiconductor laser amplifiers. IEEE J. Quantum Electron. **25**, 2297–2306 (1989)
13. E.A. Ultanir, D. Michaelis, F. Ledeerer, G.I. Stegeman, Stable spatial solitons in semiconductor optical amplifiers. Opt. Lett. **28**, 251–253 (2003)
14. R. Herrero, M. Botey, M. Radziunas, K. Staliunas, Beam shaping in spatially modulated broad-area semiconductor amplifiers. Opt. Lett. **37**, 5253–5255 (2012)
15. K. Staliunas, R. Herrero, Nondiffractive propagation of light in photonic crystals. Phys. Rev. E **73**, 016601 (2006)
16. L. Maigyte, K. Staliunas, Spatial filtering with photonic crystals. Appl. Phys. Rev. **2**, 011102 (2015)
17. M. Radziunas, M. Botey, R. Herrero, K. Staliunas, Intrinsic beam shaping mechanism in spatially modulated broad area semiconductor amplifiers. Appl. Phys. Lett. **103**, 132101 (2013)
18. M. Radziunas, R. Herrero, M. Botey, K. Staliunas, Far field narrowing in spatially modulated broad area edge-emitting semiconductor amplifiers. JOSA B **32**, 883–1000 (2015)

19. T. Winfree, Spiral waves of chemical activity. Science **175**, 634 (1972)
20. M.C. Cross, P.C. Hohenberg, Pattern-formation outside of equilibrium. Rev. Mod. Phys. **65**, 851 (1993)
21. D. Walgraef, *Spatio-Temporal Pattern Formation* (Springer, New York, 1997)
22. K. Staliunas, V.J. Sánchez-Morcillo, *Transverse Patterns in Nonlinear Optical Resonators* (Springer, Berlin, 2003)
23. S. Kumar, R. Herrero, M. Botey, K. Staliunas, Taming of Modulation Instability by Spatio-Temporal Modulation of the Potential. arXiv:1504.00236v3
24. S. Kumar, R. Herrero, M. Botey, K. Staliunas, Suppression of modulation instability in broad area semiconductor amplifiers. Opt. Lett. **39**, 5598 (2014)
25. M. Mullane, J.G. McInerney, Minimization of the linewidth enhancement factor in compressively strained semiconductor lasers. IEEE Photonics Technol. Lett. **11**, 776 (1999)
26. M. Osinski, J. Buus, Linewidth broadening factor in semiconductor lasers-An overview. IEEE J. Quantum Electron. **23**, 9 (1987)

Weakly Nonlinear Analysis and Localized Structures in Nonlinear Cavities with Metamaterials

N. Slimani, A. Makhoute and M. Tlidi

Abstract We consider an optical ring cavity filled with a metamaterial and with a Kerr medium. The cavity is driven by a coherent radiation beam. The modelling of this device leads to the well known Lugiato-Lefever equation with high order diffraction. We show that this effect alters in depth the space-time dynamics of this device. A weakly nonlinear analysis in the vicinity of the first threshold associated with the Turing instability is performed. This analysis allows us to determine the parameter regime where transition from super- to sub-critical bifurcation occurs. When the modulational instability appears subcritically, we show that bright localized structures of light may be generated in two-dimensional setting. Close to the second threshold associated with the Turing instability, dark localized structures are generated and their snaking bifurcation diagram is constructed.

1 Introduction

Dissipative structures in far from equilibrium systems can be either periodic or localized in space. They have been observed experimentally in various nonlinear chemical, biological, hydrodynamical, and optical systems (see recent overviews on this issue [1–6]). In nonlinear optics, driven resonators filled with a nonlinear medium are the basic configuration. More specifically, analytical studies have demonstrated that the

N. Slimani
Physique du Rayonnement et des Interactions Laser-Matière, Faculté des Sciences,
Université Moulay Ismail, B.P. 11201 Zitoune, Meknès, Morocco

A. Makhoute
Physique du Rayonnemet et des Interactions Laser-Matiere, Faculté des Sciences,
Département de Physique, Université Moulay Ismail, B.P. 11201 Zitoune,
Meknès, Morocco

M. Tlidi (✉)
Faculté des Sciences, Université Libre de Bruxelles (U.L.B.), CP. 231,
Campus Plaine, 1050 Bruxelles, Belgium
e-mail: mtlidi@ulb.ac.be

© Springer International Publishing Switzerland 2016   153
M. Tlidi and M.G. Clerc (eds.), *Nonlinear Dynamics: Materials,*
Theory and Experiments, Springer Proceedings in Physics 173,
DOI 10.1007/978-3-319-24871-4_11

competition between diffraction, nonlinearity and dissipation allows for the spontaneous formation of dissipative structures that can be either periodic [7] or localized
in space [8, 9]. Localized structures (LS's), often called cavity solitons, consist of
bright or dark pulses in the transverse plane orthogonal to the propagation axis. The
two dimensional spatial confinement of light was predicted [10, 11] and experimentally confirmed [12–14, 16–20]. Other mechanisms such as strong nonlocal coupling
[21, 22] or inhomogeneities of the pump could lead to the stabilisation of LS's in
regime devoid of Turing type of instability [23–25].

The However, one of the fundamental limitations of LSs is their spatial size, which is
due to the diffraction phenomenon in optical resonators. Even in the smallest soliton systems, the semiconductor microresonators, the width of LS is of the order of
15 μm. To overcome this limit, we have proposed to use left-handed metamaterials,
i.e., engineered materials with simultaneously negative permittivity and permeabil-
ity [26, 27]. These materials were first used by the scientific community to realize
imaging systems with sub-wavelength resolution then to target potential applications
including invisibility cloaks and perfect optical concentrators. They have been first
demonstrated at microwave frequencies [28, 29], and soon after in the optical domain
[30, 31]. Metamaterials are shown to exhibit novel electromagnetic phenomena such
as subwavelength imaging [32, 33] or negative diffraction, and can be used in nonlinear optical devices [6, 34–37]. In particular, the formation of LS's in a Kerr resonator
containing a metamaterial has been studied in [36, 38–44].

The combination of right-handed and left-handed materials offers the possibility
to design devices in which the effective diffraction is zero. Such systems are encountered, for example, in nonlinear optical cavities, where a true zero-diffraction regime
could lead to the formation of patterns with arbitrarily small size [41]. It has been
shown that around the zero-diffraction regime in an optical cavity containing both
left-and right-handed materials, one- and two-dimensional dark localized structures
can become stable [45]. The stabilization of these structures is the result of high-
order diffraction modeled by a bi-Laplacian term with a complex coefficient. In one
spatial dimension, the existence of a snaking bifurcation diagram has been demonstrated for these solutions, which shows a larger complexity than generally observed
in homoclinic snaking [45]. In two dimensional setting, numerical simulations have
demonstrated a similar coexistence of multiple dips in the intensity profile [45].

The aim of this contribution is twofold. First, it is to clarify analytically the formation of dissipative structures in cavity filled with left handed materials. For this
purpose we apply a weakly nonlinear analysis in the vicinity of the first threshold
associated with the Turing instability. This analysis allows us to establish the parameter range where the bifurcation appears sub-critical. Second, it is to study the
implication of high order diffraction on the stability of bright localized structures.
Throughout this contribution, we focus on two dimensional structures that occur in
the plane perpendicular to the propagation axis.

The paper is organized as follows. After briefly introducing the Lugiato-Lefever
equation with high-order diffraction describing a driven ring cavity filed with right-
handed and left-handed materials (Sect. 2), we present a weakly nonlinear analysis in

the neighborhood of the first threshold associated with the Turing instability (Sect. 3). The implications of this analysis in the formation of bright and dark localized structures are considered in Sect. 4. We conclude in Sect. 5.

2 Lugiato-Lefever Model with High-Order Diffraction

We consider a ring cavity filled with two materials having indices of refraction of opposite signs (see Fig. 2). The Kerr media with positive refractive index, while the second material consists in a linear material with a negative refractive index. The ring resonator filled with a right-handed and left-handed materials is described by the well known Lugiato-Lefever model [7], in which we incorporate the high-order diffraction effect. This model equation is valid under the following approximations: (i) at the interface separating the right-handed and left-handed materials, the reflection is assumed to be negligible, i.e., they are impedance matched; (ii) the cavity possess a high Fresnel number i.e., large-aspect-ratio system and we assume that the cavity is much shorter than the diffraction and the nonlinearity spatial scales; (iii) for the sake of simplicity, we assume a single longitudinal mode operation. Under these assumptions, the space-time evolution of the intracavity field is described by the following partial differential equation [41]

$$\frac{\partial A}{\partial t} = A_i - (1 + i\theta)A + i |A|^2 A + i\delta \nabla_\perp^2 A + i\beta \nabla_\perp^4 A, \qquad (1)$$

where A is the slowly varying envelope of light within the ring cavity, A_i is the amplitude of the injected field, and θ is the normalized cavity detuning parameter. The Laplace operator that acts on the transverse plane is $\nabla_\perp^2 = (\partial_{xx}^2 + \partial_{yy}^2)$. The effective diffraction coefficient is denoted by δ. The coefficient β accounts for the high-order diffraction effects in the cavity. In the absence of high order diffraction, we recover the well known Lugiato-Lefever (LL) model [7]. The inclusion of metamaterial slice allows us to explore the parameter regime where diffraction is negative [38]. From a practical point of view, negative diffraction could be achieved by using a self-imaging configuration [25, 46, 47]. In the absence of high order diffraction, the stabilization of localized solutions far from any pattern-forming instability has been investigated thanks to the combined influence of a negative diffraction and an inhomogeneous pumping laser beam [25]. The laser cavity can be also considered as an inhomogeneous cavity in the work by Zhang [48, 49] (Fig. 1).

The homogeneous steady states (HSS) of (1) satisfy $A_i^2 = [1 + (\theta - |A_s|^2)^2]|A_s|^2$. High-order diffraction does not affect the HSS. They are thus identical to the ones of LL model [7]. For $\theta < \sqrt{3}$ ($\theta > \sqrt{3}$) the transmitted intensity as a function of the input intensity is monostable (bistable). However, high-order diffraction plays an important role in the linear stability analysis of the HSS with respect to finite wavelength perturbations of the form $\exp(\sigma t + i\mathbf{q}.\mathbf{r}_\perp)$ with $\mathbf{q} = (q_x, q_y)$ yields the characteristic equation

Fig. 1 Schematic setup of a ring cavity filled with right-handed and left-handed materials. The cavity is driven by a coherent external injected beam. *RHM* and *LHM* denotes right-handed material and left-handed material, respectively

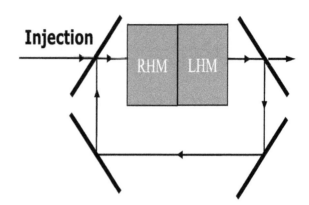

$$(\lambda + 1)^2 + 3I^2 - 4I\theta + \theta^2 + 2(2I - \theta)\left(-\delta + \beta q^2\right)q^2 + \left(-\delta + \beta q^2\right)^2 q^4 = 0 \quad (2)$$

where $I = |A_s|^2$ and $\mathbf{r}_\perp = (x, y)$. The homogeneous steady states undergo a Turing bifurcation when $\sigma = 0$ and $\partial\sigma/\partial q = 0$. These two conditions are realized at

$$|A_{T1}|^2 = 1 \text{ and } |A_{T2}|^2 = [2(\delta^2/(4\beta) + \theta) + \sqrt{(\delta^2/(4\beta) + \theta)^2 - 3/3}]$$

The first threshold may be degenerate, i.e., two marginally unstable wavelengths appear simultaneously and spontaneously in the system; q_l^2 and q_u^2; with

$$q_{l,u}^2 = (\delta \pm \sqrt{\delta^2 + 4\beta(\theta - 2)}/(2\beta). \quad (3)$$

At the second threshold I_{T2}, a new, large, critical wavelength become marginally unstable $q_{T2}^2 = \delta/2\beta$.

In contrast with the LL model where the instability domain is unbounded for large intensity values, high-order diffraction allows the Turing instability to possess a finite domain of existence delimited by two pump power values, allowing for the stationary state to stabilize at large powers.

3 Weakly Nonlinear Analysis

In what follows, we focus on the first bifurcation where both frequencies coincides, i.e., $q_l^2 = q_u^2$. This implies that $\delta^2 + 4(\theta - 2)\beta = 0$. Under this condition, the two thresholds associated with the Turing instability are $A_{1m} = 1$ and $A_{2m} = 5/3$ and the critical wavelength at both bifurcation points is $q_c^4 = (2 - \theta)/\beta$. We will not consider the situation where the first threshold is degenerate. This problem will be addressed elsewhere.

In this section we shall describe the nonlinear evolution of the system in the vicinity of the first instability point $|A_{T1}|^2 = 1$. The corresponding injected field intensity is $A_{iT1}^2 = 1 + (\theta - 1)^2$. The small-amplitude inhomogeneous stationary solutions (stripes, hexagons, rhomboids) can be calculated analytically by employing the standard theory [50]. For this purpose we first decompose the electric field into its real and imaginary parts: $A = x_1 + ix_2$ and introduce the excess variables u and v:

$$x_1 = x_{1s} + u(\mathbf{r}, t) \text{ and } x_2 = x_{2s} + v(\mathbf{r}, t) \tag{4}$$

with x_{1s} and x_{2s} are the real and the imaginary parts of the homogeneous stationary solutions of (1) given by

$$- x_{1s} + A_i - x_{1s}(x_{1s}^2 + x_{2s}^2 - \theta) = 0 \tag{5}$$

$$-x_{2s} + x_{2s}(x_{1s}^2 + x_{2s}^2 - \theta) = 0 \tag{6}$$

The solution of these equations, evaluated at the first threshold of the Turing instability, are $x_{1T} = 1/A_i$ and $x_{1T} = (1 - \theta)/A_i$.

Inserting $A = x_1 + ix_2$ in the generalized LL model (1) and using (5) and (6), we obtain the following time-space evolution equations for the real and the imaginary parts

$$\frac{\partial u}{\partial t} = -(1 + 2x_{1s}x_{1s})u - (x_{1s}^2 + 3x_{2s}^2 - \theta + \delta\nabla_\perp^2 + \beta\nabla_\perp^4)v \tag{7}$$
$$- (x_{2s}u^2 + 2x_{1s}uv + 3x_{2s}v^2 + u^2v + v^3)$$

$$\frac{\partial v}{\partial t} = - (3x_{1s}^2 + x_{2s}^2 - \theta + \delta\nabla_\perp^2 + \beta\nabla_\perp^4)u - (1 + 2x_{1s}x_{1s})v \tag{8}$$
$$+ (3x_{1s}u^2 + 2x_{1s}uv + x_{1s}v^2 + uv^2 + u^3)$$

To explore the space-time dynamics of the system, we introduce a small parameter $\varepsilon \ll 1$ which measures the distance from the critical point, and expand the input field amplitude and all variables around their critical values at the bifurcation point

$$A_i = A_{iT1} + \varepsilon A_1 + \varepsilon^2 A_2 + \cdots \tag{9}$$
$$(u, v) = \varepsilon(u_0, v_0) + \varepsilon^2(u_1, v_1) + \varepsilon^3(u_2, v_2) + \cdots \tag{10}$$
$$(x_{1s}, x_{2s}) = (x_{1T}, x_{2T}) + \varepsilon(a_1, b_1) + \varepsilon^2(a_2, b_2) + \cdots \tag{11}$$

and also introduce the slow time $\tau = \varepsilon^2 t$. Near the critical point we can approximate the solution to leading order in ε as a linear superposition of the critical modes

$$(u_0, v_0) = \left(1, \frac{\rho + 3}{1 - 3\rho}\right) \sum_{i=1}^{l} [\tilde{W}_i \exp(i\mathbf{q_i}.\mathbf{r}_\perp) + c.c.] \tag{12}$$

where the parameter ρ is $\rho = 1 - \theta$ and the *c.c.* denotes the complex conjugate. The stripes and the rhomboids are characterized by $l = 1$ and $l = 2$, respectively, and the hexagons are obtained for $l = 3$ with $\sum_{i=1}^{3} \mathbf{q_i} = \mathbf{0}$. We assume in addition that that the complex amplitude $\tilde{W}_i$ associated with the mode $\mathbf{q_i}$ depends only on the slow time and it is independent of the transverse coordinates $\mathbf{r}_\perp = (x, y)$. The quantities a_i, b_i, u_i and v_i can be calculated by inserting (10)–(11) into (7)–(9) and equating terms with the same powers of ε. The application of the solvability condition to the higher-order inhomogeneous problem leads to amplitude equations for the unstable modes. For the stripes and in terms of the unscaled amplitude ($W_1 = \varepsilon \tilde{W}_1....$) and by taking into account the slow time $\tau = \varepsilon^2 t$, we have

$$\frac{\partial W_1}{\partial t} = \mu W_1 - g(\rho)|W_1|^2 W_1, \tag{13}$$

where

$$\mu = \frac{(A_i - A_{iT1})(\rho^2 + 1)}{(1 + \rho)^2} \tag{14}$$

$$g(\rho) = \frac{8(171\rho - 85)(\rho^2 + 1)^2}{81(\rho^2 - 1)^2}, \tag{15}$$

The bifurcation is supercritical or subcritical depending on the sign of $g(\rho)$. The transition from supercritical to subcritical takes place for $\rho = 85/171 \approx 0.5$, i.e., for $\theta = 256/171 \approx 1.5$. This means that this transition occurs in the monostable regime $\theta < \sqrt{3} \approx 1.7$. However, in the absence of high-order diffraction, this transition takes place for $\theta = 41/30 \approx 1.366$. If we introduce the polar decomposition $W_1 = W(\exp i\psi)$ we see that the phase ψ is arbitrary and the real stationary amplitudes of the stripes are given by $W = 0$ and $W = \sqrt{\mu/g}$. Note that, the stripe solutions have never been observed numerically in the absence of high-order diffraction. Indeed, analytical study of LL equation through the relative stability analysis has proved that stripes are always instable with respect the hexagonale structures [51]. However, the presence of the high-order diffraction term, i.e. $\beta \nabla_\perp^4 A$, in the (1), allows to stabilize stripes structures. We have performed numerical simulations of the generalized LL model (1) in two transverse dimensions with periodic boundary conditions. The results indicate that stable stripes can occurs as shown in Fig. 2 (b_1, b_2).

Without high order diffraction, a weakly nonlinear analysis of LL model show that only the H0 hexagons are stable [51]. The H0 hexagons persists in the presence of high-order diffraction as shown in Fig. 2 (a_1, a_2). However, when taking into account high-order diffraction another type of hexagonal structures; Hπ; can be stabilized as shown in Fig. 2 (c_1, c_2).

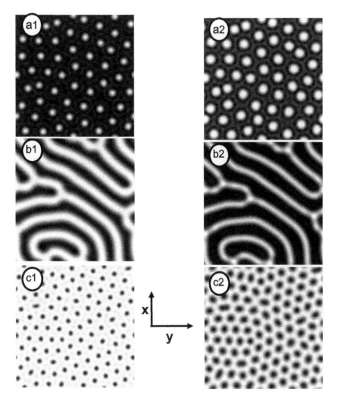

Fig. 2 As the injected beam amplitude increases, one first finds hexagonal H0, which then transform into stripes, and finally into a structure of hexagons Hπ. The three different kind of pattern are obtained for $\theta = 2$; $\delta = 0.5$; $\beta = 0.1$. (a_1, a_2) $A_i = 1.4$, (b_1, b_2) $A_i = 1.4$ and (c_1, c_2) $A_i = 1.5$. The grid is 128×128 points. *White* corresponds to the highest values of the electric field amplitude. Minima are *plain black*. Periodic boundary conditions are used in both spacial coordinates

4 Localized Structures and Localized Patterns

Many driven systems exhibit localized structures, often called localized spots, cavity solitons or localized patterns, which may be either isolated, randomly distributed, or self-organized in clusters forming a well-defined spatial pattern. Currently they attract growing interest in optics due to potential applications for all-optical control of light, optical storage, and information processing [12–20]. When they are sufficiently separated from each other, localized peaks are independent and randomly distributed in space. However, when the distance between peaks decreases they start to interact via their oscillating, exponentially decaying tails. This interaction then leads to the formation of clusters [52–55]. Localized structures (LSs) are homoclinic solutions (solitary or stationary pulses) of partial differential equations. The conditions under which LSs and periodic patterns appear are closely related. Typically, when the Turing instability becomes sub-critical, there exists a pinning domain where

localized structures are stable. In the following we discuss the formation of bright
and dark localized structures. The weakly nonlinear analysis in the vicinity of the
first threshold allows to determine the condition under which the Turing bifurcation
appears subcritically, i.e., $\theta = 256/171 \approx 1.5$. The subcritical nature of the bifurcation induces a regime where the homogeneous steady state coexist with a periodic
pattern such hexagons or stripes. The homoclinic nature of these solutions implies
that for a given set of control parameters, the number and the space distribution of
LS's immersed in the bulk of the basic homogeneous solution is determined by the
initial condition. Two-dimensional (2D) LS's are stable in nonvariational systems
such as the generalized Swift-Hohenberg equation and their existence do not require
any commutation process between homogeneous steady states [10]. Typically, LS's
arise in the regime where a homogeneous steady state solution exhibits a pattern
forming instability leading to the formation of ordered or self-organized dissipative
structures. A subcritical nature of the instability induces multistability behaviour

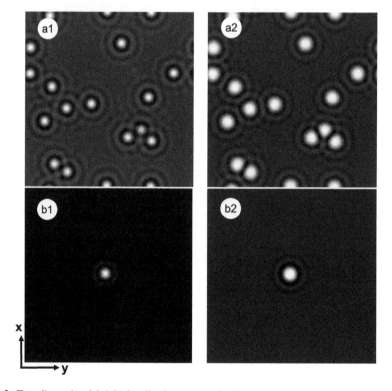

Fig. 3 Two-dimensional bright localized structures in the transverse plane (x, y). $(a_1$ and $a_2)$
corresponds to random distribution of localized bright spots $(b_1$ and $b_2)$ correspond to a single
bright localized structure. a_1 and b_1 are the spatial frofies of the real part, and a_2 and b_2 are the
profiles of the imaginary parts, respectively. The grid is 128×128 points. *White* corresponds to
the highest values of the electric field amplitude. Minima are *plain black*. Parameters are $\theta = 2$;
$\delta = 0.1$; $\beta = 0$ and $A_i = 1.45$

between the homogeneous solution and branches of self-organized periodic solutions that emerge through a pattern selection process. In that hysteresis loop there exist a so called a pinning range of parameters in which localized structures are stable. From dynamical systems point of view they are homoclinic solutions connecting the homogeneous and the periodic solutions. An example of bright LS's is shown in Fig. 3. A random distribution of bright LS's as well as a single LS's are shown in that figure. They have been obtained for the same values of parameters but with different initial conditions. The mechanism of stabilization of LS's is attributed to the nonvariational effects, i.e., the absence of a free energy or a Lyapunov functional to minimize in the vicinity of the subcritical bifurcation. Amplitude equations describing the space-time evolution of slow unstable mode derived in the weakly nonlinear regime cannot describe localized structures. This approach does not take into account the nonadibatic effects that involve fast spatial scales which are reponsable for the stabilization of LS's [56].

In the last part of this section, we focus on the second Turing instability which is generated thanks to the presence of the high-order diffraction. Depending on the initial conditions the system either evolves towards the homogeneous steady state or a periodic patterns which appear subcritically. As in the case of bright LS's, there exist a finite domain of coexistence between stable periodic structures and a stable upper HSS. In this domain, we can generate an infinite set of odd and even dark LS's, i.e., a set of stationary solutions that exhibit $n = 2p - 1$ or $n = 2p$ peaks, where p is a positive integer. The limit $p \to \infty$, corresponds to the infinitely extended periodic pattern distribution. Examples of localized structures and clusters of them having odd number of dips are shown in Fig. 4. They are obtained for the same parameter values and differ only by the initial conditions. In the pinning region, the wavelength of dark LS's is close to that of the periodic structure, i.e.,

$$\lambda \approx \pi / q_c = \pi \left(\frac{\beta}{2 - \theta} \right)^{1/4}. \tag{16}$$

Since the amplitudes of localized patterns having different number of dips are close to one another, in order to visualize the clusters properties, it is convenient to plot the "L_2 -norm"

$$\mathcal{N} = \int dx |A|^2 \tag{17}$$

as a function of A_i. This yields the two snaking curves with odd or even number dips in the spatial profile of the electric field circulating in the cavity. An example of dark localized structures having an even number of dips are shown in Fig. 4. As $\mathcal{N}$ increases, at each turning point where the slope becomes infinite, a pair of additional peaks appears in the cluster. This behavior is often called homoclinic snaking phenomenon [57–60]. These solutions are found by using appropriate initial conditions and are then continued in parameter space using a Newton method. Periodic boundary conditions are used. A typical snaking diagram consists of two snaking curves; one describes dark LS's with an odd number of dips and the other corresponds to an

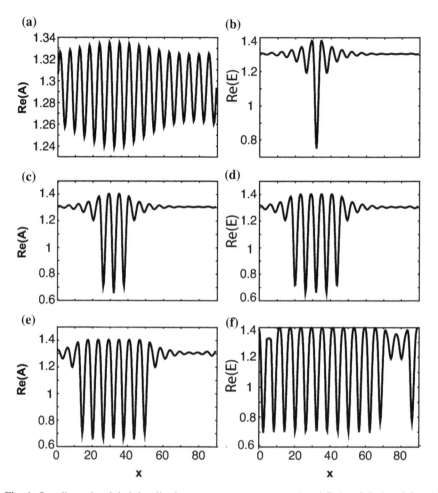

Fig. 4 One dimensional dark localized structures. parameters are $\theta = 1.7$; $\delta = 0.5$; $\beta = 0.2$ and $A_i = 1.305$

even number of dips. The branch corresponding to an odd number of dips is shown in the Fig. 5. The dark LS's bifurcate at the Turing instability of the HSS and are initially unstable. As one moves further along the snaking curve, dark LS's become better localized and acquire stability at the turning point where the slope becomes infinite. Afterwards, the dark LS begin to grow in spatial extent by adding extra dips symmetrically at either side. This growth is associated with back and forth oscillations across the pinning interval. This digram has been obtained in the case of all fiber cavity filled with a photonic crystal fiber [61]. Similar behavior has been described in cavity filled with metamaterials [45].

In two-dimensional setting, the number of dark localized structures is much larger than in 1D. Examples of 2D dark LS's are plotted in Fig. 6. Snapshots (a_1, b_1) corre-

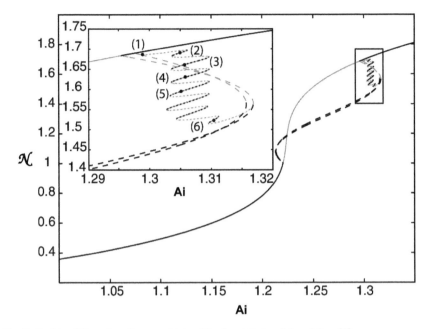

Fig. 5 Snaking bifurcation diagram obtained for $\theta = 1.7$; $\delta = 0.1$ and $\beta = 0.2$

spond to the real part, and snapshots (a_2, b_2) to the imaginary part of the intracavity field, respectively. All this figures are obtained for the same values of parameter. The system exhibits a high degree of multistability. The boundary conditions used in all our numerical simulations are periodic.

5 Conclusions

In this contribution we have investigated the formation of dissipative structures in driven resonators filled with right-handed and left-handed materials. We show that high order diffraction affects in depth the formation of dissipative structures by stabilising both stripes and Hπ hexagons. We have performed a weakly linear analysis in the vicinity of the first threshold associated with the Turing instability. We have shown that the transition to sub-critical bifurcation occurs for the detuning parameter greater than $\theta = 256/171 \approx 1.5$. We have shown also that high-order diffraction is responsible for the stabilization of both stripes and Hexagons Hπ. In the subcritical regime both bright and dark localized structures have been analysed including their snaking bifurcation diagram.

Our analysis should be applicable to all fiber resonator with high order dispersion effect. A cavity filled with photonic crystal fiber [62–65]. In this case the coupling is provided by dispersion. When dispersion and diffraction have a comparable

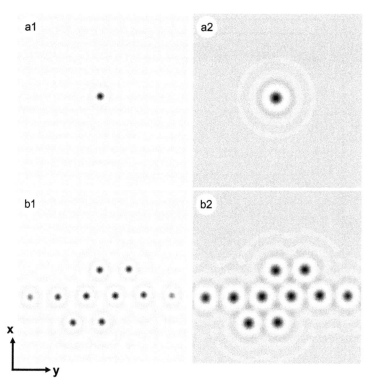

Fig. 6 Two dimensional dark localized structures. (a_1, a_2) random distribution of localized dips (b_1, b_2) single localized dips. (a_1, b_1) are the spatial profiles of the real part, and a_2 and b_2 are the profiles of the imaginary parts, respectively. The grid is 128×128 points. *White* corresponds to the highest values of the electric field amplitude. Parameters are $\theta = 2; \delta = 0.5; \beta = 0.1$ and $A_i = 1.7$

influence, three dimensional localized structures can be generated [67, 68, 70–73]. These structures consist of regular 3D lattices or localized bright light bullets traveling at the group velocity of light in the material. We plan to extend our analysis to three dimensional cavities.

Acknowledgments M.T. received support from the Fonds National de la Recherche Scientifique (Belgium). M.T. acknowledges the financial support of the Interuniversity Attraction Poles program of the Belgian Science Policy Office, under grant IAP 7-35 photonics@be.

References

1. P. Mandel, M. Tlidi, Transverse dynamics in cavity nonlinear optics (2000–2003). J. Opt. B Quantum Semiclass. Opt. **6**, R60 (2004)
2. B.A. Malomed, D. Mihalache, F. Wise, L. Torner, Spatiotemporal optical solitons. J. Opt. B Quantum Semiclass. Opt. **7**, R53 (2005)

3. M. Tlidi, M. Taki, T. Kolokolnikov, Introduction: dissipative+ localized structures in extended systems. Chaos Interdiscip. J. Nonlinear Sci. **17**, 037101 (2007)
4. H. Leblond, D. Mihalache, Models of few optical cycle solitons beyond the slowly varying envelope approximation. Phys. Rep. **523**, 61 (2013)
5. M. Tlidi, K. Staliunas, K. Panajotov, A.G. Vladimiorv, M. Clerc, Localized structures in dissipative media: from optics to plant ecology. Phil. Trans. R. Soc. A **372**, 20140101 (2014)
6. M. Lapine, I.V. Shadrivov, Y.S. Kivshar, Colloquium: nonlinear metamaterials. Rev. Mod. Phys. **86**, 1093 (2014)
7. L.A. Lugiato, R. Lefever, Phys. Rev. Lett. **58**, 2209 (1987)
8. D. McLaughlin et al., Phys. Rev. Lett. **51**, 75 (1983)
9. N. Rosanov, G. Khodova, Opt. Spektrosk. **65**, 1375 (1988)
10. M. Tlidi, P. Mandel, R. Lefever, Phys. Rev. Lett. **73**, 640 (1994)
11. A.G. Scroggie et al., Chaos Solitons Fractals **4**, 1323 (1994)
12. G. Slekys, K. Staliunas, C. Weiss, Opt. Commun. **149**, 113 (1998)
13. V.B. Taranenko, I. Ganne, R.J. Kuszelewicz, C.O. Weiss, Phys. Rev. A **61**, 063818 (2000)
14. X. Hachair, F. Pedaci, E. Caboche, S. Barland, M. Giudici, J.R. Tredicce, F. Prati, G. Tissoni, R. Kheradmand, L.A. Lugiato, I. Protsenko, M. Brambilla, Selected Topics in IEEE Journal of Quantum Electronics **12**(3), 339 (2006)
15. X. Hachair, G. Tissoni, H. Thienpont, K. Panajotov, Phys. Rev. A **79**, 011801 (2009)
16. P.V. Paulau, D. Gomila, T. Ackemann, N.A. Loiko, W.J. Firth, Phys. Rev. E **78**, 016212 (2008)
17. P. Genevet, S. Barland, M. Giudici, J.R. Tredicce, Phys. Rev. Lett. **101**, 123905 (2008)
18. V. Odent, M. Taki, E. Louvergneaux, New J. Phys. **13**, 113026 (2011)
19. F. Haudin, R.G. Rojas, U. Bortolozzo, S. Residori, M.G. Clerc, Phys. Rev. Lett. **107**, 264101 (2011)
20. E. Averlant, M. Tlidi, H. Thienpont, T. Ackemann, K. Panajotov, Opt. Exp. **22**, 762 (2014)
21. C. Fernandez-Oto, M.G. Clerc, D. Escaff, M. Tlidi, Phys. Rev. Lett. **110**, 174101 (2013)
22. D. Escaff, C. Fernandez-Oto, M.G. Clerc, M. Tlidi, Phys. Rev. E **91**, 022924 (2015)
23. A.G. Vladimirov et al., Opt. Exp. **14**, 1 (2006)
24. C. Fernandez-Oto, G.J. de Valcárcel, M. Tlidi, Panajotov, K. Staliunas, Phys. Rev. A **89**, 055802 (2014)
25. V. Odent, M. Tlidi, M.G. Clerc, P. Glorieux, E. Louvergneaux, Phys. Rev. A **90**, 011806(R) (2014)
26. V.G. Veselago, Sov. Phys. Usp. **10**, 509 (1968)
27. J.B. Pendry, Phys. Rev. Lett. **85**, 3966 (2000)
28. R.A. Shelby, D.R. Smith, S. Schultz, Science **292**, 77 (2001)
29. K. Aydin, K. Guven, M. Kafesaki, L. Zhang, C.M. Soukoulis, E. Ozbay, Opt. Lett. **29**, 2623 (2004)
30. C. Enkrich, M. Wegener, S. Linden, S. Burger, L. Zschiedrich, F. Schmidt, J.F. Zhou, T. Koschny, C.M. Soukoulis, Phys. Rev. Lett. **95**, 203901 (2005)
31. T.F. Gundogdu, I. Tsiapa, A. Kostopoulos, G. Konstantinidis, N. Katsarakis, R.S. Penciu, M. Kafesaki, E.N. Economou, T. Koschny, C.M. Soukoulis, Appl. Phys. Lett. **89**, 084103 (2006)
32. D.R. Smith, J.B. Pendry, M.C.J.K. Wiltshire, Science **305**, 788 (2004)
33. M.W. Feise, I.V. Shadrivov, Y.S. Kivshar, Phys. Rev. E **71**, 037602 (2005)
34. A.A. Zharov, N.A. Zharova, I.V. Shadrivov, Y.S. Kivshar, Appl. Phys. Lett. **87**, 091104 (2005)
35. N. Lazarides, G.P. Tsironis, Phys. Rev. E **71**, 036614 (2005)
36. G. D'Aguanno, N. Mattiucci, M. Scalora, M.J. Bloemer, Phys. Rev. Lett. **93**, 213902 (2004)
37. M. Scalora, M.S. Syrchin, N. Akozbek, E.Y. Poliakov, G. D'Aguanno, N. Mattiucci, M.J. Bloemer, A.M. Zheltikov, Phys. Rev. Lett. **95**, 013902 (2005)
38. P. Kockaert, P. Tassin, G. Van der Sande, I. Veretennicoff, M. Tlidi, Phys. Rev. A **74**, 033822 (2006)
39. P. Tassin, G. Van der Sande, N. Veretenov, P. Kockaert, I. Veretennicoff, M. Tlidi, Opt. Exp. **14**, 9338 (2006)
40. P. Tassin, L. Gelens, J. Danckaert, I. Veretennicoff, G. Van der Sande, P. Kockaert, M. Tlidi, Chaos **17**, 037116 (2007)

41. L. Gelens, G. van der Sande, P. Tassin, M. Tlidi, P. Kockaert, D. Gomila, I. Veretennicoff, J. Danckaert, Phys. Rev. A **75**, 063812 (2007)
42. A.D. Boardman, N. King, R.C. Mitchell-Thomas, V.N. Malnev, Y.G. Rapoport, Metamaterials **2**, 145 (2008)
43. A.B. Kozyrev, I.V. Shadrivov, Y.S. Kivshar, Appl. Phys. Lett. **104**, 084105 (2014)
44. A.D. Boardman, K.L. Tsakmakidis, R.C. Mitchell-Thomas, N.J. King, Y.G. Rapoport, O. Hess, Springer Series in Materials Science **200**, 161 (2015)
45. M. Tlidi, P. Kockaert, L. Gelens, Phys. Rev. A **84**, 013807 (2011)
46. J.A. Arnaud, Appl. Opt. **8**, 189 (1969)
47. P. Kockaert, JOSA B **26**, 1994 (2009)
48. H. Zhang, D.Y. Tang, L. Zhao, X. Wu, Phys. Rev. B **80**, 052302 (2009)
49. H. Zhang, D.Y. Tang, L. Zhao, X. Wu, Opt. Exp. **19**, 3525 (2011)
50. P. Manneville, Dissipative Structures and Weak Turbulance (1991)
51. M. Tlidi, R. Lefever, P. Mandel, Quantum Semiclass. Opt. **8**, 391 (1996)
52. A.G. Vladimirov et al., Phys. Rev. E **65**, 046606 (2002)
53. M. Tlidi et al., IEEE J. Quant. Electron. **39**, 216 (2003)
54. M. Tlidi, R. Lefever, A.G. Vladimirov, Lecture Notes in Physics **751**, 381 (2008)
55. A.G. Vladimirov et al., Phys. Rev. A **84**, 043848 (2011)
56. Y. Pomeau, Physica D **23**, 3 (1986)
57. P.D. Woods, A.R. Champneys, Physica D **129**, 147 (1999)
58. P. Coullet, C. Riera, C. Tresser, Phys. Rev. Lett. **84**, 3069 (2000)
59. J. Burke, E. Knobloch, Chaos **17**, 037102 (2007)
60. D.J.B. Lloyd, B. Sandstede, D. Avitabile, A.R. Champneys, SIAM J. Appl. Dyn. Syst. **7**, 1049 (2008)
61. M. Tlidi, L. Gelens, Opt. Lett. **35**, 306 (2010)
62. M. Tlidi, A. Mussot, E. Louvergneaux, G. Kozyreff, A.G. Vladimirov, M. Taki, Opt. Lett. **32**, 662 (2007)
63. M. Tlidi, L. Bahloul, L. Cherbi, A. Hariz, S. Coulibaly, Phys. Rev. A **88**, 035802 (2013)
64. L. Bahloul, L. Cherbi, A. Hariz, M. Tlidi, Philos. Trans. R. Soc. A **372**, 20140020 (2014)
65. M.J. Schmidberger, D. Novoa, F. Biancalana, P. Russell, St.J., Joly N.Y. Opt. Exp. **22**, 3045 (2014)
66. M. Tlidi, M. Haelterman, P. Mandel, EPL (Europhysics Letters) **42**, 505 (1998)
67. M. Tlidi, M. Haelterman, P. Mandel, Quantum Semiclass. Opt. J. Eur. Opt. Soc. Part B **10**, 869 (1998)
68. K. Staliunas, Phys. Rev. Lett. **81**, 81 (1998)
69. M. Tlidi, P. Mandel, Phys. Rev. Lett. **83**, 4995 (1999)
70. M. Tlidi, J. Opt. B Quantum Semiclass. Opt. **2**, 438 (2000)
71. N. Veretenov, M. Tlidi, Phys. Rev. A **80**, 023822 (2009)
72. M. Brambilla, T. Maggipinto, G. Patera, L. Columbo, Phys. Rev. Lett. **93**, 203901 (2004)
73. C.-Q. Dai, X.-G. Wang, G.-Q. Zhou, Phys. Rev. A **89**, 013834 (2014)

Secondary Instabilities and Chaos in Optical Ring Cavities

Z. Liu, F. Leo, S. Coulibaly and M. Taki

Abstract The dynamical evolution of the modulational instability is studied in a coherently driven passive optical fiber cavity. We show that periodic pattern generated close to the threshold experience convective and absolute Eckhaus instabilities. In addition to this splitting of the secondary instabilities into convective and absolute instabilities we have observed that the onset of the absolute regime is accompanied by the emergence of spatio-temporal chaos.

1 Introduction

Modulational instability (MI)—small scale periodic pattern—has been observed in almost all fields in Physics ranging from chemical reactions, biology to nonlinear optics and fluid mechanics [1, 2]. There is currently a considerable interest in understanding their dynamics in optical fiber since they have been identify as one of the mechanisms seeding the supercontinuum generation. It is known that periodic pattern can experience different generic instabilities characteristics of the reduced modeled—normal form or amplitude equation—describing their evolution: the Eckhaus instability being the famous one. In optics, MI arises naturally in many devices from the coupling of dispersion (temporal systems) or diffraction (spatial systems), nonlinearities, and dissipation. Among the possible devices, coherently driven

Z. Liu · S. Coulibaly (✉) · M. Taki
PhLAM, Université de Lille 1, Bât. P5-bis; UMR CNRS/USTL 8523,
59655 Villeneuve D'ascq, France
e-mail: saliya.coulibaly@univ-lille1.fr

Z. Liu
e-mail: zheng.liu@univ-lille1.fr

M. Taki
e-mail: abdelmajid.taki@univ-lille1.fr

F. Leo
Photonics Research Group, Department of Information Technology,
Ghent University-IMEC, 9000 Ghent, Belgium

© Springer International Publishing Switzerland 2016
M. Tlidi and M.G. Clerc (eds.), *Nonlinear Dynamics: Materials,
Theory and Experiments*, Springer Proceedings in Physics 173,
DOI 10.1007/978-3-319-24871-4_12

167

optical fiber ring cavities have recently appeared as one of the most promising systems, not only for the richness in their nonlinear dynamic [3] but also for their potential applications [4]. It has been shown that, in presence of third-order dispersion, the standard theoretical approach leading to modulation instability must be extended [5, 6]. More specifically, a nonlinear stationary state may be unstable with respect to localized perturbations, but the state that results will depend on the relative values of the amplification and the drift induced by the third-order dispersion term. This is the basis of the difference between convective and absolute regimes. In the former, the perturbation grows in time but decreases locally because it is advected away. In the latter it increases locally and not only in the moving frame, so it eventually extends over all the slow time domain. In this case, threshold values for primary absolute and convective instabilities were obtained [3]. However, as soon as the threshold is exceeded the system enters a nonlinear regime where secondary instabilities arise leading to a more complex dynamics characterized by transition from dissipative periodic solutions to non periodic and/or chaotic ones.

In this chapter we investigate these secondary instabilities in the presence of third-order dispersion and emphasize the crucial role of secondary convective and absolute instabilities in the dynamics of a coherently driven passive optical fiber cavity. Indeed, above threshold, dissipative time-modulated solutions are obtained. By increasing the incident pump power, these solutions destabilize and the system bifurcates either to a new periodic solution or enters a chaotic regime. At this stage a secondary instability threshold is reached. This threshold is crucial in the nonlinear dynamics of the system above threshold since it determines the stability range of the dissipative periodic solutions and subsequently the parameters range of their observation. An amplitude equation has been derived to describe the weakly nonlinear dynamics above the onset of instability that allows us to determine the threshold values for the different types of the secondary instability. An important results is that the threshold of absolute instability of modulated solutions determines the transition from modulated dissipative solutions to a regime of a temporal chaotic behavior [7].

2 The Models

The system under investigation depicted in Fig. 1 can be modeled by the extended nonlinear Schrdinger equation with boundary conditions. This leads to a set of two equations, usually referred to as the map equations (or mapping) that can be reduced in the mean field approximation to obtain a single equation modeling the intra-cavity field dynamics. This equation, known as the Lugiato-Lefever equation (LL model) [8], has been proven relevant for describing weakly nonlinear dynamics in cavities [9]. It reads

$$\frac{\partial \psi}{\partial t'} = S - (1 + i\Delta)\psi - is\frac{\partial^2 \psi}{\partial \tau'^2} + B_3\frac{\partial^3 \psi}{\partial \tau'^3} + i|\psi|^2\psi \qquad (1)$$

Fig. 1 Scheme of the fiber
ring cavity. BS, beam
splitter. ρ and θ are the
amplitude reflection and
transmission coefficients of
BS ($\rho^2 + \theta^2 = 1$)

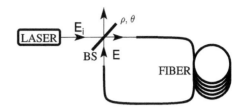

where $t' = t\theta^2/2t_R$ is a slow normalized time variable, with t the real time and
t_R the round trip time, $\tau' = T\theta/(L|\beta_2|)^{1/2}$ is a fast normalized time variable, $s = \text{sign}(\beta_2)$, $\psi = A(2\gamma L)^{1/2}/\theta$, $S = 2A_i(2\gamma L)^{1/2}/\theta^2$, $B_3 = \beta_3\theta/(3L^{1/2}|\beta_2|^{3/2})$ and
$\Delta = 2\delta_0/\theta^2$. The amplitudes A_i and A are respectively slowly varying envelopes
for the incident pump electric field and the intra-cavity electric field, δ_0 the cavity
phase detuning, L the cavity length, γ the nonlinear coefficient of the fiber and $\beta_{2,3}$
are the second-order and the third-order dispersion (TOD) terms respectively. The
steady state response ψ_s of (1) satisfies $S_s = [1 + i(\Delta - |\psi_s|^2)]\psi_s$ so that the system
is monostable (bistable) for $\Delta < \sqrt{3}$ ($\Delta > \sqrt{3}$). The linear stability analysis of
the steady state response with respect to finite frequency perturbations of the form
$\exp[-i(Kt' - \Omega\tau')]$ shows that the primary threshold is reached at $I = I_s = 1$ with
$I = |\psi_s|^2$. Hence, the system evolves toward modulated solutions with the frequency
$\Omega_c = \sqrt{(\Delta - 2I_s)/s}$ associated to the wavenumber $k_c = B_3\Omega_c^3$.

In order to describe the amplitude of linearly excited MI in the weakly nonlinear
regime, the following ansatz was introduced $\psi = \psi_s + \varepsilon a_1 + \varepsilon^2 a_2 + \varepsilon^3 a_3 + \cdots$,
where $a_1 = (1+i)[A_1 e^{i(\Omega_c\tau' + k_c t')} + A_1^* e^{-i(\Omega_c\tau' + k_c t')}]$. The parameter $\varepsilon = I^2 - 1 \ll 1$
measures the distance from the primary threshold. Next, using appropriate expansions
[9–11]: $T_0 = t, T_1 = \varepsilon t, T_2 = \varepsilon^2 t, \tau_0 = \tau$ and $\tau_1 = \varepsilon\tau$ and collecting term in ε, the
solvability condition at order $O(\varepsilon^3)$ gives:

$$\partial_{t'}S + 3B_3\Omega_c^2\partial_{\tau'}S = (\psi_s^2 - 1)S + (2\Omega_c^2 + 3i B_3\Omega_c)\partial_{\tau'}^2 S$$
$$+ (d_1 + id_2)|S|^2 S, \qquad (2)$$

where we have set $S = \varepsilon A_1$ and the parameters are defined as

$$d_1 = 24\frac{2G + 3}{G^2} + 4\frac{G^2(1 - 2G) + H^2(2G - 3)}{(G^2 - H^2)^2 + 4H^2} \qquad (3)$$

$$d_2 = \frac{4H[2(1 - 2G) + G^2 - H^2]}{(G^2 - H^2)^2 + 4H^2} \qquad (4)$$

with

$$G = 3(\Delta - 2)$$
$$H = -6B_3\Omega_c^3$$

This equation of complex Ginzburg-Landau type [12] describes the time evolution of Stokes wave above the threshold. Notice that, in absence of TOD ($\beta_3 = 0$), three terms in (2) disappear since $B_3 = 0$ and $d_2 = 0$. Moreover, the expression of d_1 greatly simplifies to

$$d_1 = \frac{30\Delta - 41}{[3/2(\Delta - 2)]^2} \tag{5}$$

and the numerator of d_1 shows clearly that the transition from super- to sub-critical bifurcation is reached at $\Delta = 41/30$ in agreement with the result in the seminal paper of Lugiato-Lefever [8]. Later, introducing $\tilde{S} = \sqrt{\frac{-d_1}{\varepsilon^2}} S, \tilde{t} = \varepsilon^2 t', \tilde{\tau} = \sqrt{\frac{\varepsilon^2}{2\Omega_c^2}}(\tau' - 3B_3\Omega_c^2 t'), b = \frac{3B_3}{2\Omega_c}, c = \frac{d_2}{d_1}$, (2) can be simplified as follows:

$$\frac{\partial \tilde{S}}{\partial \tilde{t}} = \tilde{S} + (1 + ib)\frac{\partial^2 \tilde{S}}{\partial \tilde{\tau}^2} - (1 + ic)|\tilde{S}|^2\tilde{S}. \tag{6}$$

Under this form the complex Ginzburg-Landau equation have been widely studied [10, 12] and will be the starting point of our analysis. Its simplest solution the standing wave solution takes the form:

$$\tilde{S}(\omega) = \sqrt{1 - \omega^2}e^{i(\omega\tilde{\tau} - k(\omega)\tilde{t})}, \tag{7}$$

with $k(\omega) = c(1 - \omega^2) + b\omega^2$. Solution (7) is known to be subjected to some instability that we recall in the following section.

3 Convective Versus Absolute Instability

Lets start by the linear stability analysis of the solution (7). The largest eigenvalue of the linearized system around this solution reads:

$$\lambda(q) = -|\tilde{S}_{\tilde{\omega}}|^2 - 2ibq\tilde{\omega} - q^2 \tag{8}$$

$$+ \sqrt{(1 + c^2)|\tilde{S}_{\tilde{\omega}}|^4 - (bq^2 - 2i\tilde{\omega}q + c|\tilde{S}_{\tilde{\omega}}|^2)^2}. \tag{9}$$

Based, on this latter equation, let us recall the background about this stability analysis. In the general case, when $b \neq c$ it appears a drift to modulated solutions. This drift gives rise to the appearance of two types of instability regimes of modulated solutions: a convective regime and an absolute one. The phase drift velocity of the solution (7) is given by $\tilde{v} = \tilde{k}_{\tilde{\omega}}/\tilde{\omega}$. Note that $\tilde{v}$ may be positive or negative depending on the values of b, c and $\tilde{\omega}$. The analysis of the eigenvalue λ [expression (9)] allows to understand the dynamics in different instability regimes of the solution (7). For the sake of simplicity, we develop λ for $q \to 0$ to obtain an approximate expression as follows [12, 13]

$$\lambda_a = -i V_g q - D q^2 + O(q^3) \tag{10}$$

with

$$V_g = 2(b - c)\tilde{\omega} \tag{11}$$

$$D = 1 + bc - \frac{2(1 + c^2)\tilde{\omega}}{1 - \tilde{\omega}^2} \tag{12}$$

The approximate expression of λ [expression (10)] allows us to get a good insight on the stability of (7). Indeed, the sign of D determines the stability of this solution: it is stable ($\mathrm{Re}[\lambda_a] < 0$) when $D > 0$. Using the expression (12), we can express the stability criterion as

$$\tilde{\omega}^2 < \frac{1 + bc}{3 + 2c^2 + bc} \tag{13}$$

The solution $\tilde{S}_0$ (solution corresponding to $\tilde{\omega} = 0$) is stable if the Benjamin-Feir-Newell criterion $1 + bc > 0$ is satisfied [12]. The most unstable mode (the Eckhaus mode) can be found by using the following relation

$$\left. \frac{\partial \mathrm{Re}[\lambda]}{\partial q} \right|_{q_{max}} = 0 \tag{14}$$

From the approximate expression of the eigenvalue λ_a we obtain $q_{max} = 0$. In the study of the convective and absolute instability, the relation $\tilde{\omega}^2 = (1 + bc)/(3 + 2c^2 + bc)$ determines the convective threshold. However, the approximate expression of λ_a is not sufficient to predict the absolute threshold and the most unstable mode (the Eckhaus mode) beyond the convective threshold; we then consider the exact expression of λ (9) to continue the analysis. Indeed, in the general context of convective and absolute instabilities, λ and q are both complex as $\lambda = \mathrm{Re}[\lambda] + i\mathrm{Im}[\lambda]$ and $q = \mathrm{Re}[q] + i\mathrm{Im}[q]$. By replacing these two complex terms in the exponential part of the perturbation leads to:

$$\delta a \propto e^{(\lambda \tilde{t} + iq\tilde{\tau})} = e^{(\mathrm{Re}[\lambda] - \mathrm{Im}[q]\frac{\tilde{\tau}}{\tilde{t}})\tilde{t}} e^{i(\mathrm{Im}[\lambda] + \mathrm{Re}[q]\frac{\tilde{\tau}}{\tilde{t}})\tilde{t}} \tag{15}$$

Here, we define the drift V and the total growth rate σ of the perturbation as follows [14]

$$V = \frac{\tilde{\tau}}{\tilde{t}} \tag{16}$$

$$\sigma = \mathrm{Re}[\lambda] - \mathrm{Im}[q]V \tag{17}$$

where the total gain σ represents the temporal growth rate along each drift V. The asymptotic dominant frequencies q_s are defined by the saddle-point method [14–16],

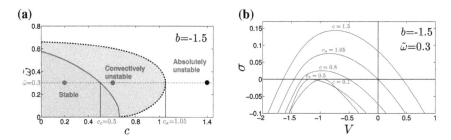

Fig. 2 (Color online) **a** The convective and absolute instabilities in the $\tilde{\omega}$-c plane with $b = -1.5$. The *solid red line* represents the limit of the Eckhaus instability $\tilde{\omega}^2 = (1 + bc)/(3 + bc + 2c^2)$ for $q \to 0$ and the *dashed black curve* shows the *marginal curve* which separates the zone of the convective and absolute instabilities, it is obtained from the relations (19) and (20). **b** The total growth rate σ (gain) versus V for 5 values of c

$$\partial_q \lambda(q_s) = V(q_s) \tag{18}$$

The absolute threshold is reached when the following conditions are satisfied [12–14]

$$\partial_q \lambda(q_{sa}) = 0 \tag{19}$$
$$\mathrm{Re}[\lambda(q_{sa})] = 0 \tag{20}$$

The frequency q_{sa} is commonly called the absolute frequency. By solving numerically (19) and (20), we can express the absolute threshold as a function of b, c and $\tilde{\omega}$. An example is given in Fig. 2a, following the same method as in [12]. The black dashed line represents the absolute threshold in the plane ($\tilde{\omega}$, c) for $b = -1.5$. A path with $\tilde{\omega} = 0.3$ (blue dashed line) is chosen to test the convective and absolute thresholds obtained. This path intersects the convective threshold curve at $c_c = 0.5$ and the absolute threshold curve at $c_a = 1.05$. Figure 2b shows the total gain versus V for five values of c with $\tilde{\omega} = 0.3$ and $b = -1.5$ to examine the transition between the different regimes. When $c \le c_c$ (stable regime), the modulated solution (??) is stable, the total gain is negative for all V except one drift V with $\sigma = 0$ which corresponds to the drift of the stable modulated solution (see the curve labeled $c = 0.1$ in Fig. 2b), it is marginally stable. By increasing c, the total gain curve does not move vertically but horizontally, this means that the modulated solution is always stable but its drift can change depending on c (see the curves labeled $c = 0.1$ and $c = 0.5$ in Fig. 2b). Just above the convective threshold, the total gain curve rises vertically with increasing c leading to a range of unstable V with σ positive (see the curve labeled $c = 0.8$ in Fig. 2b). In this regime $\sigma(V)$ is positive but only for negative values of V. This means that the perturbation grows only in one direction and the solution (7) is convectively unstable. At the absolute threshold, the limit of the positive part of the total gain curve ($\sigma = 0$) reaches $V = 0$ (see the curve labeled $c_a = 1.05$ in Fig. 2b). By increasing c, the positive part of the total gain curve extends simultaneously in the range of $V < 0$ and $V > 0$ [see the curve labeled $c = 1.3$ in Fig. 2b] and the perturbation develops in both directions so that the solution (7) is absolutely unstable.

(a)   **(b)**

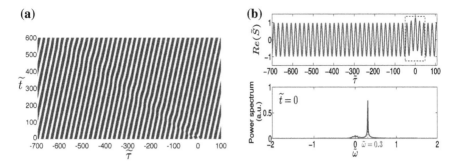

Fig. 3 a Evolution of a perturbation initially located at $\tilde{\tau} = 0$ on the modulated solution in the $\tilde{\tau}$-$\tilde{t}$ plane obtained by numerical integration of (6) in a stable case ($\tilde{\omega} = 0.3$, $b = -1.5$, $c = 0.2$). **b** Instantaneous profile and the corresponding spectrum at $\tilde{t} = 600$

We have performed three numerical simulations by integrating (6) to distinguish these different regimes: the first one in the stable regime, the second in the convective regime and the last in the absolute regime. These simulations are denoted by circles in Fig. 2a with red color ($c = 0.2$), blue color ($c = 0.6$) and black color ($c = 1.4$) respectively. In all simulations, we super imposed a localized perturbation to the solution (7) around the position $\tilde{\tau} = 0$ as an initial condition and follow its evolution. In the stable regime, the localized perturbation drifts in the $\tilde{\tau}$-$\tilde{t}$ plane as shown in Fig. 3a where the drift of the perturbation is given by $V = V_g = 2(b - c)\tilde{\omega}$. Note that the black and white fringes represent the oscillation of the modulated solution at frequency $\tilde{\omega} = 0.3$. Indeed, the slope of the fringes in the $\tilde{\tau}$-$\tilde{t}$ plane means that the modulated solution drifts itself along $\tilde{\tau}$ with a phase velocity $v = \tilde{k}_{\tilde{\omega}}/\tilde{\omega}$. In the temporal profiles and the spectrum of the signal at $\tilde{t} = 300$ (Fig. 3b), we observe that the perturbation is lessened and the system returns to the stable modulated solution (the positions of the perturbation are framed by the black dotted rectangles). In the convective regime, the evolution of the signal in the $\tilde{\tau}$-$\tilde{t}$ plane is displayed in Fig. 4a where we observe that the affected part of the signal by the initial perturbation extends and drifts only in the backward direction during the simulation, and it completely leaves the observation domain under the effect of the drift at $\tilde{t} = 500$. Note that the drift is negative as shown in Fig. 2b. The temporal profiles and the spectra show that the affected part of the signal tends to form an oscillation of the Eckhaus frequency (Fig. 4a), but this frequency eventually disappears and the spectrum find its initial state after the perturbed part of the signal leaves completely the observation domain under the effect of the drift. This is in excellent agreement with our analytical prediction about the existence of a convective regime where the system asymptotically recovers its initial state. In the absolute regime, the evolution of the signal in the $\tilde{\tau}$-$\tilde{t}$ plane is shown in Fig. 5a where the perturbation affects all the signal and it finally invades the entire observation domain. During the evolution, the initial frequency $\tilde{\omega}$ disappears, but we do not observe the amplification of the Eckhaus frequency and the affected part of the signal loses completely the modulated form (Fig. 5b). Note that a similar behavior has been reported in [7]. The authors

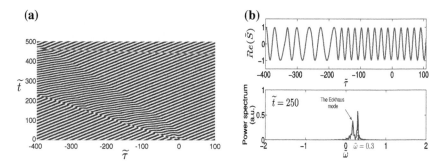

Fig. 4 **a** Evolution of a perturbation initially located at $\tilde{\tau} = 0$ on the modulated solution in the $\tilde{\tau}$-$\tilde{t}$ plane obtained by numerical integration of (6) with $\tilde{\omega} = 0.3$, $b = -1.5$, $c = 0.6$. **b** Instantaneous profile and the corresponding spectrum at $\tilde{t} = 500$

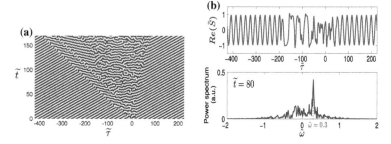

Fig. 5 **a** Evolution of a perturbation initially located at $\tilde{\tau} = 0$ on the modulated solution in the $\tilde{\tau}$-$\tilde{t}$ plane obtained by numerical integration of (6) with $\tilde{\omega} = 0.3$, $b = -1.5$, $c = 1.4$. **b** Instantaneous profile and the corresponding spectrum at $\tilde{t} = 170$

investigated transition from periodic oscillations to spatiotemporal chaos in a classic ecological system of invasion of prey population by predators. Here, in addition to the concept of convective and absolute instability, an advanced nonlinear study is necessary to explain our results by taking into account the interactions between all unstable modes and their relative stabilities.

After investigating the non-stationary modulated solutions and their stabilities in the normalized amplitude equation (6), let us return to the LL model (1). In fact, the nature of the instability depends directly on the frame of reference in which we observe the instability, it can be convective in a frame of reference and absolute in another one, the key factor here is the drift of the frame of reference we chose. In our case, we first need to reconsider the normalization which connects the time variable τ' of the LL model to the normalized variable $\tilde{\tau}$ as

$$\tau' = \sqrt{\frac{2\Omega_c^2}{\varepsilon^2}} \left(\tilde{\tau} - \frac{3B_3\Omega_c}{\sqrt{2}\varepsilon} \tilde{t} \right) \tag{21}$$

We found that the frame of reference of the LL model is moving in the frame of reference of the normalized amplitude equation with a drift $3B_3\Omega_c/\sqrt{2}\varepsilon$ that we note V_c. We also introduce the drifts of the limiting fronts of the wave packet in the normalized amplitude equation (6). These fronts are marginally stable and defined by $\sigma(V_-) = \sigma(V_+) = 0$ where V_- and V_+ represent the drift of the leading and the trailing edges, respectively. According to the concept of convective and absolute instability applied to propagating wavepackets in unstable medium [17], the instability of the modulated solution is absolute in the LL model if $(V_- - V_c)(V_+ - V_c) < 0$, otherwise, it is convective. However, with parameters in our analysis, V_+ and V_- are always smaller than V_c, leading to a convective regime in LL model.

4 Conclusion

In summary, we studied analytically and numerically the impact of the TOD on modulated solutions that appear above the onset of instability in a fiber ring cavity. We have shown that the presence of the TOD induced a drift which leads to the appearance of convective and absolute instability regimes of the modulated solution in the amplitude equation. In the convective regime, we numerically show that a localized perturbation impacts the modulated solution by shifting its modulation frequency to the Eckhaus frequency. Under the effect of the drift, the modulated solution recovers its initial state after the localized perturbation leaves completely the observation domain. In the absolute regime, the localized perturbation affects all modulated solutions, but we do not observe the appearance of the Eckhaus frequency. The modulated solution loses completely the modulated form and enters a chaotic regime. A nonlinear advanced study is needed to explain this result and it will be a prospect for the future work. Our analytical and numerical studies of the amplitude equation led us to conclude that the instability of modulated solutions in the LL model is convective according to the range of physical parameters explored in this study.

References

1. M. Tlidi, M. Taki, T. Kolokolnikov, Chaos **17**, 037101 (2007)
2. N. Akhmediev, A. Ankiewicz, *Dissipative Solitons: From Optics to Biology and Medicine* (Springer, Berlin, 2008)
3. A. Mussot, E. Louvergneaux, N. Akhmediev, F. Reynaud, L. Delage, M. Taki, Phys. Rev. Lett. **101**, 113904 (2008)
4. F. Leo, S. Coen, P. Kockaert, S.-P. Gorza, P. Emplit, M. Haelterman, Nat. Photonics **4**, 471 (2010)
5. A. Mussot, A. Kudlinski, E. Louvergneaux, M. Kolobov, M. Douay, Phys. Lett. A **374**, 691 (2010)
6. F. Leo, A. Mussot, P. Kockaert, P. Emplit, M. Haelterman, M. Taki, Phys. Rev. Lett. **110**, 104103 (2013)
7. J.A. Sherratt, M.J. Smith, J.D.M. Rademacher, Proc. Natl. Acad. Sci. USA **106**, 10890 (2009)
8. L.A. Lugiato, R. Lefever, Phys. Rev. Lett. **58**, 2209 (1987)

9. M. Haelterman, S. Trillo, S. Wabnitz, Opt. Commun. **91**, 401 (1992)
10. R. Zambrini, M. San Miguel, C. Durniak, M. Taki. Phys. Rev. E 72, 025603 (2005)
11. P. Manneville, *Instabilities, Chaos and Turbulence* (Imperial College Press, London, 1994)
12. I.S. Aranson, L. Kramer, Rev. Mod. Phys. **74**, 99 (2002)
13. I.S. Aranson, L. Aranson, L. Kramer, A. Weber, Phy. Rev. A **46**, 2992 (1992)
14. P. Huerre, P.A. Monkewitz, Annu. Rev. Fluid Mech. **22**, 473 (1990)
15. C. Bender, S. Orszag, *Advanced Mathematical Methods for Scientists and Engineers* (MacGraw-Hill, New York, 1978)
16. H. Ward, M.N. Ouarzazi, M. Taki, P. Glorieux, Phys. Rev. E **63**, 016604 (2000)
17. G.S. Triantafyllou, Phys. Fluids **6**, 164 (1993)

Part II
Mechanics, Fluids and Magnetics

Hidden High Period Accelerator Modes in a Bouncer Model

Tiago Kroetz, André L.P. Livorati, Edson D. Leonel
and Iberê L. Caldas

Abstract We characterized the influence of high period accelerator modes in the
global dynamics of a non-dissipative Bouncer model. The dynamics of the system
was investigated considering both complete and simplified approaches. Evaluating
the average of the velocity over large ensembles of initial conditions for the complete
mapping, we obtained particular ranges of the control parameter where high period
accelerating structures are located. The position, influence and shape of the accelerator modes were obtained considering the symplectic mapping. Our results, lead us
to infer that even for high period and less influent accelerator modes, the dynamics
is globally affected for long time series, causing an anomalous diffusion, in compare
with the regular Fermi acceleration.

1 Introduction

Modelling of dynamical systems, has been one of the most embracing area of interest
among physicists and mathematicians is the past decades [1–5]. Low-dimensional
systems in particular, despite the simple modelling, are very suitable to study and to
investigate chaotic properties in their phase space [1–5]. These systems can present a
very complex dynamics leading to a rich variety of nonlinear phenomena, considering
either dissipative and non-dissipative dynamics [1–5].

T. Kroetz (✉)
Universidade Tecnológica Federal do Paraná, Pato Branco, Paraná, Brazil
e-mail: kroetzfisica@gmail.com

A.L.P. Livorati · E.D. Leonel
Departamento de Física, Universidade Estadual Paulista, Rio Claro,São Paulo, Brazil
e-mail: livorati@rc.unesp.br

E.D. Leonel
e-mail: edleonel@rc.unesp.br

I.L. Caldas
Instituto de Física, Universidade de São Paulo, São Paulo, Brazil
e-mail: ibere@if.usp.br

© Springer International Publishing Switzerland 2016
M. Tlidi and M.G. Clerc (eds.), *Nonlinear Dynamics: Materials,
Theory and Experiments*, Springer Proceedings in Physics 173,
DOI 10.1007/978-3-319-24871-4_13

The Italian physicist Enrico Fermi [6] proposed in 1949, a mechanism as an attempt to explain the origin of the high energies of the cosmic rays. Fermi claimed that charged particles, which interacted with oscillating magnetic fields present in the cosmos, would in the average exhibit a gain of energy. This unlimited growth of energy is denominated Fermi acceleration (FA) and has many applications in several areas of research, as Plasma Physics [7, 8], Astrophysics [9, 10], Atom-optics [11, 12], and specially in billiard dynamics [13–18]. This unlimited energy growth is mainly associated with normal diffusion in phase space. However, FA may present distinct transport from the normal diffusion, as exponential [19–22], or where stickiness phenomenon [4, 5] plays the role of a slowing mechanism for FA [22].

The Bouncer Model [23–25] will be the focus of our study in this paper, where basically we have a free particle under the influence of a constant gravitational field suffering elastic collisions with a vibrating platform. In the non-dissipative version and depending on both control parameters and initial conditions, the bouncer ball presents FA [1]. Despite the simple dynamics, applications for this model can be found in dynamic stability in human performance, [26], vibrations waves in a nanometric-sized mechanical contact system [27], mechanical vibrations [28, 29], anomalous transport and diffusion [30], thermodynamics [31], chaos control [32, 33], granular materials [34–36], among others.

In this paper we investigate how resonances in the phase space, known as accelerator modes (AM) (or ballistic modes) [37–48], influences diffusion and transport properties of the average velocity of the Bouncer Model. We observed that in the presence of this accelerating structures, the dynamics behaves in a regular and monotonic increase of velocity, differing from the normal diffusion present by the "regular FA" [22, 49]. By exploring the the non-symplectic character of the mapping, we set a numerical search for the AM of high period and low influence on the global dynamics [49]. We found a series of hidden AM, which accelerate more the dynamics for long times than normal diffusion, where their shape and evolution were characterized by stability islands in the symplectic version of the map. We observed that for sufficient long time series, the presence of any AM, of high period or not, would affect globally the dynamics of the system.

The paper is organized as follows: In Sect. 2, we describe the dynamics of the Bouncer Model, in both symplectic and non-symplectic version. Section 3 is devoted to determine analytically the existence of period-1 accelerator mode. Also, we perform a numerical search for high period AM and characterized their period and control parameter range of stability in a table. In Sect. 4 we use the symplectic version of the model to reveal the accelerating structures in the the modulated phase space. Finally, in Sect. 5 we draw some final remarks and conclusions.

2 The Model

The dynamical system used to explore the existence of AM with high period and low influence is the Bouncer Model. The dynamical analysis of this system can be explored considering two versions of a two dimensional mapping. The complete

Fig. 1 Schematic view of
Bouncer model

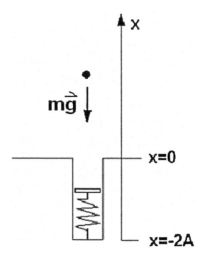

one (non-symplectic) consists in considering the full dynamics of a bouncing ball colliding elastically with a vibrating platform. The simplified version (symplectic), is set by considering the position of the platform fixed. However, when occurs a colision between the particle and the platform, they exchange momentum and energy as if the oscillatory platform where normally moving. A schematic view of the Bouncer Model is shown in Fig. 1.

In the complete version, the position of the oscillatory platform is given by $x_w(t) = A[\cos(\omega t + \varphi) - 1]$, where A is the amplitude of the platform oscillation, ω is the angular frequency and φ is the oscillation initial phase. The gravitational field acts as a return mechanism which causes repeated impacts between the particle and the platform.

The position of the particle between impacts is given by the free fall equation $x_p(t) = h_0 + vt - gt^2/2$, where h_0 is the vertical position from which the particle was previously launched by the platform, v is the launch velocity, t is the time elapsed since the last impact and g is the gravitational acceleration. The instants of impacts are obtained by equating the platform position and the particle position $x_w(t) = x_p(t)$.

2.1 Complete Model

The velocity of the particle after each impact will be given by the negative relative velocity between the particle and the platform just before the collision. Obtaining the velocity of the platform and of the particle in a recurrent way at each impact, we can write a set of discrete equations representing the particle velocity v_n and of the phase of the platform movement φ_n. The time interval t_{n+1} and the phase difference between two consecutive impacts are related by $t_{n+1} = (\varphi_{n+1} - \varphi_n)/\omega$. Therefore, the discrete map that describes the complete Bouncer Model can be written as

$$
\begin{cases}
A[\cos(\varphi_n) - \cos(\varphi_{n+1})] + v_n(\varphi_{n+1} - \varphi_n)/\omega - g(\varphi_{n+1} - \varphi_n)^2/2\omega^2 = 0 \\
v_{n+1} = -v_n + g(\varphi_{n+1} - \varphi_n)/\omega - 2A\omega\sin(\varphi_{n+1})
\end{cases} \tag{1}
$$

The first expression in (1) must be solved numerically at each collision in order to find the value of φ_{n+1}.

Defining a parameter K in terms of A, ω and g as $K = \omega^2 A/(\pi g)$ it is possible to write a new map dependent on a single control parameter. This parameter is interpreted as a ratio between accelerations of the moving platform and the gravitational field. To perform this procedure, we rewrite the map in terms of a dimensionless velocity given by $V_n = \omega v_n/(\pi g)$. So, we write the new map as

$$
\begin{cases}
K[\cos(\varphi_n) - \cos(\varphi_{n+1})] + V_n(\varphi_{n+1} - \varphi_n)(\varphi_{n+1} - \varphi_n)^2/2\pi = 0 \\
V_{n+1} = -V_n + (\varphi_{n+1} - \varphi_n)/\pi - 2K\sin(\varphi_{n+1})
\end{cases} \tag{2}
$$

To determine the non-symplectic character of this model, we need to obtain the determinant of Jacobian matrix associated with the mapping expressed in (2). The volume element in phase space varies according this determinant, whose elements are given by the partial derivatives of the mapping variables. The partial derivatives of φ_{n+1} are obtained by indirect differentiation of the first expression of the mapping (2), followed by some algebra in order to isolate the terms. These expressions are written as

$$
\frac{\partial \varphi_{n+1}}{\partial \varphi_n} = \frac{V_n - (\varphi_{n+1} - \varphi_n)/\pi + K\sin(\varphi_n)}{V_n - (\varphi_{n+1} - \varphi_n)/\pi + K\sin(\varphi_{n+1})}, \tag{3}
$$

$$
\frac{\partial \varphi_{n+1}}{\partial V_n} = \frac{-\pi(\varphi_{n+1} - \varphi_n)}{V_n - (\varphi_{n+1} - \varphi_n)/\pi + K\sin(\varphi_{n+1})}, \tag{4}
$$

$$
\frac{\partial V_{n+1}}{\partial \varphi_n} = \frac{\partial \varphi_{n+1}}{\partial \varphi_n}\left[1/\pi - 2K\cos(\varphi_{n+1})\right] - 1/\pi, \tag{5}
$$

$$
\frac{\partial V_{n+1}}{\partial V_n} = \frac{\partial \varphi_{n+1}}{\partial V_n}\left[1/\pi - 2K\cos(\varphi_{n+1})\right] - 1. \tag{6}
$$

Therefore, the Jacobian determinant is given by

$$
det(J) = \frac{V_n + K\sin(\varphi_n)}{V_{n+1} + K\sin(\varphi_{n+1})}. \tag{7}
$$

As can be seen in (7) the Jacobian has no constant value once it depends on the dynamical variables associated with the iterations. As a consequence, the system cannot be considered dissipative neither non-dissipative, since J can be greater or less than the unity for distinct regions of the phase space. The phase space volume will contract around some regions and expand around some others. This result is important for our purpose of searching small accelerating structures in the phase space, represented by high period AM.

2.2 Simplified Model

We define the simplified version considering the position of the platform as fixed, but there is an exchange of momentum and energy between the particle and the moving platform as if the platform were moving. Assuming this approximation, the time elapsed between consecutive impacts can be easily found depending only on the launch velocity of the last collision. So, considering $(\varphi_{n+1} - \varphi_n) = 2\pi V_n$, the simplified version of the bouncer model is written as

$$\begin{cases} \varphi_{n+1} = \varphi_n + 2\pi V_n \\ V_{n+1} = |V_n - 2K \sin \varphi_{n+1}| \end{cases}. \tag{8}$$

The simplified bouncer model defined by (8) is symplectic, as one can easily check. Although the simplified model does not correspond to the full dynamics of a ball bouncing in a moving floor, it can be useful to evaluate analytical calculations about the position and stability of the fixed points. Also, we use the simplified model to visualize the accelerating structures in the modulated phase space, once they will appear as periodic islands due to the symplectic character of this mapping.

3 Accelerator Modes

In the phase space, the dynamics of an AM consists of regular and repetitive jumps in V direction. This implies in a ballistic acceleration of the particle without chaotic behaviour. The simplest case happens when the map iteration leads one point in phase space to another at same value of φ and shifted in V direction by adding an integer l. We designate this kind of dynamics as period-1 AM with step-size l. If we impose an artificial periodicity along V direction by modulating the phase space, the period-1 AM is indistinguishable from a period-1 fixed point. The period of the AM refers to the number of map iterations until a repetition of coordinates in a modulated phase space.

3.1 Stability of the Period-1 Accelerator Modes Step-Size l

In this subsection we use the simplified mapping approach to obtain the position of the accelerating structure of period-1 and step-size l in the phase space as function of the control parameter K and estimate their stability. Using the symplectic map described by (8) we obtain the relation for the AM coordinates $V_{n+1} - V_n = l = -2K \sin(\varphi_{n+1})$. The position of the period-1 AM step-size l is provided by the mapping in (8) as

$$V^* = l,\tag{9}$$

$$\varphi^* = \arcsin(-l/2K).\tag{10}$$

In order to determine the stability of the period-1 AM step-size l, we linearize the system around the position (φ^*, V^*) by calculating the Jacobian matrix at this point. The calculation of the eigenvalues leads to a characteristic expression as follows

$$P(\lambda) = det \begin{pmatrix} 1 - \lambda & 2\pi \\ -2K \cos \varphi^* & 1 - 4\pi K \cos \varphi^* - \lambda \end{pmatrix},$$

$$P(\lambda) = \lambda^2 - \lambda(2 - 4\pi K \cos \varphi^*) + 1 = 0.\tag{11}$$

Since the eigenvalues are complex, the coordinate (φ^*, V^*) corresponds to an elliptical fixed point, which satisfy the stability condition of the AM. So, we obtain

$$|2 - 4\pi K \cos \varphi^*| < 4 \quad \rightarrow \quad 0 < 4\pi K \cos \varphi^* < 4.\tag{12}$$

Replacing the (10) on the expression given by (12) we can express the stability condition of the period-1 AM step-size l as a function of parameter K

$$\sqrt{l^2/4} < K < \sqrt{l^2/4 + 1/\pi^2}.\tag{13}$$

Although the interval of stability of period-1 AM step-size l was calculated by the simplified version model, the results can be extended to the complete bouncer model, where the match of their positions has a good agreement. The obtainment of analytical expressions for high order AM are extremely complicated, even using the simplified model. So, we perform a numerical approach to determine their existence, and the estimation of their stability interval and position in the phase space, as will be explained better in the next subsection.

3.2 Searching for Accelerator Modes of High Period

To determine the existence of the high period AM, we obtain numerically the average velocity of a great number of random initial condition merged in the chaotic sea during long time iterations (where each iteration corresponds to one impact) of the complete dynamics for different values of K. This procedure is efficient due to the non-symplectic character of the mapping. As a consequence, the map does not preserve the area in phase space. So, even that the location of the random initial conditions could be different from the location of accelerating structure, eventually some of them can be "attracted" to the AM [49]. If an AM exists in the non-symplectic mapping, it affects globally the dynamics of the system and the average velocity of the total ensemble will present anomalous diffusion [49].

To perform this analysis, we define the average velocity as

$$\langle V \rangle_N = \frac{1}{M} \sum_{j=1}^{M} \overline{V_j}(N, K) , \tag{14}$$

where M represents an ensemble of random initial conditions merged in the chaotic sea, and $\overline{V_j}(N, K)$ is expressed by

$$\overline{V_j}(N, K) = \frac{1}{N} \sum_{i=1}^{N} V_i , \tag{15}$$

with N representing the total iteration (impact) number,

Although the dynamics is globally affected by the existence of the AM, the high period ones, present low influence on the initial conditions far from their location. To reveal the existence of these high period AM, we set up two distinct averages over the dynamics. First the average was considered for an ensemble of $M = 5 \times 10^4$, over a time $N = 5 \times 10^3$ for (14). Then, we also perform the average over a sub-ensemble of $m = 500$ initial conditions selected in M that exhibit the highest values of the quantity expressed by (15).

The result obtained by this procedure is shown in Fig. 2, where the black line corresponds to the average over the first ensemble ($M = 5 \times 10^4$) and the red (gray) line corresponds to the average over the sub-ensemble ($m = 500$) most accelerated initial condition. The parameter K was varied with step equal to $\Delta K = 0.00005$ between $K = 0.13$ (parameter value for which the last spanning curve in phase space is destroyed [22, 49]) and $K = 0.5$ (parameter value for which occurs the first period-1 AM [49]).

As we can observe, at some values of the parameter K, the quantity $< V >_{5 \times 10^3}$ presents peaks very distinguishable from the general behavior of the curves. These peaks indicates the existence of the AM for the correspondent values of parameter K. The computation of the average over the sub-ensemble $m = 500$ of the most accelerated initial condition (red (gray) curve) highlights the effect of the AM with less influence in the global dynamics. In most of cases their existence is only revealed by the search of the $m = 500$ most accelerated initial condition. The period of each AM obtained by this procedure was numerically determined and indicated over each correspondent peak of Fig. 2. Moreover, the complete information about the AM is organized in Table 1 detailed by the period, range of stability, step-size, parameter of maximum acceleration and position of one accelerating structure.

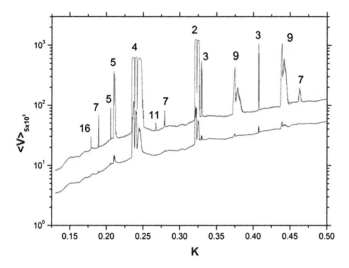

Fig. 2 Average velocity for the Bouncer model as function of K. In *black* we have the average over the first ensemble M, and in *red* (*gray*) we have the average over the sub-ensemble m. For both averages the time series we set at $N = 5 \times 10^3$. The distinct peaks in both *curves* indicate the existence of AM and the numbers indicate the correspondent period of the AM (color online)

4 Islands of Acceleration

After using the complete model to determine the existence of the AM with less influence in global dynamics, we can use the simplified model to reveal the location, shape and size of the correspondent AM of the complete version. As we discussed, the simplified mapping is symplectic and area-preserving. So, the existence of the AM in this version of the model does not affect the global dynamics of the system but only the initial conditions given inside the AM or very close to them due the effect of stickiness.

When we impose a modulation in V direction of phase space, the AM appear as periodic islands. The number of islands in modulated phase space must be equal to the period of the AM. In many cases, the size of these stability islands is too small to be visualized in the entire modulated phase space. Also, their sizes are strongly related to the influence of the AM in the global dynamics.

In Fig. 3 we illustrate the modulated phase spaces of the simplified mapping with coordinate φ at the horizontal direction and V at the vertical direction for the parameters K^* of Table 1. The red (black) circles indicate the position of each accelerating structure (islands in simplified map). As we can see, the number of accelerating structures corresponds to the period of the AM. We can also observe that most of the AM present at least one accelerating structure at $V = 1$, with exception of Fig. 3a, e, h.

Table 1 Accelerator modes for $K < 0.5$

Period	Range of stability	Step-size (l)	K^{*a}	Accelerating structure (φ, V) for $K = K^*$
16	$0.17875 < K < 0.17910$	1	0.17895	$(-1.847, 0.938)$
7	$0.18935 < K < 0.19000$	1	0.18970	$(-2.431, 1.000)$
5	$0.20570 < K < 0.20605$	1	0.20600	$(-2.520, 0.9982)$
5	$0.20965 < K < 0.21325$	1	0.21025	$(-2.483, 1.000)$
4	$0.23440 < K < 0.25065$	1	0.23600	$(-1.083, 0.815)$
11	$0.26705 < K < 0.26810$	1	0.26745	$(-0.714, 1.000)$
7	$0.27845 < K < 0.27980$	1	0.27960	$(0.865, 1.000)$
2	$0.32020 < K < 0.32720$	1	0.32100	$(-1.182, 0.797)$
3	$0.32895 < K < 0.33065$	1	0.32950	$(-1.357, 1.000)$
9	$0.37125 < K < 0.38610$	4	0.37460	$(-0.943, 1.000)$
3	$0.40725 < K < 0.40815$	2	0.40750	$(-1.398, 1.000)$
9	$0.43675 < K < 0.44810$	6	0.43920	$(-1.046, 1.000)$
7	$0.46090 < K < 0.46580$	2	0.46270	$(-1.088, 1.000)$

[a] Parameter value correspondent to the maximum acceleration in the range of stability

In Fig. 4a–n we zoomed the modulated phase spaces of Fig. 3a–n around one of the accelerating structures, respectively. We consider the modulation V mod[2] for a better view of the accelerating structure at $V = 1$. The points in red indicate the initial conditions belonging to the AM. We can observe in Fig. 4 that the most of the AM has a reduced area. This is the reason for being so difficult their detection. Also, in the Fig. 4j and l. we reveal that for the period-9 AM, we have three clusters of three accelerating structures on different locations of phase space. This fact increases the effect of stickiness around the islands of acceleration and make their influence more important to study diffusion process, besides the small size of their area on phase space.

The area of accelerating structures is not be the only aspect to determine their influence on the global dynamics in the complete model. As an example, we can compare the Fig. 4c with Fig. 4a, b. The period-5 AM shown in Fig. 4c is the smallest

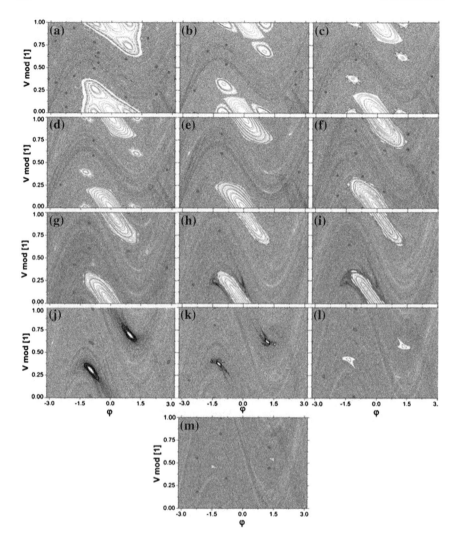

Fig. 3 Modulated phase space for the simplified model with coordinate φ on the horizontal direction and V mod[1] on the vertical direction for the respective parameters: **a** K =0.17895, **b** K =0.18970, **c** K =0.20600, **d** K =0.21025, **e** K =0.23600, **f** K =0.26745, **g** K =0.27960, **h** K =0.32100, **i** K =0.32950, **j** K =0.37460, **l** K =0.40750, **m** K =0.43920, **n** K =0.46270. The *red (black) circles* indicate the positions of the accelerating structures (color online)

accelerating structures in direct comparison with the others. However, it is more influent to a global effect in the dynamics than the period-16 AM and period-7 AM as can be seen in Fig. 2.

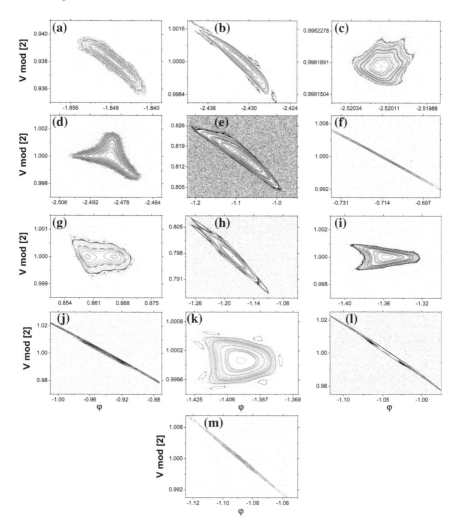

Fig. 4 Islands of Acceleration of the AM presented in Table 1 for the parameters: **a** $K = 0.17895$, **b** $K = 0.18970$, **c** $K = 0.20600$, **d** $K = 0.21025$, **e** $K = 0.23600$, **f** $K = 0.26745$, **g** $K = 0.27960$, **h** $K = 0.32100$, **i** $K = 0.32950$, **j** $K = 0.37460$, **l** $K = 0.40750$, **m** $K = 0.43920$, **n** $K = 0.46270$

5 Conclusions

The existence of high period AM and low influence on the global dynamics was investigated using the Bouncer Model. Due to the global properties of the AM in the complete version of this system, it was used to investigate the distribution of the AM for different periods as we ranged the control parameter. Until now it was a challenging task to detect the existence and location of accelerating structures in symplectic maps, since the initial conditions tested are rarely located inside the

islands of acceleration (which location is unpredictable in most of the cases). With a
model that exhibits a global behaviour for the AM, any sufficiently great ensemble of
initial conditions could be used to identify the existence of the AM and their periods.

A simplified version of the model allows us to find a correspondent symplectic mapping similar to the standard mapping. The simplified mapping was used to
visualize the accelerating structures and their positions in the phase space. When
we impose a modulation between 0 and 1 on the velocity variable, these structures
constitute a number of small islands of acceleration equal to the period of the AM.
A close view on the accelerating structures reveals that their size are not the only
aspect to determine their influence on the global dynamics of the complete model.
Other dynamical properties as: period of the AM, step-size l of the jumps, stickiness
around accelerating structures, distribution of regular periodic islands in phase space
and the chaotic saddle formed by the invariant stable and unstable manifolds should
also influence the effect of the AM on the global dynamics of the system [49].

Acknowledgments ALPL acknowledges FAPESP and CNPq for financial support. ILC thanks
FAPESP (2011/19296-1) and EDL thanks FAPESP (2012/23688-5), CNPq and CAPES, Brazilian
agencies.

References

1. A.J. Lichtenberg, M.A. Lieberman, *Regular and Chaotic Dynamics*. Applied Mathematical Sciences, vol 38 (Springer, New York, 1992)
2. R.C. Hilborn, *Chaos and Nonlinear Dynamics: An Introduction for Scientists and Engineers* (Oxford University Press, New York, 1994)
3. K.T. Alligood, T.D. Sauer, J.A. Yorke, *Chaos: An Introduction to Dynamical Systems* (Springer, New York, 1996)
4. G.M. Zaslavsky, *Physics of Chaos in Hamiltonian Systens* (Imperial College Press, New York, 2007)
5. G.M. Zaslavsky, *Hamiltonian Chaos and Fractional Dynamics* (Oxford University Press, New York, 2008)
6. E. Fermi, Phys. Rev. **75**, 1169 (1949)
7. M.A. Lieberman, V.A. Godyak, Ieee Trans. Plasma Sci. **26**, 955 (1998)
8. A.V. Milovanov, L.M. Zelenyi, Phys. Rev. E **64**, 052101 (2001)
9. A. Veltri, V Carbone, Phys. Rev. Lett. **92**, 143901 (2004)
10. K. Kobayakawa, Y.S. Honda, T. Samura, Phys. Rev. D **66**, 083004 (2002)
11. G. Lanzano et al., Phys. Rev. Lett. **83**, 4518 (1999)
12. F. Saif, I. Bialynicki-Birula, M. Fortunato, W.P. Schleich, Phys. Rev. A **58**, 4779 (1998)
13. A. Loskutov, A.B. Ryabov, L.G. Akinshin, J. Exp. Theor. Phys. **89**, 966 (1999)
14. A. Loskutov, A.B. Ryabov, L.G. Akinshin, J. Phys. A **33**, 7973 (2000)
15. R.E. de Carvalho, F.C. Souza, E.D. Leonel, Phys. Rev. E **73**, 066229 (2006)
16. F. Lenz, F.K. Diakonos, P. Schmelcher, Phys. Rev. Lett. **100**, 014103 (2008)
17. E.D. Leonel, D.F.M. de Oliveira, A. Loskutov, Chaos **19**, 033142 (2009)
18. A.L.P. Livorati, A. Loskutov, E.D. Leonel, Physica A **391**, 4756 (2012)
19. V. Gelfreich, V. Rom-Kedar, K. Shah, D. Turaev, Phys. Rev. Lett. **106**, 074101 (2011)
20. V. Gelfreich, V. Rom-Kedar, D. Turaev, Chaos **22**, 033116 (2012)
21. K. Shah, D. Turaev, V. Rom-Kedar, Phys. Rev. E **81**, 056205 (2010)

22. A.L.P. Livorati, T. Kroetz, C.P. Dettmann, I.L. Caldas, E.D. Leonel, Phys. Rev. E **86**, 036203 (2012)
23. P.J. Holmes, J. Sound Vibr **84**, 173 (1982)
24. L.D. Pustilnikov, Theor. Math. Phys. **57**, 1035 (1983)
25. R.M. Everson, Physica D **19**, 355 (1986)
26. D. Sternad, M. Duarte, H. Katsumata, S. Schaal, Phys. Rev. E **63**, 011902 (2000)
27. N.A. Burnham, A.J. Kulik, G. Gremaud, G.A.D. Briggs, Phys. Rev. Lett. **74**, 5092 (1995)
28. A.C.J. Luo, R.P.S. Han, Nonl. Dyn. **10**, 1 (1996)
29. J.J. Barroso, M.V. Carneiro, E.E.N. Macau, Phys. Rev. E **79**, 026206 (2009)
30. L. Mátyás, R. Klanges, Physica D **187**, 165 (2004)
31. E.D. Leonel, A.L.P. Livorati, Commun. Nonl. Sci. Num. Simul. **20**, 159 (2015)
32. T.L. Vincent, A.L. Mess, Int. J. Bif. Chaos **10**, 579 (2000)
33. S.K. Joseph, I.P. Mariño, M.A.F. Sanjuán, Commun. Nonl. Sci. Num. Simul. **17**, 3279 (2012)
34. F. Spahn, U. Schwarz, J. Kurths, Phys. Rev. Lett. **78**, 1596 (1997)
35. P. Müller, M. Heckel, A. Sack, T. Pöschel, Phys. Rev. Lett. **110**, 254301 (2013)
36. F. Pacheco-Vázquez, F. Ludwig, S. Dorbolo, Phys. Rev. Lett. **113**, 1108001 (2014)
37. A.J. Lichtenberg, M.A. Liberman, N.W. Murray, Physica D **28**, 371 (1987)
38. A.J. Lichtenberg, M.A. Liberman, Physica D **33**, 211 (1988)
39. Y.H. Ichikawa, Y. Nomura, T. Kamimura, Prog. Theor. Phys. Supp. **99**, 220 (1989)
40. Y.H. Ichikawa, T. Kamimura, T. Hatori, S.Y. Kim, Prog. Theor. Phys. Supp. **98**, 01 (1989)
41. G.M. Zaslavsky, B.A. Niyazov, Phys. Rep. **283**, 73 (1997)
42. V. Rom-Kedar, G.M. Zaslavsky, Chaos **9**, 697 (1999)
43. S.T. Dembinski, P. Peplowski, Phys. Rev. E **55**, 212 (1997)
44. T. Manos, M. Robnik, Phys. Rev. E **87**, 062905 (2013)
45. T. Manos, M. Robnik, Phys. Rev. E **89**, 022905 (2014)
46. M.K. Oberthaler, R.M. Godun, M.B. d'Arcy, G.S. Summy, K. Burnett, Phys. Rev. Lett. **83**, 4447 (1999)
47. S. Schlunk, M.B. d'Arcy, S.A. Gardiner, G.S. Summy, Phys. Rev. Lett. **90**, 124102 (2003)
48. S. Fishman, I. Guarneri, L. Rebuzzini, J. Stat. Phys. **110**, 911 (2003)
49. T. Kroetz, A.L.P. Livorati, E.D. Leonel, I.L. Caldas, Phys. Rev. E. **92**, 012905 (2015)

How Dissipative Highly Non-linear Solitary Waves Interact with Boundaries in a 1D Granular Medium

Lautaro Vergara

Abstract We here study the behaviour of highly non-linear solitary waves when interact with boundaries. We compare a model developed by the author with the Kubawara-Kono model. We show that this last model does not give the best results when compared with our model, proposed in Vergara (Phys. Rev. Lett. 104:118001, 2010 [1]).

Granular matter is everywhere: in agriculture, as grains; in pharmaceutical industry as pills; in soils as sand; even in outer space, as in Saturn rings! This matter could be dry or wetted and it should not come as a surprise that there is so much interest in understanding its dynamics [2]. The simplest example of this kind of matter is a line of spherical beads, which are modelled as a line of point masses connected by nonlinear springs. Nesterenko has shown that if these systems are struck on one end, highly non-linear waves are generated [3]. The interaction between non-conforming solids of elliptical shapes was derived by Hertz in 1882 [4]. Using potential theory he developed a model for the normal contact assuming a pure elastic contact. Under very restricted circumstances, this force law give reasonable results when compared with experiments (see e.g. [5, 6]). For example, when the impact between grains is such that energy dissipating phenomena become relevant Hertz theory fails. For example energy loss can be measured experimentally, at the macroscopic level, by measuring the coefficient of restitution, which has been observed to decrease with the normal component of the relative impact velocity [7], opposed to what Hertz theory establishes. Such phenomena, involving for example elasto-plastic and viscoelastic behavior [8], are so complex that it is hard to think of a closed form force law that may describe them all at once. Nevertheless, there has been approaches that have made interesting advances in modelling such phenomena.

In 1987, Kuwabara and Kono developed a model to study the restitution coefficient in a collision between two spheres, under the assumption that no plastic deformation

L. Vergara (✉)
Departamento de Física, Universidad de Santiago de Chile, USACH,
Casilla 307, Santiago 2, Chile
e-mail: lautarovergara@gmail.com

© Springer International Publishing Switzerland 2016   193
M. Tlidi and M.G. Clerc (eds.), *Nonlinear Dynamics: Materials,
Theory and Experiments*, Springer Proceedings in Physics 173,
DOI 10.1007/978-3-319-24871-4_14

is present [9]. Several years later, 1996, Brilliantov et al. have obtained a quasi-static approximation to calculate the normal force acting between colliding particles, assuming a viscoelastic force. This model happens to be the same as the one of Kubawara and Kono. One year later, Morgado and Oppenheimer [10] made a first principles derivation of the form of the instantaneous dissipative force acting between two spheres.

Since the Hertz interaction is the main one, one must add the viscoelastic term to it. The model for a set of N spheres with elastic and viscoelastic forces present is described as follows. Let $x_n(t)$ represent the displacement of the center of the nth sphere, of mass m_n, from its initial equilibrium position. The equations of motion that describe the dynamics of N beads, inclined by an angle θ, in a gravitational field are

$$m_n \ddot{x}_n = K_{n-1,n} \delta_{n-1}^{3/2} - K_{n,n+1} \delta_n^{3/2} + \eta B \left\{ \sqrt{\delta_{n-1}} \, \dot{\delta}_{n-1} - \sqrt{\delta_n} \, \dot{\delta}_n \right\} + m_n \, g \, \sin[\theta],$$
(1)

with $n = 2, \ldots, N - 1$. As known, the equations of motion for the first and last beads differ from (1) and are thus not written down here. The notation is as follows: $\dot{\delta}_n = \dot{x}_n - \dot{x}_{n+1}$ is the relative velocity of beads n and $n + 1$. The overlap between adjacent beads is $\delta_n = \max\{\Delta_{n,n+1} - (x_{n+1} - x_n), 0\}$, ensuring that the spheres interact only when in contact. For the same reason a step-function in the third term has been included, that is, $\theta_n = \theta[\Delta_{n,n+1} - (x_{n+1} - x_n)]$. $\Delta_{n,n+1} = (g \, \sin[\theta] \, n \, m_n / K_{n,n+1})^{2/3}$ appears from the pre-compression due to the gravitational interaction. The expression for the Hertz coupling $K_{i,j}$ between beads i and j is well known and depends on radii, R_i, Young moduli and Poisson ratio ν of beads [8]. Using (17) of [11] we can write the constants B and η as

$$B = \frac{1}{3} \sqrt{\frac{R_1 R_2}{R_1 + R_2}} \frac{1 - \nu}{\nu^2}, \qquad \eta = \frac{(3 \eta_2 - \eta_1)^2}{3 \eta_2 + \eta_1},$$
(2)

where η_i are the viscous constants of the particle material. Since they are unknown, we leave the constant η as a parameter.

When this force law is used to study how the coefficient of restitution changes with impact velocity theory and experiment agree well [12].

The Kubawara and Kono model was also used by Job et al. [13] to study the solitary wave reflection at a wall. They assert that they got precise quantitative agreements with their experiments. Unfortunately we have no data available from their experiment, and therefore we will use data from Daraio et al. [14]. There, among other things, they have studied the behaviour of highly non-linear solitary waves at a wall, in a line of stainless-steel beads. So, this work is rather near to what Job et al. did and it is a good stand point to make a comparison.

The set of equations in this paper have been solved using an explicit Runge-Kutta method of the 5th order with an embedded error estimator, from *Mathematica*.

Few years ago, Daraio et al. proposed a model for dissipative highly non-linear solitary waves [15] that could be thought of as an extension of a toy model proposed in [16], which incorporates a 'discrete Laplacian' in the velocities, and defined by the set of equations

$$m_n \ddot{x}_n = K_{n-1,n} \, \delta_{n-1}^{3/2} - K_{n,n+1} \, \delta_n^{3/2} + \gamma \left\{ \dot{\delta}_{n-1} - \dot{\delta}_n \right\} + m_n \, g \, \sin[\theta]. \qquad (3)$$

The equations of motion defining this model are:

$$m_n \ddot{x}_n = K_{n-1,n} \, \delta_{n-1}^{3/2} - K_{n,n+1} \, \delta_n^{3/2} + \gamma \, \mathsf{sgn}(\dot{\delta}_{n-1} - \dot{\delta}_n) \, |\dot{\delta}_{n-1} - \dot{\delta}_n|^\alpha + m_n \, g \, \sin[\theta]. \qquad (4)$$

Here γ and α are phenomenological constants that must be fitted to coincide with experiment. The signum function is included in order to coincide with the case $\alpha = 1$ of (4). The comparison of this model with experimental data seems to give good results.

When I knew about this model it was somewhat disappointing because the authors have found that, in order to fit with their experiments, they ought to use real values for the exponent α, that within the experimental errors one could agree with them that the exponent has a common value for the different materials. In my opinion, the real value of the exponent would make impossible to understand the force term from first principles. Then, I have assumed that those terms present in Model 1 must appear in a physical description of granular matter in 1-D, given that they can be obtained from first principles, under reasonable assumptions. Then I have added a phenomenological term that is quadratic in the velocities [1]. The equations of motion that define my model are:

$$m_n \ddot{x}_n = K_{n-1,n} \, \delta_{n-1}^{3/2} - K_{n,n+1} \, \delta_n^{3/2} + \eta \, B \left\{ \sqrt{\delta_{n-1}} \, \dot{\delta}_{n-1} - \sqrt{\delta_n} \, \dot{\delta}_n \right\} + m_n \, g \, \sin[\theta]$$
$$+ \lambda \left\{ \theta_{n-1} \, (\mathsf{sgn}[\dot{\delta}_{n-1}] \, \dot{\delta}_{n-1})^2 - \theta_n \, \mathrm{sign}[\dot{\delta}_n] \, \dot{\delta}_n^2 \right\}, \qquad (5)$$

where on the left term, $\mathsf{sgn}[\dot{\delta}_{n-1}]$ multiplies only the first term of $\dot{\delta}_{n-1}$, i.e. $\dot{x}_{n-1}$. In this model too we have two constants η and λ that should be fitted with the experiment.

My explanation for this Ansatz is as follows: after the impact, the dynamics becomes a multi-impact problem; this produces that the relative velocity of beads n and $n + 1$, $\dot{\delta}_n$, may change from $\dot{\delta}_n < 0$, related to an expansion phase, to $\dot{\delta}_n > 0$, corresponding to a compressional phase. The same happens for the relative velocity of beads $n - 1$ and n, $\dot{\delta}_{n-1}$. Therefore, by fixing our attention on one bead in the chain, say nth, one has a particle between two moving walls (beads n and $n + 1$) and then its dynamics depends on the dynamics of both constraints. It is worthwhile to mention that a term proportional to the square of the velocity was introduced by Pöschl [17] as an attempt to extend the Hertz theory to plastic bodies. Also notice that if beads $n - 1$, n and $n + 1$ are all in contact at a given instant, one can easily observe that the third term can be generically written as

$$\dot{x}_{n-1}^2 + 2 \, \varepsilon \, \dot{x}_n^2 + \kappa \, \dot{x}_{n+1}^2 + 2 \, \beta \, \dot{x}_n \, (\dot{x}_{n-1} - \sigma \, \dot{x}_{n+1}), \qquad (6)$$

where the constants take the values $\varepsilon = 0, 1, \kappa = \pm 1, \beta = \pm 1, \sigma = \pm 1$, depending on the sign of the relative velocities $\dot{\delta}_n$ and $\dot{\delta}_{n-1}$.

It has been demonstrated in [1] that the viscoelastic and velocity-squared terms give a contribution, in the case of interaction at the wall of the same order of magnitude, with a time interval where both force terms reinforce their contribution. Also, it has been shown that the velocity-squared force term is a continuous, although not smooth, function of time. Nevertheless, this is not crucial for reproducing the experimental data.

In [14], the effect of boundaries on solitary waves was studied. Spheres used were made of stainless steel 316 with Young modulus and Poisson ratio $E = 193 \times 10^9$ Pa and $v = 0.30$, respectively. Their diameter was 4.76 mm and mass 0.4501 g, all assembled in a vertical PTFE cylinder. This makes that the constant $B = 0.0511 \, \mathrm{m}^{1/2}$. As a wall they have used a cap of brass, with $E = 115 \times 10^9$ Pa and $v = 0.31$, in front of the piezosensor. Solitary waves were generated by impacting the column of 21 beads, each with radius 2.38 mm and mass 0.123 g, by a cylindrical alumina impactor, with Young modulus $E = 416 \times 10^9$ Pa and Poisson ratio $v = 0.23$. The cylinder had a mass 1.2 g and a velocity equal to 0.44 m/s. The force on the piezosensor was recorded, and the data presented as plots of force as a function of time.

Figure 1 shows the behaviour of the solitary wave at the wall. Because of the lack of detailed experimental information, a global time shift of 46 μs was applied to the experimental output to get time coincidence with the numerical experiment. Numerical findings from the Kubawara and Kono model are compared with the data in Fig. 1. The best value, not necessarily optimal, that approximate the data is $\eta = 6.3 \times 10^4$ kg/ms.

In Fig. 2 the same data is compared with our model. The best values for the model parameters, not necessarily the optimal one, are $\eta = 3.3 \times 10^4$ kg/ms, $\lambda = -2.2$ kg/m.

In both cases one can find a set of parameters that fit quite well the main perturbation. The difference lies in how well can each model reproduce the secondary pulse, that appear for weak dissipation. In this case, our model performs much better than

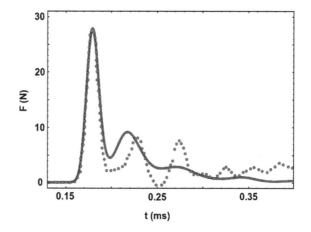

Fig. 1 Scattering of highly non-linear waves off a wall. Data is shown as *dots* and numerical results from the Kubawara and Kono model as a *solid line*

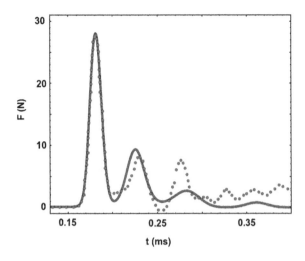

Fig. 2 Scattering of highly non-linear waves off a wall. Data is shown as *dots* and numerical results my model as a *solid line*

the Kubawara and Kono model. It is possible that the ultimate force term is not the one that we have included in our model, but given its success we are convinced that multi impacts must be included in a correct description of highly non-linear solitary waves in one-dimensional granular media.

Our model is economical, requiring only two parameters that depend on the mechanical properties of beads, one of them being well known and with clear physical origin. In addition, we have demonstrated in [1] that the parameter values do not need to be changed in order to fit simulations with experiments carried out under different conditions, and are found through a simple trial and error procedure. Of course, if one is looking for the optimal set of parameters, one should perform a more complete analysis.

Acknowledgments The author thanks Prof. Daraio for allowing me to use data that I have used to test my model. This work was partially supported by Fondecyt, Grant No. 1130492.

References

1. L. Vergara, Phys. Rev. Lett. **104**, 118001 (2010)
2. H.M. Jaeger, S.R. Nagel, Science **255**, 1523 (1992); H.J. Herrmann, J.-P. Hovi, S. Luding (eds.), *Physics of Dry Granular Media*, NATO ASI Series (Kluwer, Dordrecht, 1998); H. Hinrichsen, D. Wolf (eds.), *The Physics of Granular Media* (Wiley, Berlin, 2004); R. García-Rojo, H.J. Herrmann, S. McNamara (eds.), *Powders and Grains 2005* (Taylor and Francis, 2005)
3. V.F. Nesterenko, *Dynamics of Heterogeneous Materials* (Springer, New York, 2001)
4. H. Hertz, Journal für die reine und angewandte Mathematik **92**, 156 (1882)
5. C. Coste, E. Falcon, S. Fauve, Phys. Rev. E. **56**, 6104 (1997)
6. S. Sen, M. Manciu, J.D. Wright, Phys. Rev. E **57**, 2386 (1998); E.J. Hinch, S. Saint-Jean, Proc. R. Soc. A **455**, 3210 (1999); Hascoet, H.J. Herrmann, Eur. Phys. J. B **14**, 183 (2000);

M. Manciu, S. Sen, A.J. Hurd, Physica D **157**, 2386 (2001); S. Sen et al. Phys. Rep. **46**, 21 (2008), and references therein

7. W. Goldsmith, Impact (Edward Arnold Publications, London, 1960); R. Sondergaard, K. Chaney, C.E. Brennen. J. Appl. Mech. **57**, 694 (1990)
8. K.L. Johnson, *Contact Mechanics* (Cambridge University Press, London, 1992)
9. G. Kuwabara, K. Kono, Jpn. J. Appl. Phys. **26**, 1230 (1987)
10. W.A.M. Morgado, I. Oppenheim, Phys. Rev. E. **55**, 1940 (1997)
11. N.V. Brilliantov, F. Spahn, J.-M. Hertzsch, T. Pöschel, Phys. Rev. E. **53**, 5382 (1996)
12. R. Ramirez, T. Pöschel, N.V. Brilliantov, T. Schwager, Phys. Rev. E. **60**, 4465 (1999)
13. S. Job, F. Melo, A. Sokolow, S. Sen, Phys. Rev. Lett. **94**, 178002 (2005)
14. C. Daraio, V.F. Nesterenko, E.B. Herbold, S. Jin, Phys. Rev. E **73**, 026610 (2006)
15. R. Carretero-González, D. Khatri, M.A. Porter, P.G. Kevrekidis, C. Daraio, Phys. Rev. Lett. **102**, 024102 (2009)
16. A. Rosas, A.H. Romero, V.F. Nesterenko, K. Lindenberg, Phys. Rev. E **78**, 051303 (2008)
17. T. Pöschl, Zeit. f. Phys. **46**, 142 (1927)

Inviscid Transient Growth on Horizontal Shear Layers with Strong Vertical Stratification

Cristóbal Arratia, Sabine Ortiz and Jean-Marc Chomaz

Abstract We report an investigation of the three-dimensional stability of an horizontal free shear layer in an inviscid fluid with strong, vertical and constant density stratification. We compute the optimal perturbations for different optimization times and wavenumbers. The results allow comparing the potential for perturbation energy amplification of the free shear layer instability and the different mechanisms of transient growth. We quantify the internal wave energy content of the perturbations and identify different types of optimal perturbations. Intense excitation of gravity waves due to transient growth of perturbations is found in a broad region of the wavevector plane. Those gravity waves are eventually emitted away from the shear layer.

1 Introduction

During the last years, there has been an increased interest in the effects of horizontal shear in geophysical flows. In that context, a horizontal mixing layer with vertical stratification is one idealized flow model presenting the effect of shear and also the free shear or Kelvin-Helmholtz (KH) instability. The two dimensional (2D) modes of the KH instability are not affected by stratification. The vertical structure developing in such a flow is however strongly affected by stratification, as shown by the direct

C. Arratia (✉)
Departamento de Física FCFM, Universidad de Chile, Blanco Encalada 2008,
Santiago, Chile
e-mail: cristobal.arratia@gmail.com

C. Arratia · S. Ortiz · J.-M. Chomaz
LadHyX, CNRS-École Polytechnique, 91128 Palaiseau Cedex, France

S. Ortiz
e-mail: ortiz@ladhyx.polytechnique.fr

J.-M. Chomaz
e-mail: chomaz@ladhyx.polytechnique.fr

S. Ortiz
UME/DFA, ENSTA, Chemin de la Hunière, 91761 Palaiseau Cedex, France

© Springer International Publishing Switzerland 2016
M. Tlidi and M.G. Clerc (eds.), *Nonlinear Dynamics: Materials,
Theory and Experiments*, Springer Proceedings in Physics 173,
DOI 10.1007/978-3-319-24871-4_15

numerical simulations (DNS) performed by Basak and Sarkar [7]. They show the appearance of a layered structure consisting of a stack of 'pancake' vortices with a vertical correlation length of the order of the buoyancy length. Despite the non applicability of Squire's theorem to this flow, it has been shown in [10] that the most unstable mode is still 2D. This implies that modal stability analysis on the parallel shear layer does not select a specific vertical lengthscale. For strong stratification, however, the range of unstable vertical wavenumbers widens proportionally to the inverse of the Froude number, which means that stronger stratification destabilizes smaller vertical lengthscales [10]. This is a consequence of the self-similarity of strongly stratified inviscid flows found by Billant and Chomaz [8].

Another important aspect of horizontal shear flows with vertical stratification concerns internal gravity wave generation and emission. The generation and emission of internal waves has been studied, among others, by Vanneste and Yavneh [12] in the rotating case and by Bakas and Farrell [6] in the non-rotating case. Both studies focus on linearized dynamics of perturbations of plane constant shear flow, which allows finding analytic expressions for the generated wave amplitude in different asymptotic regimes. This simplified model has no unstable modes nor intrinsic horizontal lengthscale.

Here we consider linearized perturbations of a horizontal mixing layer with a tanh velocity profile and vertical stratification on a non-rotating frame. We study the sensitivity to initial conditions by computing the optimal perturbations, which are those that maximize the energy growth up to an optimization time T. We compute the optimal perturbations for a broad range of streamwise and spanwise wavenumbers and for different optimization times T. In this way we can determine, for different times, whether the KH instabilty or other transient growth mechanisms are more efficient in extracting energy from the basic flow. We then quantify the vortex and gravity wave energy content of the perturbations by means of a Craya-Herring, or poloidal-toroidal, decomposition [11]. This decomposition helps identifying different types of optimal perturbations according to their wave/vortex energy content. We show that some of the optimal perturbations involve the generation of waves that are eventually radiated away from the shear layer.

2 Formulation and Methods

We consider the evolution of the perturbative velocity $\mathbf{u}(\mathbf{x}, t)$ and density $\rho(\mathbf{x}, t)$ fields according to the linearized incompressible Euler equations in the Boussinesq approximation. Here $\mathbf{x} = (x, y, z)$ is the cartesian coordinate vector with z increasing upwards, and t is the time coordinate. Our base state is $\mathbf{U} = \tanh(y)\mathbf{e}_x$ with a stable linear density stratification $\rho_B(z) = \rho_0(1 - N^2 z/g)$, where ρ_0 is a reference density, g is the acceleration of gravity and $N = \sqrt{-g/\rho_0 d\rho_B/dz}$ is the Brunt-Väisälä frequency. With this variables, the horizontal Froude number corresponding to the ratio between the buoyancy and advective time scales is $F_h = N^{-1}$.

Linearity and homogeneity in x and z of the evolution equations allow us to rewrite the fields as $[\mathbf{u}, \rho](x, y, z, t) \longrightarrow [\mathbf{u}, \rho](y, t)e^{i(k_x x + k_z z)}$, and to consider independently the evolution of the different streamwise-spanwise wavenumbers. Because of the latter, for each point in the (k_x, k_z)-plane we can define the optimal gain as

$$G(T) = \max_{[\mathbf{u}, \rho](y, 0)} \left(\frac{E(T)}{E(0)} \right), \tag{1}$$

where the total energy E is given up to a constant by

$$E = \int \left(\mathbf{u}^2 + N^2 \rho^2 \right) \, dy. \tag{2}$$

Thus, $G(T)$ is the maximum attainable increase in energy up to the optimization time T. We span the (k_x, k_z)-plane for each time T at which we compute the optimal perturbation. We characterize the transient growth by the optimal mean growth rate of the perturbation

$$\sigma_m(k_x, k_z, T) = \frac{1}{2T} \int_0^T \frac{\partial}{\partial t} \log\left[E(k_x, k_z, t)\right] \, dt, \tag{3}$$

where $E(k_x, k_z, t)$ is the energy of the optimal perturbation. Using definition (3) we can directly compare the amplification of the optimal perturbations with the growth rate of the most unstable KH mode, $\sigma_{KH} = 0.1897$ [10].

The optimal perturbations are computed by the iterative procedure described in [9], whereby the succesive numerical integration of the direct and the time reversed adjoint equations is performed until convergence is achieved. The adjoint equations and optimization algorithm have been adapted on a pseudo-spectral DNS code, same as in [3].

Throughout this paper we will use $F_h = 0.1$, which is small enough so that the strongly stratified similarity described in [8] holds to a very good approximation [2], as is also the case for the eigenmodes [10] and for the viscous transient growth after correcting by the viscous damping factor $e^{\nu k_z^2 T}$, where ν is the viscosity [4]. In this context, the main consequence of this scaling law is that, for $F_h \lesssim 0.1$, we can plot the results as a function of $k_z F_h$ to capture the dependence on both k_z and F_h. The results are not significantly different for $F_h < 1$ [2].

3 Results

Figure 1 shows $\sigma_m(k_x, F_h k_z)$ for $T = 4, 7, 10$ and 15. The colorbars indicate that σ_m decreases as T increases. The dashed line on each of the figures indicates roughly the ratio $r = F_h k_z / k_x$ at which the optimal perturbations are less amplified. The value of r corresponding to the dashed lines, say r_{min}, increases as T increases. For

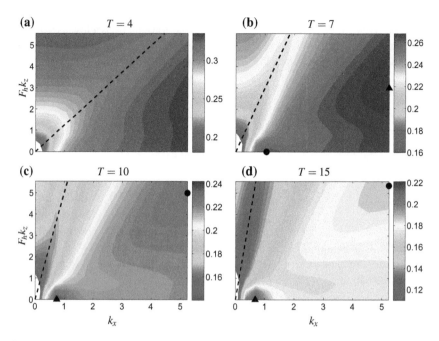

Fig. 1 Optimal mean growth rate $\sigma_m(k_x, F_h k_z)$ for optimization times $T = 4$ (**a**), $T = 7$ (**b**), $T = 10$ (**c**), and $T = 15$ (**d**). The *colorbar* is indicated next to each figure. The *horizontal* and *vertical axis* are the same on all figures, as shown in the *lower left figure* (**c**). The values of the ratio $F_h k_z / k_x$ indicated by the *dashed lines* are $r_{min} = 5/4, 3, 5$ and 8 for $T = 4, 7, 10$ and 15, respectively. In (**b–d**), the ▲ indicates the maximum of the computed σ_m (at the boundary in **c**) and the • indicates a secondary local maximum (at the boundary in **c** and **d**)

$T = 4$, σ_m increases as $k = \sqrt{k_x^2 + (F_h k_z)^2}$ increases and shows no clear maximum. For $T = 7$, 10 and 15, the global maximum of the computed σ_m is indicated by a ▲ and a second, local maximum of the computed σ_m is marked by a •; these are located at large k and around the KH unstable region with $k_z = 0$. For $T = 7$, the maximum occurs for large k at the ▲ around $r = 3/5$, and the maximum close to the KH unstable region is only a local maximum. For $T = 10$ and 15, the global maximum is 2D and located in the KH unstable region, while the secondary maximum appears at large k, at the boundary of the domain.

3.1 Craya-Herring Decomposition

Following [11], we define the Craya-Herring basis $(\mathbf{e}_1, \mathbf{e}_2, \mathbf{e}_3)$ as $\mathbf{e}_1 = \mathbf{k} \times \mathbf{e}_z / |\mathbf{k} \times \mathbf{e}_z|$, $\mathbf{e}_2 = \mathbf{k} \times (\mathbf{k} \times \mathbf{e}_z) / |\mathbf{k} \times (\mathbf{k} \times \mathbf{e}_z)|$ and $\mathbf{e}_3 = \mathbf{k} / |\mathbf{k}|$, where $\mathbf{k} = (k_x, k_y, k_z)^T$ is the wave vector. In this orthonormal basis, the Fourier transform of the velocity field becomes $\hat{\mathbf{u}} = \hat{\phi}_1 \mathbf{e}_1 + \hat{\phi}_2 \mathbf{e}_2$ because $\hat{\mathbf{u}} \cdot \mathbf{e}_3 = 0$ due to the incompressibility condition.

When $\sqrt{k_x^2 + k_y^2} \neq 0$, the energy density in spectral space $\varepsilon(\mathbf{k})$ is given by

$$\varepsilon(\mathbf{k}) = |\hat{\phi}_1|^2 + |\hat{\phi}_2|^2 + |\hat{\phi}_3|^2, \tag{4}$$

with

$$\hat{\phi}_1 = \frac{i}{\sqrt{k_x^2 + k_y^2}} \hat{\omega}_z, \tag{5a}$$

$$\hat{\phi}_2 = -\sqrt{\frac{k_x^2 + k_y^2 + k_z^2}{k_x^2 + k_y^2}} \hat{u}_z, \tag{5b}$$

$$\hat{\phi}_3 = N\hat{\rho}, \tag{5c}$$

where $\hat{\omega}_z$ is the Fourier transformed vertical vorticity. The total energy may be expressed as $E = \int \varepsilon \, \mathrm{d}k_x \mathrm{d}k_y \mathrm{d}k_z$.

The $\hat{\phi}_1 \mathbf{e}_1$ part of the velocity field is purely horizontal so it does not directly affect the disturbance density field. For the linear dynamics in the absence of a basic flow, decomposition (5) provides a separation of the fields in which the $\hat{\phi}_1 \mathbf{e}_1$ part of the velocity field decouples from the internal wave dynamics given by $\hat{\phi}_2$ and $\hat{\phi}_3$. Despite the fact that the wave and vortical parts of the velocity field can not be unambiguously split in a general case, this decomposition provides an objective, physically motivated way of quantifying the wave content of the perturbative field. We will thus refer to the different components as the wave and vortex parts of the flow fields.

Figure 2 shows the fraction of the total energy contained in the vortex part $\hat{\phi}_1$ of the velocity field, for each of the optimal perturbations of Fig. 1. Figure 2 shows the energy fractions of the optimal initial condition ($t = 0$) and the optimal response ($t = T$). Also shown are the dashed lines of Fig. 1 indicating r_{min}, the value of $r = F_h k_z / k_x$ where the optimal perturbations have lower growth. Blue contours indicate that most of the perturbation energy corresponds to gravity waves. In the left column we observe that the wave content of the optimal initial condition depends mainly on r. The energy of the optimal initial condition is given by the vortex part for $r = 0$ and the energy of the wave part becomes increasingly important as r increases. It can be seen that the dashed lines at r_{min} coincide, roughly but consistently for all T, with the region where the energy of the optimal initial condition changes from being mostly vortex to mostly wave. The energy of the optimal response, on the other hand, is mostly given by the vortex part for $k_x \sim 0$ and for $F_h k_z \sim 0$, while for most of the domain where k_x, $F_h k_z \gtrsim 1$, the energy of the optimal response is mainly in the wave part.

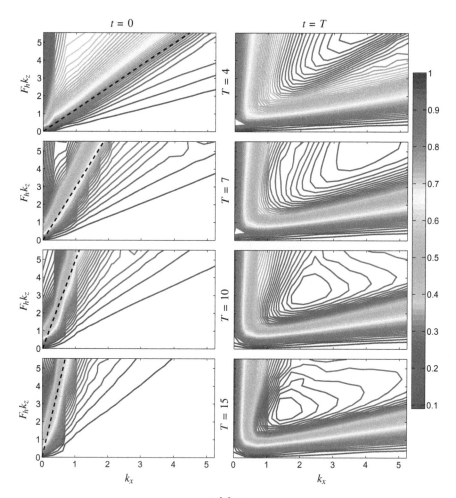

Fig. 2 Energy fraction in the vortical part $\frac{\int |\hat{\phi}_1|^2 \, dk_y}{E}$ for the optimal perturbations at $t = 0$ (*left column*) and optimal response at $t = T$ (*right column*). The optimization time for each row, starting from above, are $T = 4$, $T = 7$, $T = 10$ and $T = 15$, as indicated on the figure. Here, for $k_x = 0$, the energy of the horizontal mean flow (which corresponds to $\sqrt{k_x^2 + k_y^2} = 0$) has been added to the energy of $\hat{\phi}_1$

4 Discussion

Figures 1 and 2 help in distiguishing three main regions on the $(k_x, F_h k_z)$ plane. First, for $F_h k_z \sim 0$, the dynamics is dominated by the vertical vorticity (Fig. 2) and is essentially 2D. In this region, the Orr mechanism dominates the inviscid transient growth for short time (Fig. 1a) and for large k_x, while the KH instability becomes increasingly important as T increases (Fig. 1b–d). Second, for $k_x \sim 0$, the optimal transient growth consists of waves generating large horizontal velocity (Fig. 2), giving large transient growth for short times (Fig. 1a, b). The mechanism involved here is the transient generation of streamwise streaks during the passage of waves through shear [2, 5]. This transient growth mechanism is similar to the lift-up mechanism, it produces streamwise velocity as a result of cross-stream transport. This mechanism is perhaps related to the cross-stream transport of KH billows that is linked to the layered structure reported in [7], see also [1]. Third, for k_x, $F_h k_z \gtrsim 1$ and $r \lesssim r_{min}$, the optimal perturbations result mainly in the generation of wave energy from the vortical part (Fig. 2). This is an efficient mechanism of perturbation energy growth, the most efficient indeed for intermediate times ($T = 7$ in Fig. 1b). In the absence of viscosity, this mechanism does not seem to reach a maximum of σ_m for finite (k_x, k_z).

An important aspect concerns whether the generated wave energy remains in the wave part after T. Figure 3 shows spatio-temporal diagrams of the vortex and wave energy density for an optimal perturbation at $T = 7$, for (k_x, k_z) in the region of wave generation and largest energy growth (close to the ▲ in Fig. 1b). From $t = 0$ to 7, the perturbation energy grows by a factor of 43.6, more than any unstable mode during that time. Initially, most of the energy growth occurs on the vortical part. Wave energy starts being noticeable around $t = 5$ and then increases quickly when the vortex energy is transferred into the wave part of the flow. The generated wave is then radiated away after the optimization time. This qualitative behaviour is the same for different optimization times.

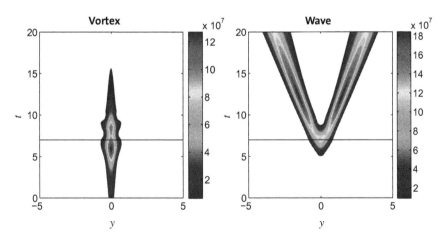

Fig. 3 Spatio-temporal diagram of the optimal perturbation for $T = 7$ (*horizontal line*), $(k_x, F_h k_z) = (4.89, 2.96)$. *Colorbar* shows the energy density of the vortex (*left*) and internal gravity wave (*right*) components of the flow. The vortex energy density is obtained by taking the inverse Fourier transform in the y direction of $|\hat{\phi}_1|^2$ evaluated at the corresponding (k_x, k_z). The same is done with $|\hat{\phi}_2|^2 + |\hat{\phi}_3|^2$ for the energy of the wave part

5 Conclusion

We have computed the optimal perturbations on a horizontal mixing layer with strong vertical stratification. For short optimization times (up to T between 7 and 10), the most amplified perturbations occur for large wavenumbers, away from the KH unstable region that dominates for large T. We have done a Craya-Herring decomposition to quantify the wave and vortex energy content of the optimal perturbations. Using this decomposition we can distinguish 3 main different types of optimal perturbations: quasi 2D for $F_h k_z \sim 0$, streamwise streaks generated by waves for $k_x \sim 0$, and waves generated by vortical motion for k_x, $F_h k_z \gtrsim 1$ and $r \lesssim r_{min}$.

An important result is the fact that, for moderate times ($T \sim 7$), the most amplified optimal perturbations produce waves that are eventually radiated away from the shear layer. This wave generation and emission mechanism remains efficient in the large wavenumber region even for larger T, when the KH instability becomes the most efficient mechanism.

References

1. E. Arobone, S. Sarkar, The statistical evolution of a stratified mixing layer with horizontal shear invoking feature extraction. Phys. Fluids **22**(11), 115108 (2010)
2. C. Arratia, Non-modal instability mechanisms in stratified and homogeneous shear flows. PhD Thesis, Ecole Polytechnique (2011). http://tel.archives-ouvertes.fr/pastel-00672072/
3. C. Arratia, C.P. Caulfield, J.-M. Chomaz, Transient perturbation growth in time-dependent mixing layers. J. Fluid Mech. **717**(1), 90–133 (2013)
4. C. Arratia, S. Ortiz, J.M. Chomaz, Transient evolution and high stratification scaling in horizontal mixing layers, in *Advances in Turbulence XII*, ed. by B. Eckhardt, pp. 183–186. Springer, Heidelberg (2009)
5. N.A. Bakas, B.F. Farrell, Gravity waves in a horizontal shear flow. Part I: growth mechanisms in the absence of potential vorticity perturbations. J. Phys. Oceanogr. **39**(3), 481–496 (2009a)
6. N.A. Bakas, B.F. Farrell, Gravity waves in a horizontal shear flow. Part II: interaction between gravity waves and potential vorticity perturbations. J. Phys. Oceanogr. **39**(3), 497–511 (2009b)
7. S. Basak, S. Sarkar, Dynamics of a stratified shear layer with horizontal shear. J. Fluid Mech. **568**(1), 19–54 (2006)
8. P. Billant, J.-M. Chomaz, Self-similarity of strongly stratified inviscid flows. Phys. Fluids **13**(6), 1645–1651 (2001)
9. P. Corbett, A. Bottaro, Optimal linear growth in swept boundary layers. J. Fluid Mech. **435**(1), 1–23 (2001)
10. A. Deloncle, J.-M. Chomaz, P. Billant, Three-dimensional stability of a horizontally sheared flow in a stably stratified fluid. J. Fluid Mech. **570**(1), 297–305 (2007)
11. F.S. Godeferd, C. Cambon, Detailed investigation of energy transfers in homogeneous stratified turbulence. Phys. Fluids **6**(6), 2084–2100 (1994)
12. J. Vanneste, I. Yavneh, Exponentially small inertia-gravity waves and the breakdown of quasi-geostrophic balance. J. Atmos. Sci. **61**(2), 211–223 (2004)

Numerical Solution of a Novel Biofilm Growth Model

Patricio Cumsille, Juan A. Asenjo and Carlos Conca

Abstract In this work we simulate biofilm structures ("finger-like", as well as, compact structures) as a result of microbial growth in different environmental conditions. At the same time, the numerical method that we use in order to carry out the computational simulations is new to the biological community, as far as we know. The use of our model sheds light on the biological process of biofilm formation since it simulates some central issues of biofilm growth: the *pattern formation of heterogeneous structures, such as finger-like structures*, in a substrate-transport-limited regime, and the formation of more compact structures, in a growth-limited-regime.

1 Biofilm Growth Model

In this work we aim at solving and validating a novel biofilm growth model which has been derived in our previous work [3]. The equations, variables and parameters of the model read as follows:

P. Cumsille (✉)
Group of Applied Mathematics, Department of Basic Sciences, University of Bío-Bío,
Campus Fernando May, Av. Andrés Bello S/n, Casilla 447, Chillán, Chile
e-mail: pcumsill@gmail.com

P. Cumsille
Centre for Biotechnology and Bioengineering, University of Chile, Beauchef 850,
Santiago, Chile

J.A. Asenjo
Centre for Biochemical Engineering and Biotechnology, Centre for Biotechnology and
Bioengineering, University of Chile, Beauchef 850, Santiago, Chile
e-mail: juasenjo@ing.uchile.cl

C. Conca
Department of Mathematical Engineering (DIM), Centre for Mathematical
Modelling (CMM) (UMI CNRS 2807), and Centre for Biotechnology
and Bioengineering (CeBiB),University of Chile, Beauchef 851,
P.O. Box 170-3, Santiago, Chile
e-mail: cconca@dim.uchile.cl

© Springer International Publishing Switzerland 2016   207
M. Tlidi and M.G. Clerc (eds.), *Nonlinear Dynamics: Materials,*
Theory and Experiments, Springer Proceedings in Physics 173,
DOI 10.1007/978-3-319-24871-4_16

$$-\nabla^2 p = g(U)\chi_{\Omega_2(t)} \quad \text{in } \Omega, \tag{1}$$

$$[p] = d_0\kappa \quad \text{on } \Gamma(t), \tag{2}$$

$$[p_n] = 0 \quad \text{on } \Gamma(t), \tag{3}$$

$$p = 0 \quad \text{on } z = 1, \tag{4}$$

$$p_z = 0 \quad \text{on } z = 0, \tag{5}$$

$$p(0, z) = p(L_X, z) \quad \forall z \in (0, 1), \tag{6}$$

$$S_t + \nabla \cdot (S\mathbf{u}) - \frac{D_S T}{L_Z^2}\nabla^2 S = -U(S)\chi_{\Omega_2(t)} \quad \text{in } \Omega, \tag{7}$$

$$[S] = 0 \quad \text{on } \Gamma(t), \tag{8}$$

$$[S_n] = 0 \quad \text{on } \Gamma(t), \tag{9}$$

$$S = 1 \quad \text{on } z = 1, \tag{10}$$

$$S_z = 0 \quad \text{on } z = 0, \tag{11}$$

$$S(0, z) = S(L_X, z) \quad \forall z \in (0, 1). \tag{12}$$

The variables of the model are: $\mathbf{u}$ (velocity field), p (pressure field), g (biomass volumetric flow rate), $U(S)$ (substrate uptake rate), S (substrate concentration). Variables $\mathbf{u}$, $g(U)$ and $U(S)$ are given by:

$$\mathbf{u} = -\nabla p, \quad g(U) = \frac{\nu\delta_g\mu_m S_m}{U_{S_m}}\left[(1+\mu)\frac{S}{K+S} - \mu\right] \tag{13}$$

$$U(S) = \frac{T\delta_U U_{S_m}}{S_m}(1+\mu)\frac{S}{K+S}. \tag{14}$$

In (3), $\Gamma(t)$ represents the interface between the liquid and the biofilm at time t, $\kappa(x, z, t)$ is the mean curvature at the point (x, z) on the interface $\Gamma(t)$ at time t. $[f]$ denotes the jump of a function f across the interface $\Gamma(t)$.

In (1) and (7) we have denoted by $\chi_{\Omega_2(t)}$ the characteristic function of the biofilm compartment at time t. Likewise, the notation p_n stands for the normal derivative of p, n is the normal point outwards $\partial\Omega$, and p_z is the derivative of p with respect to z. The same notations are also used for S.

The key parameters of the dimensionless model are d_0 the *amalgamated* surface tension coefficient G the growth number of the biofilm and ν, which are defined by:

$$d_0 = \frac{\lambda\gamma\nu}{D_S G}, \tag{15}$$

$$G = \frac{L_Z^2 U_{Sm}}{D_S S_m}, \tag{16}$$

$$\nu = \frac{T U_{Sm}}{S_m}. \tag{17}$$

The growth number G is a dimensionless quantity representing, in one parameter, the factors that many researchers have found to affect the biofilm structure: the vertical length, L_Z, since this is the direction in which the substrate diffuses, the concentration of soluble nutrient in the bulk, S_m; its diffusion coefficient D_S, and the maximum substrate consumption rate, U_{Sm}. High G makes the biofilm structure more heterogeneous. At low G, the resulting structure is more compact and homogeneous (see [5]). In the numerical section, we refer these two scenarios as substrate transport limited regime (high G or low d_0) and growth limited regime (low G or high d_0) respectively. Another important parameter, involved in the definition of the pressure p and of d_0, is ν which measures the ratio between the time scale of biofilm growth, T, and the substrate consumption time, S_m/U_{Sm}, near the top of the biofilm.

The other parameters of the model are: L_X, the width of the domain; λ, a parameter which varies in form inversely proportional to the viscosities. Although, by simplicity we have assumed in our simulations that λ is a small constant, in order to take into account the fact that the biofilm behaves as a fluid with large viscosity; γ, the surface tension coefficient in the liquid; μ_m, the maximum biomass growth rate; $K = K_S/S_m$, the (dimensionless) saturation constant for substrate; μ, the maintenance coefficient; and δ_g, δ_U scaling parameters introduced for numerical convenience.

2 Numerical Methods

In this section, we describe the numerical methods required to solve our model.

2.1 Numerical Method for the Interface Evolution

The method we have used to compute the interface evolution is the level set method. The central idea of the level set method is to define a smooth function $\phi(x, z, t)$ that represents the unknown interface $\Gamma(t)$ as the set where $\phi(x, z, t) = 0$. The motion of the interface is analyzed by convecting the values of ϕ (level sets of ϕ) with the velocity field $\mathbf{u}$. This is done by means of equation

$$\phi_t + \mathbf{u} \cdot \nabla\phi = 0. \tag{18}$$

We add an initial condition, which implicitly represents the initial interface.

The numerical method we have used to solve (18) is a combination of the second-order TVD RK approximation [9] to update ϕ from time t_n to t_{n+1} with the second-order ENO approximation to $\nabla\phi$, as devised in [2]. Of course, in order to guarantee the stability of the numerical method, we have taken the usual CFL time-step restriction:

$$\Delta t < \frac{h}{\max |\mathbf{u}|}.$$

We have also carried out a *re-initialization method* as devised by Peng et al. [4], each 15 time iterations.

2.2 Numerical Method for Solving the Substrate Equation

In order to solve (7), together with the boundary conditions (10)–(12), we have used a fractional step method. Thus, we split (7) into three equations:

$$S_t - \frac{D_S T}{L_Z^2} \nabla^2 S = 0 \quad \text{in } \Omega, \tag{19}$$

$$S_t + \nabla \cdot (S\mathbf{u}) = 0 \quad \text{in } \Omega, \tag{20}$$

$$S_t = -U(S)\chi_{\Omega_2(t)} \quad \text{in } \Omega. \tag{21}$$

The numerical solution of (7) is then achieved in three steps. First, we solve (19) for initial data S^n, next, we solve (20) whose initial data is the numerical solution obtained in step one, and finally, we solve (21) whose initial data is the numerical solution obtained in step two.

2.3 Numerical Method for Solving the Pressure Equation

In this subsection we explain how the biofilm balance equation (1) has been solved:

$$\nabla^2 p = -\delta_g \nu \frac{\mu_m S_m}{U_{Sm}} \left[(1 + \mu) \frac{S}{S + K} - \mu \right] \chi_{\Omega_2(t)} \quad \text{in } \Omega. \tag{22}$$

In order to solve (22) together with the transmission/boundary conditions (2)–(3), and (4)–(6), respectively, a suitable numerical method is needed.

In practice, the approximation of the derivatives of p is the same as for S, except that at those grid points which are close to the interface, the finite difference formula must be adjusted. This is done by adding a corrector term on the right-hand side of the numerical equation, so as to impose the transmission conditions that satisfy p at the interface, in such a way that a high accuracy is achieved in the numerical reconstruction of the solution. This is the main principle of the immersed interface method (IIM). This method was first introduced to deal with partial differential equations whose solution or coefficients, as well as its gradients in the normal direction, may have a jump of discontinuity at the interface.

Next, a short description of the IIM is given. The numerical approximation for p^{n+1} is based upon the classical centered five-point finite difference scheme. Equation (22) is solved by means of the centered five-point finite difference scheme:

$$\nabla_h^2 p_{i,j}^{n+1} = -\delta_g v \frac{\mu_m S_m}{U_{Sm}} \left[(1+\mu) \frac{S_{i,j}^n}{S_{i,j}^n + K} - \mu \right] \chi_{\Omega_2^n} + C_{i,j}^n, \qquad (23)$$

$\forall i = 0, \ldots, N_X$, $\forall j = 0, \ldots, N_Z$. Here above, $\nabla_h^2 p^{n+1}$ is the discrete Laplace operator, computed by using the centered five-point finite difference scheme, $\chi_{\Omega_2^n}$ is the characteristic function of Ω_2^n (the biofilm compartment at time t_n), $S_{i,j}^n$ is the numerical solution to (7), and $C_{i,j}^n$ is a corrector term which acts only on the *irregular grid points* (i, j), that is, the grid points (i, j) whose standard five-point stencil centered around it, is located on both sides of the interface. The grid points whose standard five-point stencil centered around it is located only on one side of the interface are called *regular grid points*. Thus,

$$C_{i,j}^n = \begin{cases} 0 & \text{if } (i, j) \text{ is a regular grid point,} \\ \neq 0 & \text{if } (i, j) \text{ is an irregular grid point.} \end{cases}$$

Since the irregular grid points are adjacent to the interfacial curve and form a lower-dimensional set, it turns out to be sufficient to require an $O(h)$ truncation error at these points.

Note that the matrix of the resulting system for (23) does not change with the application of the IIM, because we just have to change the right-hand side, and moreover it does not change with the time iterations. Thus, we compute the solution by means of the *LU* factorization of the matrix system, which is only computed once before of the time iterations. Once p is calculated the velocity of the medium, has to be computed, which is given by $\mathbf{u} = -\nabla p$. Again, this is computed by using classical finite difference formulae at regular grid points, and by adding a corrector term as devised by the IIM, at irregular grid points.

2.4 Summary of the Algorithm

We can summarize our algorithm as follows:

1. $n \leftarrow 0$. Initialize the initial interface, i.e., initialize the level-set function $\phi(x, z, t = 0)$. Moreover, at time zero, substrate is at its maximum concentration and uniformly distributed in the space $S(x, z, t = 0) = 1$ for all $(x, z) \in \Omega$.
2. Solve (22) with transmission and boundary conditions (2)–(3) and (4)–(6), respectively, and compute $\mathbf{u} = -\nabla p$ based on the IIM, at time $t_n = n \Delta t$, as explained in Sect. 2.3. The time-step Δt is chosen in such a way that the level-set (18) be numerically stable, as explained in Sect. 2.1.
3. Solve (18) at time $t_n = n \Delta t$, with $\mathbf{u} = -\nabla p$, and re-initialize ϕ as explained in Sect. 2.1. The zero level-set of ϕ gives us the new position of the interface.
4. Solve (7) coupled with the boundary conditions (10)–(12) at time $t_n = n \Delta t$, as explained in Sect. 2.2.
5. Change $n \leftarrow n + 1$. Repeat steps (2)–(5) to let the system evolve.

3 Numerical Results and Discussion

We have done numerical experiments with the same initial biofilm-liquid interface and different growth numbers (or surface tensions). The crucial parameter which affects the stability in the Hele-Shaw flow is the amalgamated surface tension coefficient d_0, and thus the growth number G [see (15)]. The bigger is G, the more unstable is the Hele-Shaw flow. Our main results are presented below. In all the simulations the time is in hours.

3.1 Grid Refinement Analysis

In this subsection we present a grid refinement analysis of our numerical method. To demonstrate convergence of our method, we carry out our analysis for two different simulations. The first one consists of an initial condition representing an already formed homogeneous biofilm layer. To do this we take an interface which is a perturbed horizontal straight line, defined implicitly by the level set function:

$$\phi_0(x, z) = (z - L_Z/4) + 0.02\cos(4\pi x/L_X). \tag{24}$$

We start our study with a uniform grid of 30 intervals in the x-direction and 45 in the z-direction (refered as the coarse grid in the rest of this section), and double it twice to conduct the grid refinement analysis (the other two grids are refered as the intermediate and fine grid in the rest of this section).

If a re-initialization is applied too many times for a coarse grid, then the biofilm is not allowed to grow, because this procedure plays the role of a geometric regularization. Thus, starting from a homogeneous biofilm, the re-initialization prevents the development of irregular forms in the biofilm structure, which is observed in real biofilms and also in biofilm model simulations for a transport-limited regime (high metabolic rates, that is, high G) [8]. The stopping criterion we have used leads to impose the condition

$$|\nabla\phi| = 1 + O(h^2),$$

for the reinitialized level set function ϕ, on the grid points located in a tube about the interface, where h is the grid size.

We present our first grid refinement analysis for

$$G = 395, \quad S_m = 4 \times 10^{-3} \text{ kg m}^{-3},$$

for a transport-limited regime (high metabolic rates). The values for the parameters S_m, D_S, K_S and μ_m were obtained from [8]. The parameter U_{Sm} was estimated in order to obey relation (16) for a biofilm of characteristic length $L_Z = 2 \times 10^{-3}$ m. The time scale of biofilm growth was taken as $T = 1000$ s [12]. The parameter

Table 1 Common parameters to all simulations

Parameter	Symbol	Value	Units
Time scale of biofilm growth	T	1000	s
Monod half-saturation constant	K_S	3.5×10^{-4}	kg m^{-3}
Diffusion coefficient	D_S	2.3×10^{-9}	m^2 s^{-1}
Surface tension coefficient in water	γ	72.8×10^{-3}	kg s^{-2}
Maximum specific growth rate	μ_m	1.5×10^{-5}	s^{-1}

γ is the surface tension of the water (see p. 408 in [1]). The parameters ν and d_0 were derived from the values of the other parameters according to (17) and (15), respectively. The parameter μ was estimated in the order of a usual maintenance coefficient [6], λ was estimated in order to take into account that the biofilm behaves as a fluid with large viscosity and particularly to obtain d_0 in a reasonable physical range. Finally the scaling parameters δ_U and δ_g were estimated in order to reproduce the desired behavior (mushroom-shaped biofilm/liquid interface for high G and low S_m, and more compact and homogeneous biofilm structure for low G and high S_m). The values for the parameters D_S, K_S, μ_m and T are the same through all the simulations, whereas the values for G, S_m, U_{Sm}, d_0, μ, ν, δ_U and δ_g were varied in order to simulate other behaviors (such as biofilm growth-limited regime). See Table 1 for a list of the common parameters. See also Table 2 for a list of the parameters used in the first simulation.

The way to calculate the errors is explained below. We estimate the error on each grid by using the next finer grid as the reference solution, rather than using the same reference solution for both coarser grids. More precisely, we compute the errors at each time t as follow:

$$E_h(t) = \|\phi_h(t) - \phi_{h/2}(t)\|_{L^1(\mathrm{T}_h)}/M_h,$$
$$E_{h/2}(t) = \|\phi_{h/2}(t) - \phi_{h/4}(t)\|_{L^1(\mathrm{T}_{h/2})}/M_{h/2},$$

where M_h (resp. $M_{h/2}$) stands by the number of grid points in $\mathrm{T}_h = \{|\phi_h| \leq \alpha_h\}$ (resp. $\mathrm{T}_{h/2} = \{|\phi_{h/2}| \leq \alpha_{h/2}\}$), where α_h (resp. $\alpha_{h/2}$) is the tube width T_h (resp. $\mathrm{T}_{h/2}$) for the re-initialization procedure corresponding to the coarse (resp. intermediate) grid. Errors E_h and $E_{h/2}$ measure the difference of the computed level-set functions in (averaged) L^1-norm in the tube where we test convergence of our re-initialization procedure (see [10]), that is, we measure the errors in a vicinity of the interface, because its location and shape are the best indicators of convergence of our method. In order to determine the order of convergence, we examine the ratio $E_{h/2}/E_h$, which should be at most 0.25 if the method is second-order accurate. See Table 3 for a list of the numerical parameters α_h used in each simulation.

Table 2 Parameters used in the first simulation

Parameter	Symbol	Value	Units
System dimensions			
Length	L_X	1.333×10^{-3}	m
Height	L_Z	2×10^{-3}	m
Substrate concentration in the bulk liquid	S_m	4×10^{-3}	kg m^{-3}
Maximum substrate consumption rate	U_{Sm}	9.088×10^{-4}	kg m^{-3} s^{-1}
Ratio between biofilm growth and substrate consumption	ν	4.4014×10^{-3}	Dimensionless
Proportionality coefficient in pressure equation	λ	4.4014×10^{-11}	kg^{-1} m^3 s
Maintenance coefficient	μ	1.3204×10^{-4}	Dimensionless
Amalgamated surface tension	d_0	8.0106×10^{-4}	m
Scaling parameter for U	δ_U	8×10^{-3}	Dimensionless
Scaling Parameter for g	δ_g	6.6021×10^{-4}	Dimensionless

Table 3 Tube width used in each simulation

Grid	First	Second	Third
Coarse	1/350	1/40	1/230
Intermediate	1/340	1/400	1/305
Fine	1/1250	1/250	1/720

Figure 1 qualitatively demonstrates convergence of our method as we refine the mesh. The figure depicts the zero level set of ϕ (the biofilm/liquid interface) at time $t = 17.6056$. Table 4 depicts quantitatively the grid refinement analysis at three different times, by showing the computed errors as explained before. The results in Table 4 clearly indicate second-order accuracy at three different times.

Our second simulation consists in the same initial condition as for the first simulation [see (24)], but with different parameters. See Table 5 for a list of the parameters used in the second simulation. The results in Table 6 clearly show second-order accuracy at three different times. Figure 2 qualitatively shows convergence for the second simulation at time $t = 0.1956$.

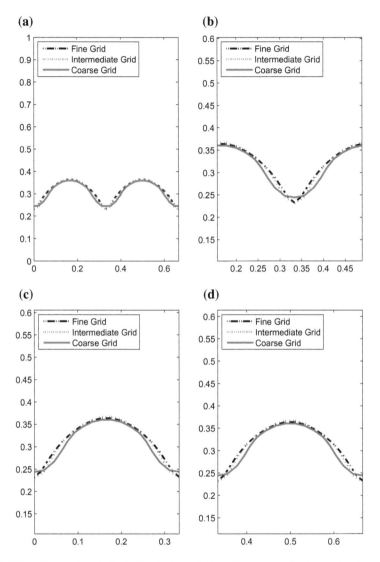

Fig. 1 Grid refinement analysis for the first simulation. Convergence is observed. **a** The whole interface, **b** zoom of the middle of the interface, **c** zoom of the left of the interface, **d** zoom of the right of the interface

3.2 Further Experiments and Analysis

Below we compare our results with those obtained in the literature. We compare the first two simulations shown in the grid refinement analysis: the first one for $G = 395$ and $S_m = 4 \times 10^{-3}$ kg m^{-3}, for a transport-limited regime, and the second one for

Table 4 Grid refinement analysis for the first simulation

Time	E_h	$E_{h/2}$	$E_{h/2}/E_h$
5.8685	4.1248×10^{-6}	1.7253×10^{-7}	0.0418
11.7371	4.6715×10^{-6}	4.9488×10^{-7}	0.1059
17.6056	4.3121×10^{-6}	3.1629×10^{-7}	0.0733

Table 5 Parameters used in the second simulation

Parameter	Symbol	Value	Units
System dimensions			
Length	L_X	1.1926×10^{-4}	m
Height	L_Z	1.7889×10^{-4}	m
Substrate concentration in the bulk liquid	S_m	1×10^{-1}	kg m^{-3}
Maximum substrate consumption rate	U_{Sm}	1.136×10^{-1}	kg m^{-3} s^{-1}
Ratio between biofilm growth and substrate consumption	ν	8.8028×10^{-4}	Dimensionless
Amalgamated surface tension	d_0	2.00×10^{-2}	m
Proportionality coefficient in pressure equation	λ	8.8028×10^{-12}	kg^{-1} m^3 s
Maintenance coefficient	μ	2.6408×10^{-5}	Dimensionless
Scaling parameter for U	δ_U	7.0423×10^{-2}	Dimensionless
Scaling Parameter for g	δ_g	7.0423×10^{-3}	Dimensionless

Table 6 Grid refinement analysis for the second simulation

Time	E_h	$E_{h/2}$	$E_{h/2}/E_h$
0.1467	2.5060×10^{-6}	5.4183×10^{-7}	0.2162
0.1956	5.0392×10^{-6}	8.7318×10^{-7}	0.1733
0.2445	7.7270×10^{-6}	1.8515×10^{-6}	0.2396

$G = 16$ and $S_m = 1 \times 10^{-1}$ kg m^{-3}, for a growth-limited regime, starting from the same initial condition (24).

The evolution of a biofilm (in hours) is presented in Figs. 3 and 4. In these figures, we have plotted the negative level sets of the level function ϕ (with a variation

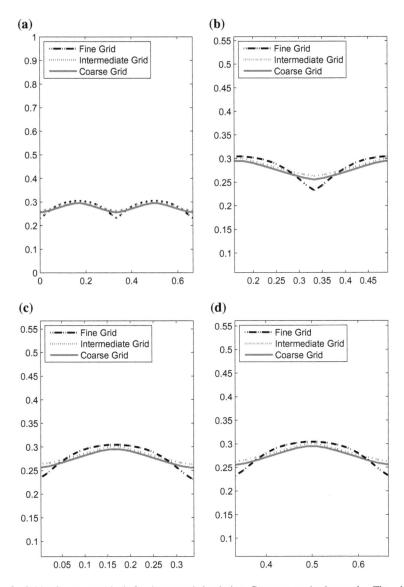

Fig. 2 Grid refinement analysis for the second simulation. Convergence is observed. **a** The whole interface, **b** zoom of the middle of the interface, **c** zoom of the left of the interface, **d** zoom of the right of the interface

of $h = 0.0111$ between lines), i.e., those level sets corresponding to the biofilm compartment. The thicker line represents the zero level set of ϕ, and it corresponds to the biofilm/liquid interface.

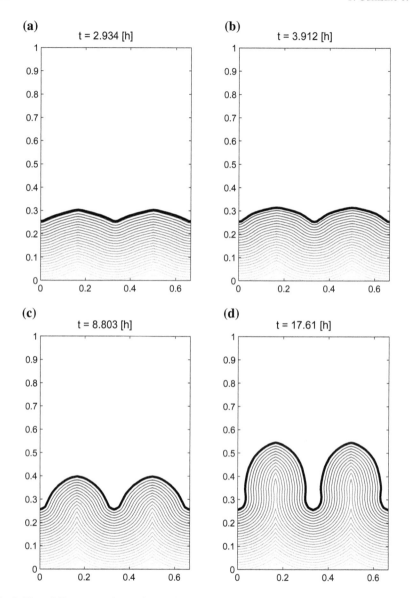

Fig. 3 Four different snapshots of a mushroom-shaped biofilm, simulated with the intermediate grid (first simulation)

At the beginning, when there is sufficient substrate in the environment (i.e., there is no important nutrient limitation), the biofilm grows in all directions (see Figs. 3a, b, and 4a, b). As the biofilm gets thicker, two situations can occur. If there is still no substrate limitation (i.e., low G) the biofilm grows forming a relatively compact or smooth structure (see Fig. 4c, d). The other possible scenario is when the

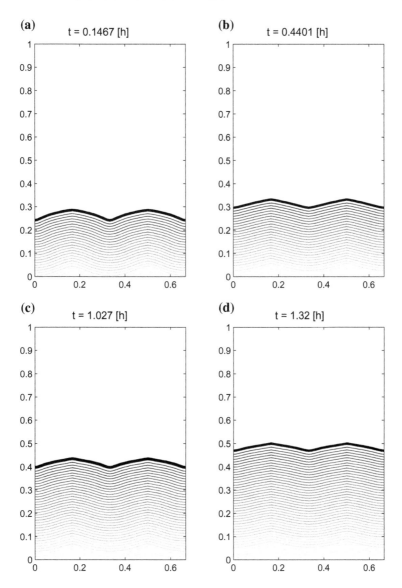

Fig. 4 Four different snapshots of a compact-shaped biofilm, simulated with the intermediate grid (second simulation)

nutrient is depleted in the biofilm depth. In this case, there is almost no flux of substrate to the cells situated in the "valleys". Since only microbes in the top regions are active, dividing and creating new biomass, the biofilm grows forming "finger" or "mushroom" like structures toward the liquid bulk (see Fig. 3c, d). In general,

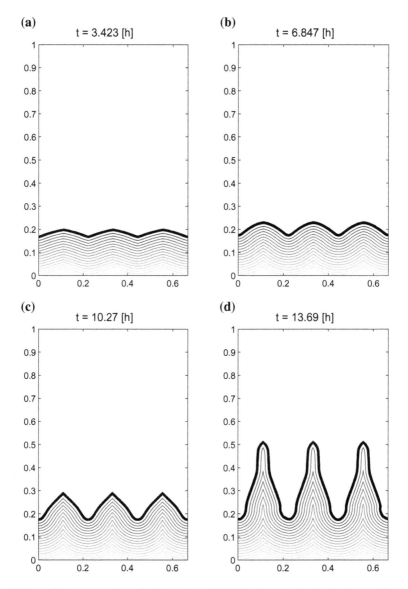

Fig. 5 Four different snapshots of a finger-shaped biofilm, simulated with the intermediate grid (third simulation)

for homogeneously distributed biomass on the carrier (Fig. 4), the preferential growth direction is perpendicular to the carrier surface.

We can observe the formation of finger-like structures after longer periods of time (see Fig. 3c, d), which is a common pattern observed in biofilm growth (see [5–8, 11]). In the second simulation (Fig. 4), starting from the same initial interface,

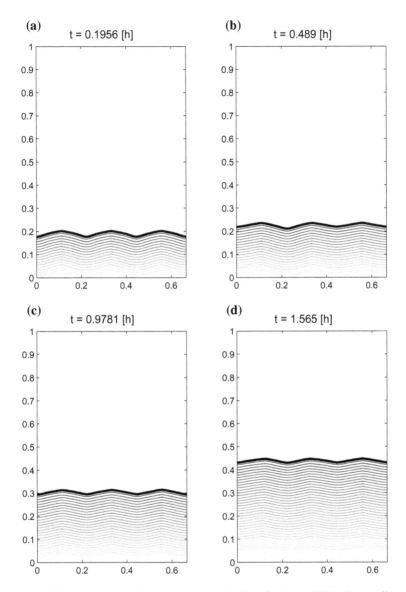

Fig. 6 Four different snapshots of a compact-shaped biofilm, simulated with the intermediate grid (fourth simulation)

but using different values for the parameters than those used in the first simulation, it can be observed that the biofilm arrives to the top of the computational domain (where the source of substrate is located) faster than in the previous simulation. This is because of the growth-limited regime (low G), in whose case there is not an important limitation of substrate. The behavior shown in the two simulations, carried out in this subsection, show that our model can follow more smooth development of biofilms structures in the shorter time periods as depicted in Fig. 4 in the second simulation, whereas in longer time periods more pronounced "finger" like structures as depicted in Fig. 3d. Since this is a numerical solution in which parameters may be adjusted to the specific biofilm that needs to be simulated, it clearly shows that the model proposed in this paper has the potential to simulate a range of behaviors that have been observed in biofilm models and their practical behavior.

Finally, the same analysis done before for simulations 1 and 2 also holds true for simulations 3 and 4, which are shown in Figs. 5 and 6 respectively.

Acknowledgments This work was partially supported by PIA-CONICYT grant PFBasal-01. The work of first author was also founded DIUBB 121909 GI/C and DIUBB 122109 GI/EF. Third author thanks partial support from Ecos-Conicyt C13E05. He is also partially supported by PFBasal 01 and PFBasal 03 projects and by Fondecyt 1140773.

References

1. G.W. Castellan, *Physical Chemistry*, 3rd edn. (Addison-Wesley, 1983)
2. Y.C. Chang, T.Y. Hou, B. Merriman, S. Osher, A level-set formulation of eulerian interface capturing methods for incompressible fluids flow. J. Comput. Phys. **124**(72), 449–464 (1996)
3. P. Cumsille, J.A. Asenjo, C. Conca, A novel model for biofilm growth and its resolution by using the hybrid immersed interface-level set method. Comput. Math. Appl. **67**(1), 34–51 (2014)
4. D. Peng, B. Merriman, S. Osher, H. Zhao, M. Kang, A pde-based fast local level set method. J. Comput. Phys. **155**, 410–438 (1999)
5. C. Picioreanu, M.C.M. van Loosdrecht, J.J. Heijnen, A new combined differential-discrete cellular automaton approach for biofilm modeling: application for growth in gel beads. Biotechnol. Bioeng. **57**(6), 718–731 (1998)
6. C. Picioreanu, M.C.M. van Loosdrecht, J.J. Heijnen, Mathematical modeling of biofilm structure with a hybrid differential-discrete cellular automaton approach. Biotechnol. Bioeng. **58**(1), 101–116 (1998)
7. C. Picioreanu, M.C.M. van Loosdrecht, J.J. Heijnen, A theoretical study on the effect of surface roughness on mass transport and transformation in biofilms. Biotechnol. Bioeng. **68**(4), 355–369 (2000)
8. C. Picioreanu, M.C.M. van Loosdrecht, J.J. Heijnen, Effect of diffusive and convective substrate transport on biofilm structure formation: a two-dimensional modeling study. Biotechnol. Bioeng. **69**(5), 504–515 (2000)
9. C.-W. Shu, S. Osher, Efficient implementation of essentially nonoscillatory shock-capturing schemes. J. Comput. Phys. **77**(2), 439–471 (1988)
10. M. Sussman, P. Smereka, S. Osher, A level set approach for computing solutions to incompressible two-phase flow. J. Comput. Phys. **114**, 146–159 (1994)
11. O. Wanner, H.J. Eberl, E. Morgenroth, D.R. Noguera, C. Picioreanu, B.E. Rittmann, M.C.M. van Loosdrecht, *Mathematical Modeling of Biofilms* (IWA Publishing, Alliance House, 2006)
12. T. Zhang, N.G. Cogan, Q. Wang, Phase field models for biofilms ii. 2-d numerical simulations of biofilm-flow interaction. Commun. Comput. Phys. **4**(1), 72–101 (2008)

Dynamics of a One-Dimensional Kink in an Air-Fluidized Shallow Granular Layer

J.E. Macías, M.G. Clerc and C. Falcón

Abstract We report on the observation and characterization of the dynamics of one-dimensional granular kinks in shallow granular layer subjected to an air flow oscillating in time. We characterize experimentally the properties of this extended solution and present results of the appearance of an effective drift as a function of the inclination of the experimental cell, which can be understood using a simple phenomenological amplitude equation to describe the onset of these solutions, their morphology, dynamical properties and the pinning-depinning transition of one-dimensional granular kinks.

1 Introduction

Macroscopic systems under the influence of injection and dissipation of quantities such as energy and momentum usually exhibit coexistence of different states, which is termed *multistability* [1–3]. Heterogeneous initial conditions–usually caused by the inherent fluctuations–generate spatial domains which are separated by their respective interfaces. These interfaces are known as fronts in the nonlinear science community [2]. The evolution of these solutions can be regarded as a particle-type one, i.e. they can be characterized by a set of continuous parameters such as the position, width, charge, and so forth. In the particular case where fronts separate symmetric states, these front solutions are termed kinks solutions. Kinks have been a central element in classical and quantum field theory to understand the dynamics of several physical systems [4]. Parametrically driven systems exhibit instabilities which

J.E. Macías · M.G. Clerc · C. Falcón (✉)
Departamento de Física, Facultad de Ciencias Físicas y Matemáticas,
Universidad de Chile, Casilla 487-3, Santiago, Chile
e-mail: cfalcon@ing.uchile.cl

J.E. Macías
e-mail: juanmacias@ug.uchile.cl

M.G. Clerc
e-mail: marcel@dfi.uchile.cl

© Springer International Publishing Switzerland 2016
M. Tlidi and M.G. Clerc (eds.), *Nonlinear Dynamics: Materials,
Theory and Experiments*, Springer Proceedings in Physics 173,
DOI 10.1007/978-3-319-24871-4_17

lead to the emergence of symmetric states, which are out of phase in half the period of the forcing [5]. Kink solutions naturally have been observed in two-dimensional vertically vibrofluidized granular layers (see [6]) and, recently, in one-dimensional systems where their properties have been studied [7–10]. Although progress has been done in the study of kink properties and features, there are still several unanswered questions on how to control these properties affecting their short term and long term dynamics, which would affect strongly the theory of pattern-forming systems, and would be of straightforward application in technology [11].

In this chapter we present a study on the dynamics of kink solutions which appear at the surface of a one-dimensional shallow granular layer fluidized by periodic air flow as a control parameter (the inclination angle of the experimental cell ϕ) is changed (cf. Fig. 1). Experimentally, we show that granular kinks display a pinning-depinning transition controlled by ϕ. This transition is explained with a simple model for the position of the kink $x_o(t)$ arising from a phenomenological amplitude equation close to a focusing-defocusing transition which displays a good agreement with experimental findings.

2 Experimental Setup

The experimental setup under study is displayed in Fig. 1. A cell (200 mm wide, 200 mm tall and 3.5 mm in depth) made out of two large glass walls with an horizontally placed thick-band like sponge (6 mm thick, 200 mm wide and 15 mm tall) acts as a porous floor where approximately 40,000 monodisperse bronze spheres (diameter $d = 350\,\mu m$) are deposited. In grain diameter units, the granular layer is $570d$ wide, $10d$ deep, and $8d$ tall. The excitation system of the granular layer is similar to the one described in [9, 10], where a time-modulated air flow is generated by an air compressor and regulated by an electromechanical proportional valve. The valve aperture is set by a variable voltage signal controlled by the first output of a 2-channel function generator through a power amplifier. A symmetrical triangular signal with frequency f_o and a non-zero offset is used to generate through the air flow a time-modulated controllable pressure signal. Pressure oscillations are measured 50 cm before the flow enters the cell with a dynamic pressure sensor and a signal conditioner. We have checked experimentally the linearity between the peak voltage delivered by the function generator and the peak pressure fluctuations P_o at the forcing frequency f_o. The experimental cell can be tilted with respect to the horizontal (x-axis) with an angle ϕ, measured from images of the experimental cell acquired with a CCD camera in a 1080×200 px spatial window (0.19 cm/px sensitivity in the horizontal direction and 0.18 cm/px in the vertical direction). Variations of the off-plane inclination angle on the x-axis are forbidden (using a plate to constrain the motion on a plane) to ensure that only in-plane movements of the cell are allowed. Hence, the control parameters are the excitation frequency f_o, the peak pressure P_o and the inclination angle ϕ.

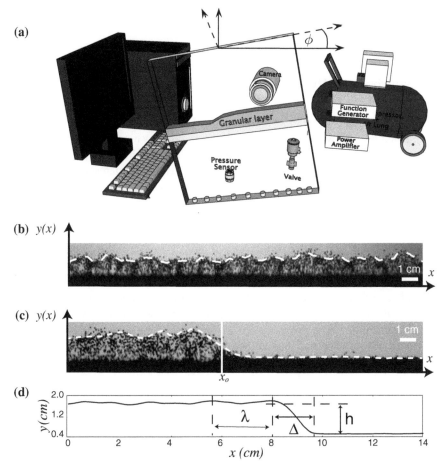

Fig. 1 **a** Experimental setup as described in the text. **b** Typical image of the excited granular layer at $\phi = 0°$. The *dashed white line* corresponds to the numerically calculated granular interface $y(x)$. **c** Granular surface kink on shallow granular layers at $\phi = 0°$. *Vertical continuous line* depicts where the position of the kink x_o is calculated. **d** Average kink for 10^4 images for $P_o = 7.0 \pm 0.2$ kPa at $f_o = 14$ Hz. Here h, Δ, and λ are the granular kink height, width and wavelength respectively

Images of the granular bed motion are acquired over a 100 s time window. For each experimental configuration, two image sequences are taken. The first one, acquired at high frame rate (100 fps), is used to study the typical oscillation frequencies of the granular layer. The second one, set at the subharmonic frequency $f_o/2$ using the second output of the function generator as a trigger, is used to ensure a stroboscopic view of the oscillating layer. The granular interface $y(x, t)$ is tracked for every point in space x at each time t using a simple threshold intensity algorithm, as it is shown in Fig. 1. To do this, white light is sent through a diffusing screen from behind the granular layer as images are taken from the front, enhancing contrast and thus, surface tracking algorithms. Figure 1 shows a snapshot of the granular layer and a kink solution using the above mention tracking algorithm.

3 Experimental Results

For $f_o = 14$ Hz and $\phi = 0°$, and above a critical peak pressure $P_o^c = 5.1 \pm 0.2$ kPa, the granular layer shows a parametric instability, as already reported in [9, 10], where the layer oscillates at half the forcing frequency. The amplitude of this oscillation grows roughly as $[(P_o - P_o^c)/P_o^c]^{1/2}$ for $P_o \gg P_o^c$ and no hysteresis is experimentally observed, which makes the instablity supercritical in nature. For $P_o \sim P_o^c$, an effective amplitude curve has been found [9, 10] following the noisy bifurcation theory of Agez et al. [12]. Therefore, the granular layer presents an effective parametric resonance as a consequence of the forcing: the periodic air flow is responsible for inducing the oscillatory behavior of the layer and its respective parametric resonance [13].

As the flat layer undergoes this parametric instability, it develops sporadically spatial fluctuations with a characteristic wavelength and frequency that appear with a typical lifetime (of the order of 5–10 oscillating periods) and disappear randomly. This phenomenon, called *precursor* is a consequence of the balance between energy injection, caused by internal fluctuations of the granular layer, and the dissipation of the slowest decaying spatial mode of the uniform steady state of the layer interface. The typical wavelength of the spatial precursor $\lambda \sim 1.5$ cm.

The parametric oscillation allows the system to display bistability, i.e., two stable extended states (one in phase with the forcing and another one in anti-phase with the forcing) that can be connected spatially anywhere in the spatial domain via a kink solution (see Fig. 1). That is, there is a height jump as we go from one side of the experimental cell to the other through a finite region of the layer where this jump occurs. Thus, a kink can appear spatially at any given point of the experimental cell where, at any given time, on one side of the region the granular layer is moving upwards and on the other side it is moving downwards. The spatial features of the kink solution, that is, its height h, wavelength λ and width of the kink Δ, displayed in Fig. 1, depend on P_o. It has been reported elsewhere [9, 10] that h grows linearly with P_o while Δ and λ are independent of P_o.

At $f_o = 14$ Hz, fixing now $P_o = 7.0 \pm 0.2$ kPa, we change ϕ from $-2.2°$ to $1.4°$. In this configuration, the stationary kink solution suffers a secondary instability as it begins to drift to the left (right) for positive (negative) ϕ with a well-defined mean velocity $\langle v \rangle$. Thus it presents a pinning-depinning transition. It must be noticed that this drift does not occur for any non-zero angle. There is range of ϕ where, although an non negligible inclination has been given to the experimental cell, no average motion of the kink in the experimental time window is observed. This range in ϕ-space is called pinning region. Outside this pinning region, the kink moves. For ϕ larger than the presented values, the granular layer is too shallow to sustain a kink (on the shallower side) and displays a bifurcation towards a granular patterns (as the one

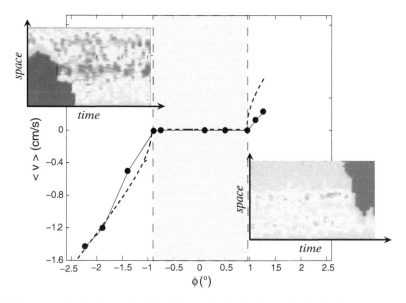

Fig. 2 Average velocity $\langle v \rangle$ as a function of the angle of inclination ϕ. The *light blue zone* represents the region where the kink rests pinned. Outside this pinning region, the kink moves with a well defined mean velocity $\langle v \rangle \neq 0$. The *insets* are typical colormaps showing a kink moving to the left for $\phi = -1.0°$ or to the right for $\phi = 1.3°$. *Dashed lines* are the theoretical fit using the model described in the next section for ϕ. Note the asymmetry of the acquired data for positive and negative ϕ, both in $\langle v \rangle \neq 0$ and in the pinning region

studied in [7, 8]), thus no data of these measurements is presented in this work. The measured velocity is computed as follows. For a given ϕ and a peak amplitude P_o, a set of images is acquired at $f_o/2$ for 2000 s. From each image acquired at a time t, the position of the kink $x_o(t)$ is computed (as shown in Fig. 1d), and then its derivative computed and smoothed, which gives $\langle v \rangle$. Therefore, fluctuations of v are small with respect to $\langle v \rangle$. Figure 2 shows $\langle v \rangle$ for different values of ϕ at $P_o = 7.0 \pm 0.2$ kPa.

4 Theoretical Description

A major difficulty to describe granular media is that they lack a well-established continuum description [6]. Then, the strategy we adopt here to describe the granular kink is to use amplitude equations based on symmetry arguments and bifurcation theory used in systems showing the same type of behavior. It is important to note that, increasing a small number of layers of the fluidized granular bed, the parametric instability displayed by the layer leads to the pattern formation [7]. Such a type of transition has already been observed in parametrically amplified surface waves on a newtonian fluid in a rectangular container. A channel with water vertically driven exhibits standing subharmonic waves, through the well-known Faraday instability [14], for layer depths greater than the channel width. However, when the water height is smaller than the channel width, kink solutions are observed on the fluid surface [15]. This transition from surface kinks to patterns can be understood as a focusing to defocusing one [16], as we will explain below.

Due to the similarity of the dynamical behavior between these systems, we consider that the height of the granular layer, close to the parametrical instability as an order parameter which can be described by $y(x, t) = h_{f_o}(t) + \psi(x, y)e^{i\pi f_o t} + c.c.$, where $\{x, t\}$ denote, respectively, the spatial and temporal coordinates, $h_{f_o}(t)$ is a periodic function of period f_o^{-1} which accounts for the uniform harmonic motion of the granular layer. $\psi(x, t)$ is the envelope of subharmonic mode, which we assume varies slowly in space and time, near the focusing-defocusing transition. Based on the amplitude equation theory, the evolution of $\psi(x, t)$ satisfies parametrically driven Ginzburg-Landau equation [2, 3]

$$
\begin{aligned}
\partial_t \psi = {} & -(\mu + i\nu)\psi - (\alpha + i\delta)|\psi|^2\psi + (\beta + i)\partial_{xx}\psi \\
& + \gamma\bar{\psi} - i\eta|\psi|^4\psi + \Gamma\partial_x\psi + \sqrt{\kappa}\zeta(x, t).
\end{aligned}
\tag{1}
$$

The variable $\bar{\psi}$ stands for the complex conjugate of ψ. ν is the detuning parameter, which is proportional to the difference between half of the forcing frequency and the natural frequency of the oscillator field, $\{\mu, \alpha\}$ are the damping parameters which account, respectively, for the linear and nonlinear dissipation, β accounts for diffusion process and γ is the amplitude of the parametric forcing. δ and η represent the nonlinear response of the envelope frequency as a function of the modulus. $\zeta(x, t)$ accounts for random fluctuations of granular surface, which is modeled by gaussian white noise with zero mean value and correlation $\langle \zeta(x, t)\zeta(x', t') \rangle = \delta(x - x')\delta(t - t')$, and κ stands for the intensity of the noise. Finally, Γ describes the drift induced by the tilt, which in this weakly nonlinear case is proportional to ϕ.

In the case of $\phi = 0°$, $\Gamma = 0$, we have kink solutions. Usually δ is of order one, and then the term proportional to η is negligible, which corresponds to the parametrically driven Ginzburg-Landau equation ($\eta = 0$). When δ is positive (negative), the system is focusing (defocusing), i.e. the envelope frequency decreases (increases) with the amplitude. Hence, when δ is small enough near the focusing-defocusing transition, the first correction term is given by the quintic term proportional to η. Then, (2) accounts for a parametric instability simultaneously with a focusing-defocusing transition.

In focusing media, model (2) exhibits patterns and dissipative solitons [17, 18]. In contrast, in defocusing media it shows kink solutions [15, 19], usually called dark solitons in the framework of nonlinear optics. Kink solutions typically connect monotonically the uniform states in a parametrically driven system. Even analytically, these solutions are well described by hyperbolic tangent functions [19], which is inconsistent with our experimental observations (see Fig. 1). This scenario changes drastically when one considers the influence of the quintic term ($\eta \neq 0$). This term modifies the value of uniform states and the shape of the kink profile, which now exhibits spatially d'amped oscillations as illustrated in Fig. 3. These damped

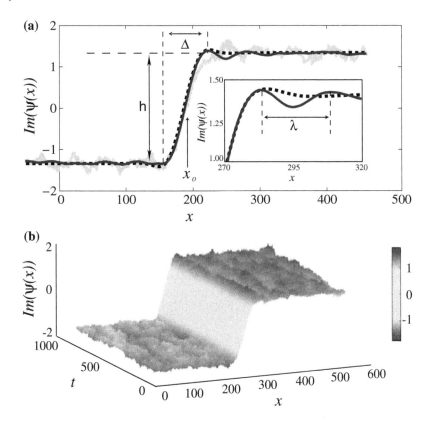

Fig. 3 Kink solution obtained using the parametrically driven Ginzburg-Landau equation (2), with $v = 0.06, \mu = 0.2, \alpha = 0.005, \delta = -0.01, \gamma = 0.25, \beta = 0.2, \eta = 0.01$ and $\kappa = 1.0$. **a** Profiles of $Im(\psi)$. Instantaneous (*light gray*) and averaged (*dark gray*) kink solution of model (2) with noise. *Dotted line* depicts the kink solution without noise. Here $\kappa = 0.0$. Here h, Δ, λ and x_o are the granular kink height, width, wavelength and position. *Inset* Zoom of the local amplitude jump near the position of the kink for the averaged kink solution with (*continuous line*) and without (*dotted line*) noise. Here λ is the wavelength of the kink. **b** Spatiotemporal diagram of $Im(\psi)$

oscillations are a manifestation that the slowest spatial mode of the uniform state has a wavelength different from zero. Therefore, when one considers random fluctuations, it induces an effective pattern–*a precursor*–which is highly fluctuating (see Fig. 3). Note that the kink exhibits an overshooting amplitude oscillation around x_o, similar to the one observed experimentally (cf. Fig. 1). Moreover, the emergence of these kinks is due to the quiescent state ($\psi = 0$) which shows a supercritical bifurcation when one increases the amplitude of the forcing or decreasing the detuning. That is, the kink height increases monotonically with the amplitude of the forcing. However, the laws found for the kink height and width as a function of the forcing amplitude are different that those found experimentally (see Fig. 1). Note that the characteristic features of kink solutions, both experimental and theoretical, following qualitatively a similar behavior. Thus, the proposed model (2) describes qualitatively the dynamics observed in a quasi-one-dimensional fluidized shallow granular bed driven by a periodic air flow in a small cell.

In the case of nonzero $\phi \propto \Gamma$, as already described in [8], all linear coefficients of (2) become inhomogeneous (they have a spatial dependence on coordinates). Without loss of generality, we can assume that they are linear functions of the spatial coordinate, i.e., a ramp. Using this simple model, we can predict the evolution of $\psi(x, t)$, which displays a two-fold effect as in [8]: the kink solution is deformed and it propagates (drifts) for large enough Γ. Following the same scheme as in [10], we can compute a simple model for the position of the kink $x_o(t)$ from (2) assuming the ansatz $\psi(x, t) = \psi_o(x - x_o(t)) + w(x_o(t), t, x)$ with $\psi_o(x - x_o)$ the kink solution for (2) which connects the homogeneous states arising from $\partial_t \psi = \partial_x \psi = 0$ and $w(x_o(t), t, x)$ is a small correction of order $\mu^{3/2}$. Introducing this ansatz in (2), an equation for $x_o(t)$ can be found as a solvability condition [8]. Although the calculations are not straightforward because of the cumbersome algebra related to finding the stationary states and latter on the front solution (even in the case of constant coefficients), we can infer a simple evolution equation for $x_o(t)$ from the observed dynamics and previous experiments (see [10]). The proposed Langevin equation for $x_o(t)$ is

$$\dot{x}_o = \Gamma_o + \delta_o \cos\left(\frac{\pi x_o}{\lambda} + \varphi\right) + \sqrt{\eta_o}\zeta_o(t), \tag{2}$$

where $\Gamma_o \propto \phi$ is a constant drift related to the linear inhomogeneity of the system, δ_o is the amplitude of a periodic forcing stemming from the precursor of the kink with periodicity π/λ, κ_o is the effective intensity of the noise term and φ is a phase mismatch related to the initial position of the kink in the spatial domain. Both δ_o and κ_o have already been studied in [10] as functions of P_o. The calculation of their dependence on parameters of (2) is still lacking. Γ_o changes the typical dynamics of $x_o(t)$: for $|\varepsilon| = |\Gamma_o/\delta_o| \leq 1$ there is a pinning region (already described above) where the dynamics of x_o is a hopping one with zero mean displacement in the experimental time window, but as $|\varepsilon| > 1$ it becomes a drifting (directed) one. Integrating directly the above model we find that the mean velocity $\langle \dot{x}_o \rangle$ is 0 when $|\varepsilon| \leq 1$ and $-\text{signc}(\varepsilon)\sqrt{\varepsilon^2 - 1}$ when $|\varepsilon| > 1$. This theoretical prediction is roughly observed in Fig. 2. The mismatch between these curves lies in the inhomogeneities of the experimental cell, which perturb the symmetry in $\langle \dot{x}_o \rangle$ but not the net effect: the generation of a pinning range and thus a pinning-depinning transition for one-dimensional kinks.

5 Conclusion

We have studied the pinning-depinning bifurcation of kink solutions in a shallow one-dimensional fluidized granular layer subjected to a periodic air flow when the experimental cell is inclined. We have proposed a suitable model that accounts for the main properties of granular surface kinks. The inherent noise of the system simultaneously induces fluctuations on the position of the position of the kink, and sustains an effective pattern over the extended homogeneous state. Therefore, one expects a highly complex and intriguing dynamics of these granular kink solutions as they evolve in time or interact with other structures or between themselves. Using a simple reduced Langevin equation for the evolution of the position of the kink $x_o(t)$, the mean velocity can be computed, showing that a pinning range appears. This phenomenological prediction agrees roughly with the acquired data. Further work on the statistics of $x_o(t)$ and how to control the pinning range with external fluctuations is already in progress.

Acknowledgments This work was possible with the financial support of FONDECYT grant 1130354.

References

1. G. Nicolis, I. Prigogine, *Self-organization in Non Equilibrium Systems* (Wiley, New York, 1977)
2. L.M. Pismen, *Patterns and Interfaces in Dissipative Dynamics*, Springer Series in Synergetics (Springer, Berlin, 2006)
3. M.C. Cross, P.C. Hohenberg, Pattern formation outside of equilibrium. Rev. Mod. Phys. **65**, 851 (1993)
4. N. Manton, S. Sutcliffe, *Topological Solitons* (Cambridge University Press, Cambridge, 2004)
5. L.D. Landau, E.M. Lifshiftz, *Mechanics*, vol. 1, 1st edn. (Pergamon Press, New York, 1976)
6. I. Aronson, L. Tsimring, *Granular Patterns* (Oxford University Press, New York, 2009)
7. I. Ortega, C. Falcon, M.G. Clerc, N. Mujica, Subharmonic wave transition in a quasi-one-dimensional noisy fluidized bed. Phys. Rev. E **81**, 046208 (2010)
8. J. Garay, I. Ortega, M.G. Clerc, C. Falcón, Symmetry-induced pinning-depinning transition of a subharmonic wave pattern. Phys. Rev. E (R) **85**, 035201 (2012)
9. J.E. Macías, M.G. Clerc, C. Falcón, M.A. Garca-Nustes, Spatially modulated kinks in shallow granular layers. Phys. Rev. E **88**, 020201(R) (2013)
10. J.E. Macías, C. Falcón, Dynamics of spatially modulated kinks in shallow granular layers. New J. Phys. **16**, 043032 (2014)
11. R. Tomasello, E. Martinez, R. Zivieri, L. Torres, M. Carpentieri, G. Finocchio, A strategy for the design of skyrmion racetrack memories. Sci. Rep. **4**, 6784 (2014)
12. G. Agez, M.G. Clerc, E. Louvergneaux, Universal shape law of stochastic supercritical bifurcations: theory and experiments. Phys. Rev. E **77**, 026218 (2008)
13. M.G. Clerc, C. Falcon, C. Fernandez-Oto, E. Tirapegui, Effective-parametric resonance in a non-oscillating system. EPL **98**, 30006 (2012)
14. M. Faraday, On a peculiar class of acoustical figures; and on certain forms assumed by groups of particles upon vibrating elastic surfaces. Philos. Trans. R. Soc. London **121**, 299 (1831)
15. B. Denardo, W. Wright, S. Putterman, A. Larraza, Observation of a kink soliton on the surface of a liquid. Phys. Rev. Lett. **64**, 1518 (1990)
16. J.W. Miles, Parametrically excited solitary waves. J. Fluid Mech. **148**, 451 (1984)
17. P. Coullet, T. Frisch, G. Sonnino, Dispersion-induced patterns. Phys. Rev. E **49**, 2087 (1994)
18. M.G. Clerc, S. Coulibaly, D. Laroze, Localized states and non-variational IsingBloch transition of a parametrically driven easy-plane ferromagnetic. Phys. Rev. E **77**, 056209 (2008)
19. I. Barashenkov, S. Woodford, E. Zemlyanaya, Parametrically driven dark solitons. Phys. Rev. Lett. **90**, 54103 (2003)

Measurements of Surface Deformation in Highly-Reflecting Liquid-Metals

Pablo Gutiérrez, Vincent Padilla and Sébastien Aumaître

Abstract We present an experimental study of surface deformation in a liquid metal. The investigation has two main parts. First we present an optical setup allowing to obtain the surface profile along a line, for a highly-reflecting liquid-metal. We track the diffusion of a laser sheet on the surface from two opposite angles, avoiding saturations due to specular reflection. In the second part, the technique is used to study some aspects of the wave-vortex interaction problem. Special attention is given to the surface deformation produced by a flow composed by moving vortices. Asymmetric height statistics are observed and discussed. Some indications of wave emission are presented. Finally, the attenuation of propagating waves by the same flow is briefly discussed.

1 Introduction

It is difficult to think about a world without waves. Waves are everywhere: sometimes they are visible to the eye and sometimes not. This chapter deals with waves in fluids. And even if these waves are visible to our eyes, they can be elusive to quantification. Such a challenge is usually overcome with a thorough use of light (which is, in fact, electromagnetic waves), as it will be discussed here from an experimental point of view.

P. Gutiérrez (✉)
Departamento de Física, Facultad de Ciencias Físicas y Matemáticas,
Universidad de Chile, Casilla 487-3, Santiago, Chile
e-mail: pagutier@gmail.com

P. Gutiérrez · V. Padilla · S. Aumaître
Service de Physique de l'Etat Condensé, DSM, CEA-Saclay, UMR-CNRS 3680,
91191 Gif-sur-Yvette, France

S. Aumaître
Laboratoire de Physique, ENS de Lyon, UMR-CNRS 5672, 46 allée d'Italie,
69007 Lyon, France

© Springer International Publishing Switzerland 2016
M. Tlidi and M.G. Clerc (eds.), *Nonlinear Dynamics: Materials,
Theory and Experiments*, Springer Proceedings in Physics 173,
DOI 10.1007/978-3-319-24871-4_18

Fig. 1 Flows under consideration, exhibiting wave-like motion on the surface. A liquid metal is driven with the Lorentz force (see Sect. 3). *Left* highly fluctuating motion when the forcing is strong ($I = 400$ A). *Right* with a gentle forcing ($I = 80$ A), a linear wave is excited at 5 Hz from a corner of the container

More generally, we are interested in the motion of the free surface of a liquid. It happens both because of the influence of an underlying turbulent motion, or because of the external perturbation of the surface. Theoretical studies of the generation of waves from turbulent motion can be traced back to Lighthill [15], when he studied the generation of sound by a turbulent flow. More recently, the analysis was extended to water waves by exploiting the formal analogy between (non-dispersive) sound waves and waves in shallow water [4, 9]. Other approach, probably also influenced by Lighthill's work, was proposed by Phillips to study the interaction of a turbulent flow with surface water waves in an oceanographical setting [23].

Two important problems come from the interaction between turbulence and a restoring mechanism (as gravity) in the surface of a flow [4, 9, 23]: (a) the generation of waves; (b) the scattering of propagating waves. Both situations are presented in Fig. 1: left panel show a wavy surface which appears as a consequence of the underlying turbulent motion; right panel present a similar setup, where the surface is perturbed both by an underlying flow (although the forcing is almost ten times smaller compared to the left panel) and by a wave that is excited at the bottom-right corner. More details are given in Sects. 3.2 and 3.3, respectively.

In many situations it is important to measure the deformation of a liquid surface after a perturbation. From ripple tank experiments and its pedagogical interest of visualizing concepts of sound waves [14, 16, 29]; to more sophisticated experiments relating the statistical features of subsurface water motion with the one on the surface [25]. In the fifties, Cox and Munk pioneered the use of light to obtain spatially-resolved measurements of the deformation field of a fluid surface [6]. However, last years were prolific in the development of new measurement techniques to study water motion in the laboratory [2, 5, 21, 24, 31], taking advantage of different optical properties of water. Liquid metals, on the other hand, continue to be more elusive, since they are opaque and highly reflecting.

Liquid metals are relevant because they appear in planetary flows (see, for instance, [22]) and in industrial applications [7]. Among industrial problems, improvements in the efficiency of aluminum reduction cells largely depend on reliable measurements and analysis of surface deformation. Here, the surface of liquid alumina (surmounted by lighter liquid electrolytes) exhibits instability as a consequence of a complex interaction between strong currents and magnetic fields. Control of the interfacial instability would allow to reduce the thickness of the electrolyte layer, considerably increasing the efficiency of the process [7].

This chapter organizes as follows. The measurement technique is described in Sect. 2. It consists on a classical geometrical arrangement to measure the profile of a surface, which is cleverly extended to a highly-reflecting liquid metal flow. We recall the optical arrangement first (Sect. 2.1). Then, it is discussed our particular setup and its practical advantages in the study of liquid metals (Sects. 2.2 and 2.3). Finally, we discuss the associated image processing (Sect. 2.4). Section 3 concerns the applications that motivated our study. We first describe the setup we use to study magnetohydrodynamical (MHD) flows (Sect. 3.1). Then we focus on the surface deformation induced by this flow (Sect. 3.2). Finally we give a brief account of what happens with a wave that propagates on the surface of the turbulent flow (Sect. 3.3). We end this chapter with the conclusions.

2 Measurement of Surface Deformation

The aim of this section is to present our method to perform surface level measurements along a line in a liquid metal.

2.1 Oblique View

The idea behind the method is very simple. It is conceptually the same as looking a surface profile with a camera perpendicular to it. This *perpendicular view* is particularly useful when the surface moves roughly on a single plane. Examples of this situation are quasi-unidimensional experiments, as those of solitary waves in water [10] or in granular layers [20]. The advantage of these methods is that one can directly obtain the surface profile, as naturally as in human vision.

However, we are interested in a surface which moves vertically in a large horizontal area. To do so, we perform observations on a single isolated line. This can be done, for instance, by illuminating the surface with a laser sheet from the top, and by looking to the light diffused on it, which is always moving jointly with the surface. In that case, the view of the line may be obstructed. It usually happen if part of the surface deforms strongly enough to cover the light path between the illuminated line and the observation point (at the camera position). To avoid obstructions, one can change the observation angle, going from a perpendicular view to an oblique one.

The geometry introduced for the oblique view is still very simple and it is shown in Fig. 2 for a single point (rather than for a line). We consider the paraxial approximation $\beta_{max} \approx H/L \ll 1$, as the distance L between the camera and the measurement area,

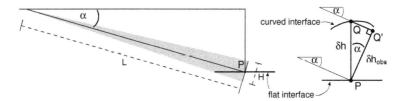

Fig. 2 Geometry linking the observed displacement δh_{obs} and the vertical displacement δh, in the paraxial approximation

is much larger than the maximal measured deformation H. In this approximation, one can consider a reference point P, lets say on the flat surface. Then, when the surface deforms, the point in the same vertical, Q will be perceived as being Q'. The vertical displacement $\overline{PQ} \equiv \delta h$ and the observed displacement $\overline{PQ'} \equiv \delta h_{obs}$ are related by

$$\delta h = \delta h_{obs}/\cos\alpha, \tag{1}$$

where α is the angle between the camera orientation direction and the horizontal (see Fig. 2). This argument is valid for a whole line. It is defined by the direction $\hat{r}$, and gives $\delta h(r) = \delta h_{obs}(r)/\cos\alpha$. The angle α lays between zero and $\pi/2$ (see (1) or Fig. 2). If α is close to zero, the measured δh_{obs} is very similar to the real δh, and the amplitude of motion can be directly obtained. However, in this case the view can be easily obstructed by other vertical deformations happening closer to the camera. For larger α on the other hand, obstructions are less probable, but to the detriment of the measured amplitude δh_{obs}. Despite δh is reconstructed geometrically, larger αs imply a lower resolution in the measurement. Therefore, α is chosen having in mind these two constrains.

2.2 Application to Highly Reflective Surfaces

A major challenge comes when using this technique in a highly reflecting surface, like the one of a liquid metal. This is because most of the light coming from the laser sheet is reflected specularly (i.e. in the vertical direction when the surface is flat). The consequence is twofold:

1. Only a small amount of light can be seen by diffusion (from any angle α). It is observed that the light diffusion properties are related to the cleanliness of the surface: the more impurities cumulate on the surface, the better the diffused line is seen. Therefore, the purest is the liquid metal, the more difficult will be to track the diffused line, although there is always possible to carry out the analysis in practical situations.
2. The reflected part of the beam coming to the surface (i.e. most of the light) is projected always specularly. As the surface deforms erratically, the reflected line

does as well. Eventually, the slope of the surface may be such that the reflected beam is in the observer direction, saturating the camera sensor. As expected, if the camera is placed with an angle α close to $\pi/2$, saturations occur frequently, even if the interfacial deformation is gentle. When α is reduced, saturations are less. Therefore, a third constrain is added to the choice of α.

2.3 Stereoscopic Extension

As it can be seen in Fig. 1, the flows under consideration have strong deformation. Thus, we expect to deal with saturations sooner or later. However, one can easily overcome the problem by putting a second camera on the opposite angle with respect to the vertical (they can be identified with M2 and M2$'$ in Fig. 3). If the surface is smooth, it has a well-defined slope at each point on the line. Therefore, if the reflection of a ray saturates a camera sensor, the same ray cannot saturate the sensor on the other camera at the same time.

Then, in the post-processing stage, one can reconstruct all the line by discarding the saturated parts of both simultaneous records. In practice, as the vertical deformation of the line could be registered within a few pixels, we decided to use a set of mirrors in order to register both views in a single picture (see Fig. 3). This trick has also the advantage of avoiding camera synchronization. As it is actually a stereoscopic measurement and it is performed with a single camera, we may call it *single camera stereoscopic measurement*.

A realization of the setup is schematized in Fig. 3. A laser sheet is obtained by placing a cylindrical lens Le in the path of a laser beam La. With the help of the mirror $M1$, the sheet is sent vertically to the liquid-metal surface, where part of the light diffuses along a line. The laser we use is a continuous green (wavelength of 532 nm). With an array of mirrors ($M2$, $M2'$ and P), images of the diffused line are

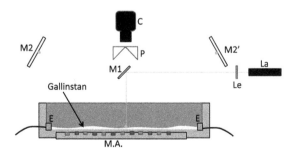

Fig. 3 Stereoscopic setup. A 1 cm layer of Gallinstan is placed between two electrodes (E), over a magnet array ($M.A.$) in a container of 50×40 cm^2. A laser La, a cylindrical lens Le and a mirror $M1$ produce a light sheet. The diffused line is tracked with two opposite angles by a single camera C, mirrors $M2$, $M2'$ and prisms P

Fig. 4 Example of line reconstruction scheme. **a** and **c** are the two registered views. They are binarized in (**b**) and (**d**), respectively. The matrix multiplication is presented in (**e**) after noise removal. The flow under study is strong, produced with $I = 500$ A. A saturation is seen in (**a**)

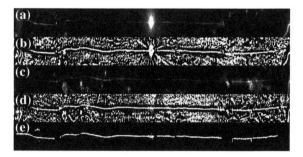

sent to the camera C. The associated α angles are $\pm 50°$. We take images at 60 Hz with a high resolution Dalsa camera that gives 2000×1700 pixels^2 images. When taking pictures of a line of 40 cm, one gets 5 pixels/mm as horizontal resolution. Vertical resolution depends on the angle α, and here it is found to be 3 pixels/mm. It was obtained by measuring the profile of a well calibrated stairway-shaped object.

2.4 Image Processing

Once images were obtained, the goal is to track the diffused line, converting it to a continuous and derivable function $\eta(x, t)$. The main steps for the analysis can be summarized as follows:

1. To remove any distortion between both views, induced by residual misalignments in the optical configuration. This can be achieved by performing a transformation (which may include translation, rotation and elongation) of one view. Once it is found for a single image-pair, it could be applied to the whole set of images. Further steps assume that both views are equivalent, with the only exception of saturated regions.
2. To filter the saturated spots. Here again we take advantage of the stereoscopic acquisition of images. It allows to filter one view with the complementary one. Among the several ways to do it, perhaps the simplest is to average the light intensity of both views. A slightly more complex procedure is shown in Fig. 4: each view is binarized and then both views are multiplied, eliminating saturated spots (see caption). After this step, we end up with a single image merging both views.
3. To extract a continuous line from the intensity map. Several methods are available, including very sophisticated global procedures [13, 32]. We computed the convolution with a Gaussian intensity profile for each line on the image.

Typical examples of the obtained deformation fields $\eta(x, t)$ are given in Sect. 3, in the form of spatiotemporal diagrams. These measurements also helped to characterize the flow obtained in our MHD setup [11].

3 Applications

Now we have a tool to study how a liquid-metal-surface is deformed. It can be either as a consequence of an underlying flow, or because of external perturbations. But before discussing our experimental results, we first present the setup used to produce the flow, together with its main characteristics.

3.1 Magnetohydrodynamical Setup

To stir a fluid in a constrained situation, we use a classical setup [3]. We apply the Lorentz force $\mathbf{F}_L = \mathbf{J} \times \mathbf{B}$, as depicted in Fig. 3. A vertical magnetic field $\mathbf{B}$ is produced with bands of strong permanent magnets with alternating polarities (named $M.A.$ in Fig. 3). The spatial structure of $\mathbf{B}$ fixes the one of the forcing. A horizontal density of current $\mathbf{J}$ is obtained by imposing an electrical current I between two electrodes placed on the container ends (named E in Fig. 3). The efficiency of this forcing depends on the fluid conductivity, justifying our choice of working with a liquid metal. We use Gallinstan, an alloy made of gallium, indium and tin, which is liquid at room temperature.[1] Thus, we can impose electrical currents I going from few Amperes to 600 with negligible heat losses. The control parameter of the experiment will be I. The characteristics of the flow were given elsewhere [11]. However, we can underline that the forcing is principally horizontal, and it has a fixed length scale. This produces several vortices of a size comparable with the one of the forcing (around 4 cm), which move and interact with each other, in a rather unpredictable way.

3.2 Deformation of the Surface of a Vortical Flow

We now focus on the surface deformation η produced by the underlying turbulent flow. An example of spatiotemporal evolution is shown in Fig. 5a, after the mean level is subtracted. Several features can be emphasized: (i) as expected for the motion of interacting vortices, surface deformation is highly fluctuating. (ii) The more visible deformation comes in large patches, and negative events (in blue) seem to have a characteristic width, despite it is not fixed. (iii) Negative events (*vortical depletions*) cover a smaller area in the diagram, although their magnitude is higher (the color-code is not symmetrical around zero). (iv) As a consequence of the conservation of mass, a larger area in the diagram has a level η larger than zero. (v) One can observe tenuous straight lines, specially in higher (red) zones.

---

[1]From the safety datasheet acc, Guideline 93/112/EC of Germatherm Medical AG, the Gallinstan is made of 68.5 % of Gallium, 21.5 % of indium, 10 % of Tin. Its density is $\rho = 6.440 \times 10^3$ kg/m^3, its kinematic viscosity is $\nu = 3.73 \times 10^{-7}$ m^2/s, its electrical conductivity $\sigma = 3.46 \times 10^6$ S/m.

Fig. 5 Surface deformation induced by a vortex-dominated flow. Panel **a** present 5 s of the spatiotemporal evolution of η after the mean level $\bar\eta$ is subtracted. The MHD flow is produced here with an electrical current of $I = 350\,A$. Panel **b** present the PDF of $(\eta - \bar\eta)/\sigma_\eta$, which is the deformation normalized by the standard deviation σ_η. The *color code* represent here the current I. The *black dashed line* correspond to a Gaussian distribution. The *inset* present σ_η as a function of I. Panel **c** show two examples of *vortex depletions*, where the hight η has been previously normalized. The position is given by x_0

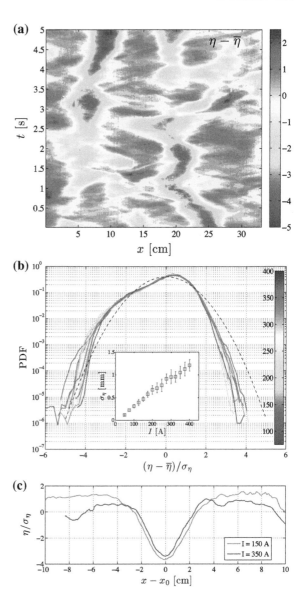

From the whole measurement (of 60 s), we compute the statistics of the level η. The obtained probability distribution function (PDF) is shown in Fig. 5b for different intensities in the MHD forcing. Colors go from a gentle forcing in blue to a strong one in red. Remarkably, all these PDFs show a fair collapse when rescaled in terms of the standard deviation (shown as an inset in Fig. 5b), suggesting a common origin. However, they considerably differ from the Gaussian shape shown as a dashed black line. The most probable value in the PDF is above zero, consistent with the observations made before.

Random linear waves—as those encountered in a calm ocean—show Gaussian PDFs [17, 26]. For gravity waves, when nonlinear effects become important, it appears an asymmetry between crests and troughs. It imposes an asymmetry in the PDF through positive values [18, 26, 30]. However, it was shown that the asymmetry can be removed by geometrical constrains for capillary waves, and Gaussian statistics are recovered [8]. The asymmetry of the PDFs is quantified with the skewness: it is zero for symmetric distributions (for instance for random linear waves) and positive for random nonlinear waves [27].

Our PDF, on the other hand, show an asymmetry through negative values. This is consistent with our previous observations of Fig. 5a, as we noticed that vortical depletions have an important contribution in η. Depletions are produced by a lower pressure at the vortex-core, and its magnitude depends on its vorticity, which in turbulence is expected to follow a Gaussian distribution. Thus, vortex profiles are intrinsically asymmetric through negative values with respect of the mean level of η (see two examples in Fig. 5c, where η has been normalized). This asymmetry is expected to impose the asymmetry in the distribution of η, expressed in a negative skewness. Note that, even if the PDF in Fig. 5b looks slightly similar to a bimodal distribution, we check that there is no second stable value for the surface level. Therefore, the PDF is shaped mainly by the statistical nature of vortical depletions.

A theoretical analysis inspired in this flow will be developed elsewhere. The idea is to consider the hydrodynamics of a single vortex without discharge (but inspired in those analysis [1, 19, 28]), combined with statistical arguments as in the study of random sea-waves.

We conclude that the main contribution to η comes from vortical structures interacting in a erratic way. However, as we noticed in Fig. 5a, one may also distinguish tenuous straight lines. As traveling waves appear as straight lines in spatiotemporal diagrams (see also Sect. 3.3), we may identify these features as the subtle signature of waves. The frequency spectrum of a velocity signal is less steep than the one of the position. Thus any wave-like motion present in our data should be emphasized by taking the temporal derivative. Indeed, in the example of $\partial\eta/\partial t$ shown in Fig. 6a, straight-line features appear much more clear than in the original signal.

From velocity signals, we computed the fast Fourier transform (FFT), which is shown in Fig. 6b. It presents a map of the frequencies and wavelengths involved in the motion. As the surface is perturbed from the bulk, the response in wavelength to a given frequency (the dispersion relation) is far from being known. Indeed, one can see that there is a large patch concentrating surface energy (from 0.2 to 3 Hz and from 0 to 0.25 cm^{-1}). As a reference, we plotted—in a thick white line—the usual dispersion relation for linear waves:

$$\omega^2 = \left(\frac{\rho - \rho'}{\rho + \rho'} gk + \frac{\gamma}{\rho + \rho'} k^3 \right) \cdot \tanh(kH) \tag{2}$$

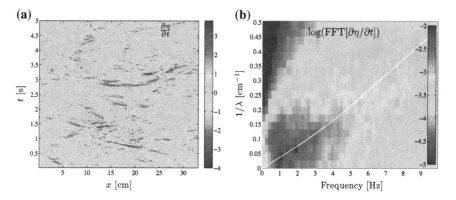

Fig. 6 Spatio-temporal diagram of the vertical velocity of the surface $\partial\eta/\partial t$ (*left*) and its fast Fourier transform (*right*) in log scale. The associated linear dispersion relation is added as a *thick white line* on the *right panel*. The forcing here corresponds to $I = 400$ A

where $k = 2\pi/\lambda$ is the wavenumber, λ is the wavelength, $\omega = 2\pi f$ is the angular frequency, H is the fluid depth, g is the gravity acceleration, γ is the surface tension, ρ is the liquid metal density and ρ' is the density of acidified-water used to prevent surface oxidation. We see that part of the surface energy lies on linear dispersion relation, despite it not confined to it at all. These result represent another experimental observation of waves generated by unstationary flows [5, 25]. However, a more proper identification of the structures containing surface energy is another perspective of this work.

3.3 Waves Propagating over a Turbulent Flow

We are interested now in what happens to a wave that propagates on the surface of the turbulent flow. The extensive study is described elsewhere [12]. Here we only explain our procedure, and we show typical deformations fields obtained with the measurement technique described before.

We excite waves on the liquid metal surface by means of an electromagnetic shaker. A vertical sinusoidal vibration is applied to the liquid surface by a cylindrical paddle. This *source* of waves is placed on one corner of the working area (bottom-right corner in Fig. 1 and near the x-origin in Fig. 7), and measurements are performed along the central line in the container. Figure 7 present our typical observations. In Fig. 7a, there is only the mechanically excited wave (here at $f_0 = 5$ Hz). Wave propagation appears very clear as oblique lines, despite the wave pattern is not completely trivial. Figure 7b present the deformation of the surface when there are both a wave (again at $f_0 = 5$ Hz) and an underling fluid motion produced by MHD forcing at $I = 80$ A. In this case, the contributions to surface deformation are easy to identify: waves correspond to oblique lines (as in left panel), and the big red and blue

(a)   **(b)**

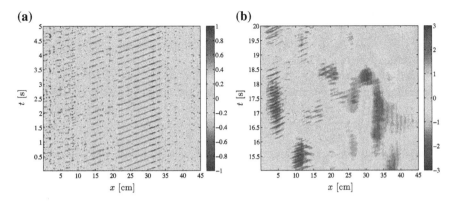

Fig. 7 Spatio-temporal diagrams of the surface displacement η. They show the influence of fluid motion on the propagation of waves. *Left* a wave is excited externally at 5 Hz with a shaker. *Right* the same wave is propagating over a fluctuating background produced by a forcing of $I = 80$ A

patches correspond to larger scale fluid motion. Interestingly, one may appreciate here that the amplitude of the propagating wave gradually decreases. It is visible only until around $x \sim 25$ or 30 cm, as a consequence of an enhanced dissipation produced by fluid motion.

To go further in the study, we measure η for different flow strengths (controlled by the current I) and for excitation frequencies f_0 between 3 and 9 Hz. Knowing that I and the flow velocity U_o are related as $U_o \sim \sqrt{I}$ [11], we can define the Froude number $Fr = U_o/C_w$, where C_w is the phase velocity of the wave. Two main results come from our observations: (a) Wave dissipation is enhanced by the flow, as observed in Fig. 7b. This enhanced dissipation is found to scale linearly with Fr. (b) A shift in the wavelength is found for frequencies between 6 and 9 Hz. The shift can also be described as a function of Fr. A detailed account of these results is given in [12].

4 Conclusions

In this chapter we present a conceptually simple technique to perform measurements of the surface deformation in the highly-reflecting surface of a liquid metal. The idea is to send a laser sheet on the surface of the liquid-metal and to track the diffused light with a camera from a given angle. When the deformation of the surface is strong, specular reflections of the laser saturates the camera sensor. This problem is overcome by using an optical arrangement allowing to record two complementary views of the diffused line in the same picture.

The technique is used to study two problems of the interaction between surface waves and turbulent flows. First we considered the influence of turbulence on surface deformation. The statistics of the height level are presented. They are asymmetric, with a negative asymmetry coefficient (skewness), in opposition to steep gravity waves, were the skewness is positive. This is interpreted as a statistical signature of vortices. Then, we discussed the presence of wavy motion in our signals, highlighted in the vertical velocity of the surface.

Finally, we present some observations about the influence of turbulence on wave propagation. Here we considered simultaneously our turbulent MHD flow and externally excited waves. Wave dissipation by turbulence is visible in a presented spatiotemporal diagram. A more complete study is given elsewhere [12].

Acknowledgments We would like to thanks C. Wiertel-Gasquet for helping with the automation of the experiment. F. Daviaud, M. Bonetti and G. Zalczer for helpful discussions. PG received support from the Triangle de la Physique and CONICYT/FONDECYT postdoctorado N 3140550.

References

1. A. Andersen, T. Bohr, B. Stenum, J. Juul Rasmussen, B. Lautrup, The bathtub vortex in a rotating container. J. Fluid Mech. **556**, 121–146 (2006)
2. A. Benetazzo, Measurements of short water waves using stereo matched image sequences. Coast. Eng. **53**, 1013–1032 (2006)
3. N.F. Bondarenko, M.Z. Gak, F.V. Dolzhanskiy, Laboratory and theoretical models of plane periodic flow. Izv. Atmos. Ocean. Phys. **15**, 711–716 (1979)
4. E. Cerda, F. Lund, Interaction of surface waves with vorticity in shallow water. Phys. Rev. Lett. **70**, 3896–3899 (1993)
5. P.J. Cobelli, A. Maurel, V. Pagneux, P. Petitjeans, Global measurement of water waves by Fourier transform profilometry. Exp. Fluids **46**, 1037–1047 (2009)
6. C. Cox, W. Munk, Measurement of the roughness of the sea surface from photographs of the sun's glitter. J. Opt. Soc. Am. **44**, 838–850 (1954)
7. P.A. Davidson, Magnetohydrodynamics in materials processing. Annu. Rev. Fluid Mech. **31**, 273–300 (1999)
8. G. Düring, C. Falcón, Symmetry induced four-wave capillary wave turbulence. Phys. Rev. Lett. **103**, 174503 (2009)
9. R. Ford, Gravity wave generation by vortical flows in a rotating frame. Ph.D. thesis, University of Cambridge, 1993
10. L. Gordillo, T. Sauma, Y. Zárate, I. Espinoza, M.G. Clerc, N. Mujica, Can non-propagating hydrodynamic solitons be forced to move? Eur. Phys. J. D **62**, 39–49 (2010)
11. P. Gutiérrez, S. Aumaître, Clustering of floaters on the free surface of a turbulent flow: an experimental study. (2015)
12. P. Gutiérrez, S. Aumaître, Surface waves propagation on a turbulent flow forced electromagnetically (2015)
13. M. Kass, A. Witkin, D. Terzopoulos, Snakes: active contour models. Int. J. Comput. Vis. **1**, 321–331 (1988)
14. G. Kuwabara, T. Hasegawa, K. Kono, Water waves in a ripple tank. Am. J. Phys. **54**, 1002–1007 (1986)
15. M.J. Lighthill, On sound generated aerodynamically. I. General theory. Proc. R. Soc. Lond. A **211**, 564–587 (1952)
16. M.J. Lighthill, Waves in Fluids (Cambridge University Press, Cambridge, 1978)
17. M.S. Longuet-Higgins, The statistical analysis of a random, moving surface. Phil. Trans. R. Soc. Lond. A **249**, 321–387 (1957)
18. M.S. Longuet-Higgins, On the distribution of the heights of sea waves: some effects of nonlinearity and finite band width. J. Geophys. Res. **85**, 1519–1523 (1980)

19. T.S. Lundgren, The vortical flow above the drain-hole in a rotating vessel. J. Fluid Mech. **155**, 381–412 (1985)
20. J.E. Macías, C. Falcón, Dynamics of spatially modulated kinks in shallow granular layers. New J. Phys. **16**, 043032 (2014)
21. F. Moisy, M. Rabaud, K. Salsac, A synthetic Schlieren method for the measurement of the topography of a liquid interface. Exp. Fluids **46**, 1021–1036 (2009)
22. R. Monchaux, M. Berhanu, M. Bourgoin, M. Moulin, P. Odier, J.F. Pinton, R. Volk, S. Fauve, N. Mordant, F. Pétrélis, A. Chiffaudel, F. Daviaud, B. Dubrulle, C. Gasquet, L. Marie, F. Ravelet, Generation of a magnetic field by dynamo action in a turbulent flow of liquid sodium. Phys. Rev. Lett. **98**, 044502 (2007)
23. O.M. Phillips, The scattering of gravity waves by turbulence. J. Fluid Mech. **5**, 177–192 (1959)
24. R. Savelsberg, A. Holten, W. van de Water, Measurement of the gradient field of a turbulent free surface. Exp. Fluids **41**, 629–640 (2006)
25. R. Savelsberg, W. van de Water, Turbulence of a free surface. Phys. Rev. Lett. **100**, 034501 (2008)
26. H. Socquet-Juglard, K. Dysthe, K. Trulsen, H.E. Krogstad, J. Liu, Probability distributions of surface gravity waves during spectral changes. J. Fluid Mech. **542**, 195–216 (2005)
27. M.A. Srokosz, M.S. Longuet-Higgins, On the skewness of sea-surface elevation. J. Fluid Mech. **164**, 487–497 (1986)
28. Y. Stepanyants, G. Yeoh, Stationary bathtub vortices and a critical regime of liquid discharge. J. Fluid Mech. **604**, 77–98 (2008)
29. B. Ströbel, Demonstration and study of the dispersion of water waves with a computer-controlled ripple tank. Am. J. Phys. **79**, 581–589 (2011)
30. M.A. Tayfun, Narrow-band nonlinear sea waves. J. Geophys. Res. **85**, 1548–1552 (1980)
31. W.B. Wright, R. Budakian, S.J. Putterman, Diffusing light photography of fully developed isotropic ripple turbulence. Phys. Rev. Lett. **76**, 4528–4531 (1996)
32. C. Xu, J.L. Prince, Snakes, shapes, and gradient vector flow. IEEE Trans. Image Process. **7**, 359–369 (1998)

Parametric Phenomena in Magnetic Nanostripes

Alejandro O. León

Abstract Systems with a time-modulated injection of energy self-organize into patterns and solitons. Magnetic systems forced by a direct spin-polarized current and a constant external field are equivalent to parametrically driven systems. This parametric equivalence implies that both systems are described by the same model, the *parametrically driven, damped nonlinear Schrödinger* equation, and that they exhibit the same parametric phenomena, which includes patterns and solitons. We review here recent literature on the topic, and we investigate the case of long and narrow magnets. This configuration reduces significantly the critical currents at which self-organization emerges, and it allows the formation of bound states of solitons.

1 Introduction

In the presence of external forcing, magnetic materials exhibit a wide range of dynamical responses which include self-oscillations [1, 2], chaos [3, 4], patterns [5], solitons [6, 7], skyrmions [8] and rogue-waves [9]; this versatility renders magnetic materials an ideal framework to study nonlinear dynamics and self-organization. A particularly promising scenario occurs at nanoscales, where the magnetization is manipulated by effects which are negligible in larger systems, such as the *spin-transfer torque*. This effect occurs when the spins of an electric current interact with the spins of a ferromagnetic medium [10–12]. When the spin of conduction electrons have a preferred orientation—*spin-polarized current*—the spin-transfer torque can generate self-oscillations with microwave frequencies and magnetic reversions [13].

From a dynamical systems viewpoint, spin-transfer torques are non-variational effects that can inject or dissipate the ferromagnetic energy. The first case is important for technological applications since energy injections can induce permanent behaviors such as limit cycles [2, 13] and chaos [14]. In addition, spin-polarized currents

A.O. León (✉)
Departamento de Física, Facultad de Ciencias Físicas y Matemáticas,
Universidad de Chile, Casilla 487-3, Santiago, Chile
e-mail: alejandroleonvega@gmail.com

© Springer International Publishing Switzerland 2016
M. Tlidi and M.G. Clerc (eds.), *Nonlinear Dynamics: Materials,*
Theory and Experiments, Springer Proceedings in Physics 173,
DOI 10.1007/978-3-319-24871-4_19

can switch the magnetization from one equilibrium to another, which is particularly important for the development of memory technologies [12].

The presence of a dissipative spin-transfer torque permits to generate stationary patterns [5], domain walls and dissipative solitons [6]. This rich variety of magnetization textures can be understood because a magnetic medium in presence of spin-polarized direct currents obeys the same equations of macroscopic systems with a time-dependent forcing [6], known as parametrically driven systems [15]. These systems are characterized by oscillating at half the forcing frequency and exhibiting sub-harmonic resonances [15]. In several cases, these resonances induce patterns or Faraday-type waves [16] and solitons [17–20]. A few examples of parametrically driven systems are vibrating fluids [17], nonlinear lattices [21], light pulses in optical fibers [22], optical parametric oscillators [23], and ferromagnetic materials under oscillatory magnetic fields [24, 25]. Parametrically driven systems can be described in a unified manner using the *parametrically driven, damped nonlinear Schrödinger equation* (PDNLS) [26], this model is valid when the forcing frequency is close to twice the natural frequency of oscillations, and the injection and dissipation of energy are small. The PDNLS equation admits several analytic solutions for one-dimensional systems, in particular, expressions for solitons [17, 27] and patterns [28] are known. Hence, the equivalence between parametrically driven systems and magnets forced by a direct spin-polarized current, has permitted to predict a large variety of states in nanomagnetism, and to use the PDNLS analytic solutions as approximations for the magnetic states [6].

We review here the parametric equivalence, and apply it to large and narrow magnets driven by spin-polarized currents. Some advantages of this configurations are the stabilization of multiple localized states for long enough magnets, the reduction of critical currents at which patterns and solitons emerge, and the simplicity of one-dimensional systems.

The chapter is organized as follows. In next section we describe the theoretical model of a ferromagnetic medium under spin-polarized currents. In Sect. 3 we review the parametric equivalence and its most important predictions. The discussions and conclusions are left to the final section.

2 Theoretical Model for Magnetization Dynamics

In the continuum approach, magnetic nanostripes are described by their magnetization vector $\mathbf{M}(x', t)$ [1]. Time $t' = \gamma_0 M_s t$ and the space coordinates $x' = l_{ex} x$, where x is the direction of the stripe, are written in terms of the following material properties: the gyromagnetic ratio γ_0, the saturation magnetization M_s and the exchange length l_{ex}. At nanoscales, the magnetization norm $||\mathbf{M}|| = M_s$ is constant within the material, and therefore the dynamical variable is the unitary vector $\mathbf{m} = \mathbf{M}/M_s$. The following phenomenological model, known as *Landau-Lifshitz-Gilbert* (LLG) dimensionless equation, describes the magnetization evolution [1]

$$\partial_t \mathbf{m} = -\mathbf{m} \times \mathbf{h}_{\textit{eff}} + \alpha \mathbf{m} \times \partial_t \mathbf{m}, \qquad (1)$$

where the first term of (1) accounts for precessions around the effective field $\mathbf{h}_{\textit{eff}}$ defined by

$$\mathbf{h}_{\textit{eff}} \equiv -\frac{L_0}{\mu_0 M_s^2} \frac{\delta E_M}{\delta \mathbf{m}} = \partial_{xx}\mathbf{m} + \mathbf{h}_0 + \mathbf{h}_D + \beta_{x0} m_x \mathbf{e}_x, \qquad (2)$$

and

$$E_M = \frac{\mu_0 M_s^2}{L_0} \int_0^{L_0} \left[\frac{1}{2} |\partial_x \mathbf{m}|^2 - \mathbf{m} \cdot \mathbf{h}_0 - \frac{1}{2}\mathbf{m} \cdot \mathbf{h}_D - \frac{\beta_{x0}}{2} m_x^2 \right] dx, \qquad (3)$$

the vectors $\partial_{xx}\mathbf{m}$ and $\mathbf{h}_0$ stand for the exchange and external fields, respectively. The magnetostatic field $\mathbf{h}_D$ is usually a nonlocal function of the magnetization, and it is obtained by solving Maxwell equations inside the material. The magnetocrystaline anisotropy is given by $\beta_{x0} m_x \mathbf{e}_x$, and it favors orientations along the x axis. E_M is the magnetic energy of the material. When the last term of (1) is zero, the magnetization is a Hamiltonian oscillator and the energy is conserved.

The term proportional to α in (1) models dissipation mechanisms [1]. If the external field $\mathbf{h}_0$ is independent of time, then the magnetic energy converges monotonically to its minima $dE_m/dt = -(\alpha \mu_0 M_s^2 / V_0) \int |\partial_t \mathbf{m}|^2 \le 0$ at a rate proportional to α. Hence, magnetic materials are damped oscillators, and after transients the magnetization reaches its stable equilibrium. One possibility to break the relaxation-type dynamic of the LLG equation is by using a time-dependent external field $\mathbf{h}_0 \to \mathbf{h}_0(t)$. Under this parametric injection, ferromagnetic media exhibit Faraday-type waves [28], fronts and several types of localized states [24, 26, 27, 29–33]. The case of oscillatory fields has been extensively studied in the literature, and then we will focus on another physical effect that permits to manipulate the magnetization, namely the spin-transfer torque.

Spin-transfer torques emerge in multilayer nanopillars (see Fig. 1). This type of metallic device is composed by two conducting ferromagnetic layers separated by a nonmagnetic metal. The electric current is uniform and it flows perpendicular to the plane of the layers. One magnetic film has *fixed* magnetization and it is used

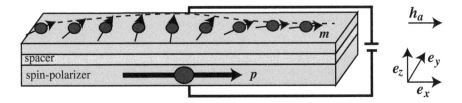

Fig. 1 Nanopillar device. This conducting structure is composed by two magnetic nanostripes and a nonmagnetic spacer. The magnetization is fixed for one ferromagnet, and it is used to polarize the electric current that flows perpendicular to the layers. We are interested on the magnetization dynamics of the second material, which are affected by the electric current and an external field $\mathbf{h}_a$

as a filter (polarizer) for the spins of the electric current. We are interested on the magnetization dynamics of the second material, which is *free* and evolves according to the LLG equation. Spin-polarized currents can be modeled as an additional torque τ_{stt} in the (1). At leading order, this term reads

$$\tau_{stt} = g'\mathbf{m} \times (\mathbf{m} \times \mathbf{p}),\tag{4}$$

where $\mathbf{p}$ is the direction of the spins of the electric currents. The function $g'(\mathbf{m} \cdot \mathbf{p})$, known as *spin-transfer efficiency*, accounts for the details of the transport processes of conduction electrons. There are several expressions in literature [10, 35–40] for g', the simplest one is to approximate $g'(\mathbf{m} \cdot \mathbf{p}) \approx g'(1)$, this approach is known as *sine-approximation*. In this chapter we use this approximation. The parameter $g \equiv g'(1)$ depends on the intensity of the applied electric current and device properties [34]

$$g = \frac{\hbar}{2|e|d_z} \frac{\eta_0 J}{\mu_0 M_s^2},$$

where J is the current density of electrons, d_z the thickness of the free layer and $e < 0$ the electric charge. The current density of electrons J and the parameter g are negative when the electrons flow from the fixed to the free layer. All other transport properties of the materials are summarized in the η_0 coefficient.

2.1 Simplified Model for Magnetization Dynamics in Nanostripes

Let us focus on the nanostripe magnet of Fig. 1. A nanostripe is a large and narrow thin film, satisfying $d_x \gg d_y \gg d_z$ for the lateral dimensions along the respective Cartesian axis. We approximate the magnetostatic field by shape anisotropy terms β_{x1} and β_z, where β_{x1} (β_z) favors (disfavors) configurations along the x axis (z axis). We set both the external field direction $\mathbf{h}_0 = h_0\mathbf{e}_x$ and the spin-current direction $\mathbf{p} = \mathbf{e}_x$ along the x axis. Under these assumptions, the LLG model simplifies to

$$\partial_t\mathbf{m} = -\mathbf{m} \times \left[(h_0 + \beta_x m_x)\mathbf{e}_x - \beta_z m_z\mathbf{e}_z + \partial_{xx}\mathbf{m}\right] + \alpha\mathbf{m} \times \partial_t\mathbf{m} + g\mathbf{m} \times (\mathbf{m} \times \mathbf{e}_x),\tag{5}$$

where $\beta_x \equiv \beta_{x0} + \beta_{x1}$. This model admits two homogeneous stationary states: $\mathbf{m} = \pm\mathbf{e}_x$ in which the magnetization points parallel ($+$) and antiparallel ($-$) to the polarization direction of the current. We concentrate on the first state, however the dynamics of the antiparallel state are equivalent in the appropriate region of the parameter space, due to the symmetry $(\mathbf{h}_a, g) \rightarrow -(\mathbf{h}_a, g)$ of (5).

It is worth noting that we study time-independent forcing mechanisms, namely a direct electric current and a constant magnetic field, and then this systems is not parametrically driven. Even so, using an appropriate change of variables, the system can be described by the PDNLS equation (as we will see in next section) and then it exhibits the phenomena usually found in parametrically driven systems, such as patterns and solitons [6].

3 Parametric Equivalence

3.1 Stereographic Projection

The Landau-Lifshitz-Gilbert equation conserves the magnetization norm $\partial_t |\mathbf{m}| = (\mathbf{m} \cdot \partial_t \mathbf{m})/|\mathbf{m}| = 0$, and then it can be written in different coordinate systems such as spherical [1, 5], canonical [1], and stereographic variables [1, 41]. In stereographic representation, the equations for magnetization dynamics take the form of a generalized *complex Ginzburg-Landau* model, which is a paradigmatic description of nonlinear oscillators. Figure 2 shows the stereographic representation, which is obtained by projecting the phase space—spherical surface—over the equatorial plane $m_x = 0$, according to [41]

$$a(x, t) = \frac{m_y + im_z}{1 + m_x}. \tag{6}$$

The complex field a represents deviations from the parallel state $\mathbf{m} = \mathbf{e}_x$. Replacing formula (6) in (5) and after straightforward calculations one obtains (the generalized *complex Ginzburg-Landau* model)

$$(i + \alpha) \partial_t a = (ig - h_a) a - \frac{\beta_z}{2} (a - \bar{a}) \frac{1 + a^2}{1 + |a|^2} - \beta_x a \frac{1 - |a|^2}{1 + |a|^2} + \partial_{xx} a - 2 \frac{\bar{a} (\partial_x a)^2}{1 + |a|^2}, \tag{7}$$

(a)   **(b)**

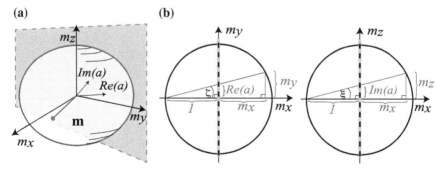

Fig. 2 Stereographic representation. **a** The unitary spherical surface $m_x^2 + m_y^2 + m_z^2 = 1$ is mapped to the equatorial plane defined by $m_x = 0$. **b** The projections on the complex field $a = (m_y + im_z)/(1 + m_x)$

where the term proportional to β_x is a nonlinear saturation. The term proportional to β_z breaks the phase invariant $a \rightarrow ae^{i\phi_0}$, and it is usually associated to parametric forcing. The coefficients α and g break the temporal reversion invariance $(t, a) \rightarrow (-t, \bar{a})$. Notice that in the absence of spin transfers ($g = 0$ but $\alpha \neq 0$) the (7) has the structure of a variational (relaxation type) wave equation.

3.1.1 PDNLS-Limit

A notable limit of (7) is obtained for small amplitude magnetic motions and gradients $|\partial_x a| \ll |a| \ll 1$ and small anisotropy $\beta_z \ll 1$; in this case we can replace $A \equiv ae^{i\pi/4}/\sqrt{2\beta_x + \beta_z}$ into (7) and after simplifications one obtains the *parametrically driven, damped nonlinear Schrödinger* equation

$$\partial_t A = -i\left(\nu A + A|A|^2 + \partial_{xx}A\right) - \mu A + \gamma \bar{A}, \tag{8}$$

where $\nu \equiv -h_a - \beta_x - \beta_z/2$ is the detuning between half the forcing frequency and the response frequency in usual parametrically driven systems. The dissipation and parametric injection are given by $\mu \equiv -g - \alpha\nu$ and $\gamma \equiv \beta_z/2$, respectively. The equation scales as $\nu \sim \mu \sim \gamma \sim |A|^2 \sim \partial_{xx} \sim \partial_t$, and all the higher order terms have been neglected.

It is worth noting that in the present case the dissipation is an experimental control parameter, while the injection $\gamma \equiv \beta_z/2$ is not. The nanostripe geometry is between two natural limits: the nanowire and the two-dimensional cross-section device. For the nanowires we have $d_y \sim d_z \ll d_x$ and the anisotropies are $\beta_z \approx 0$ and $\beta_x \equiv \beta_{x0} + \beta_{x1} \approx \beta_{x0} + 1$. In this case the injection γ is small and the required currents are also small. On the other hand, for two-dimensional cross-section nanopillars, we have $d_z \ll d_y \sim d_x$, and $\beta_z \approx 1$ and $\beta_x \approx \beta_{x0}$. Then, the two-dimensional magnets have a bigger parametric injection and higher electric currents are necessary to obtain the scaling $\gamma \sim \mu$.

3.2 Interpretation in Terms of Equivalent Systems

The simple stereographic projection presented above permitted us to obtain the PDNLS model in magnetic media forced by direct currents. Another approach to derive this equation is to notice that the spin-transfer torque has the form of a pseudo torque (non-inertial effect), and then the LLG equation can be transformed under an appropriate change of reference frames into a rotating system.

Let us consider a stripe rotating with constant angular velocity along the x axis direction $\Omega = \Omega_0 e_x$, in presence of an external field $\mathbf{h}'_a = (h_a + \Omega_0)e_x$, as shown in Fig. 3. For this system, the magnetization equations can be written in two frames: the inertial coordinate systems S' defined by $\{e'_x, e'_y, e'_z\}$ and the co-movil coordinate system S given by the unit vectors $\{e_x, e_y, e_z\}$, which is fixed to the stripe. In the first

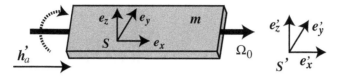

Fig. 3 Rotating magnetic stripe. The magnetization equations describe a parametrically driven device in the inertial frame S', on the other hand the magnetic equations in the co-movil frame S are the same as (5)

frame we have

$$\left(\frac{\partial \mathbf{m}}{\partial t}\right)\Bigg|_{S'} = -\mathbf{m} \times \mathbf{h}'_{eff}(t) + \alpha \mathbf{m} \times \left(\frac{\partial \mathbf{m}}{\partial t}\right)\Bigg|_{S'}, \tag{9}$$

$$\mathbf{h}'_{eff}(t) = (h_0 + \Omega_0 + \beta_x m_x)\mathbf{e}_x - \beta_z (\mathbf{m} \cdot \mathbf{e}_z(t)) \mathbf{e}_z(t) + \partial_{xx}\mathbf{m}. \tag{10}$$

Note that the projection over the fixed axis of the inertial frame produces a temporal dependence of the form

$$(\mathbf{m} \cdot \mathbf{e}_z(t)) \mathbf{e}_z(t) = \beta_z(-\sin(\Omega_0 t)m'_y + \cos(\Omega_0 t)m'_z)(-\sin(\Omega_0 t)\mathbf{e}'_y + \cos(\Omega_0 t)\mathbf{e}'_z).$$

Let us remark that the anisotropy field is a function of time, then, rotating stripes are parametrically driven systems. Moreover, the frequency ω of the anisotropy vector oscillations is close to twice the rotating frequency $\omega = 2(\Omega_0 + \nu)$, where ν is the detuning parameter defined in (8). Hence, the oscillation envelope of the magnetization obeys the PDNLS equation (8).

The equation (9) can be written in the co-movil system S. After transforming the temporal derivatives of vectors according to $\partial_t|_{S'} = \partial_t|_S + \Omega \times$ [15], we obtain

$$\left(\frac{\partial \mathbf{m}}{\partial t}\right)\Bigg|_{S} = -\mathbf{m} \times \mathbf{h}_{eff} + \alpha \mathbf{m} \times \left(\frac{\partial \mathbf{m}}{\partial t}\right)\Bigg|_{S} - \alpha \Omega_0 \mathbf{m} \times (\mathbf{m} \times \mathbf{e}_x), \tag{11}$$

$$\mathbf{h}_{eff} = (h_0 + \beta_x m_x)\mathbf{e}_x - \beta_z m_z \mathbf{e}_z + \partial_{xx}\mathbf{m}, \tag{12}$$

where the last term of (11) is the non-inertial torque originated in the transformation of the damping torque. The transformation of the left hand side of (9) is only a shift of the external field. Note that (11) and (12) are exactly the same LLG model of the spin-transfer driven stripe (5), where the intensity of the spin-transfer is proportional to the angular velocity $g = -\alpha \Omega_0$.

In brief, magnetic stripes driven by spin-transfer torques are equivalent to rotating stripes without spin-transfer torques. Moreover, in an appropriate reference frame both physical systems obey the same equations. Since rotating stripes are parametrically driven systems, they exhibit sub-harmonic instabilities, solitons, and patterns.

The PDNLS amplitude equation has been used to predict self-organization in several contexts. In the next subsection we review some of the solutions of this equation and compare them with the phenomena observed in (5).

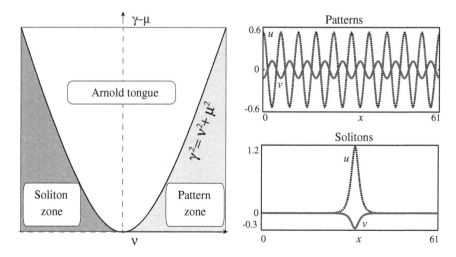

Fig. 4 Bifurcation diagram of the PDNLS model. The *right panel* shows dissipative solitons and patterns obtained from (8). Parameter values are $\gamma = 0.5$, $\mu = 0.45$, and $\nu = -0.5$ and $\nu = 1$ for solitons and patterns, respectively

3.3 Parametric Phenomena in Magnetic Nanostripes

We start writing the PDNLS equation in terms of the real and imaginary parts of the amplitude $A = u + iv$,

$$\partial_t u = (\gamma - \mu)u + (\nu + \partial_{xx})v + v(u^2 + v^2),$$
$$\partial_t v = -(\nu + \partial_{xx})u - (\gamma + \mu)v - u(u^2 + v^2). \qquad (13)$$

The stability of the trivial state $u = v = 0$ is determined by its eigenvalues

$$\lambda_{\pm} = -\mu \pm \sqrt{\gamma^2 - (\nu - k^2)^2}, \qquad (14)$$

which predict several bifurcations. The first one, is an Andronov-Hopf instability for $\mu \geq 0$. When this instability saturates, the system exhibits uniform self-oscillations [13] with frequency $\omega_0 = \sqrt{\nu^2 - \gamma^2}$. When the bifurcation is not saturated, the magnetization switches to another equilibria. In this regime, the magnetization behaves as a nonlinear oscillator with negative damping [2]. Note that the control parameter of this bifurcation requires is negative dissipation, which cannot be obtained in parametrically driven systems.

Another bifurcation predicted by expression (14) is a stationary instability when $\gamma \geq \sqrt{\nu^2 + \mu^2}$, this region is the well-known Arnold tongue (see Fig. 4). The critical curve describing the Arnold tongue in the magnetic system is $g^2 + [h_a - (\beta_x + \beta_z/2)]^2 = \beta_z^2/4$. Inside this region several states appear, such as localized states, patterns, and domain walls [6, 26].

3.3.1 Pattern Formation

The eigenvalues of expression (14) reveal a third bifurcation of the magnetic system, which occurs for $\gamma \geq \mu$ and positive detuning $\nu \geq 0$. This instability is characterized by the emergence of spatially periodic patterns with intrinsic wavenumber $\sqrt{\nu}$. At the onset of this bifurcation, a reduced description in terms of the pattern amplitude $T(t)$ is obtained using standard techniques of weakly nonlinear analysis. Moreover introducing the ansatz

$$\begin{pmatrix} u \\ v \end{pmatrix} = Te^{i\sqrt{\nu}x} \begin{pmatrix} 1 \\ 0 \end{pmatrix} - \frac{3}{2\mu}T|T|^2 e^{i\sqrt{\nu}x}\begin{pmatrix} 0 \\ 1 \end{pmatrix} + \frac{T^3}{8\nu}e^{3i\sqrt{\nu}x}\begin{pmatrix} 1 \\ 0 \end{pmatrix} + c.c. + \cdots, \quad (15)$$

in (13), one gets the following Solvability condition for the envelope $T(t)$ [28]

$$\partial_t T = (\gamma - \mu)T - \frac{9}{2\mu}T|T|^4. \quad (16)$$

This equation predicts the formation of a stable pattern for $\gamma \geq \mu$. Figure 4 shows this solution, which at leading order reads $A \approx 2(2\mu(\gamma - \mu)/9)^{1/4}\cos(\sqrt{\nu}x)$. This bifurcation was studied in detail in [5] for both the one and two-dimenssions devices. In the first case, the full solution of patterns obtained from (5) is [5]

$$m_y \approx 2\sqrt[4]{\frac{4\beta_z\,(g + \beta_z/2)}{(6\beta_x + 3\beta_z - 2\nu^2)^2}}\cos(\nu x), \quad (17)$$

where $m_y \approx -m_z$ and $m_x \approx 1 - (m_y^2 + m_z^2)/2$. This solution is exactly the same as the PDNLS solution in the limit $\nu \ll 1$.

Let us remark that the critical current of the spatial instability is controlled by the perpendicular anisotropy of the stripe $g = \beta_z/2$. The perpendicular anisotropy is fixed for square cross-section devices, but it can be significantly decreased for stripes by adjusting the d_y length.

3.3.2 Dissipative Solitons

The PDNLS equation (8) can be written in the terms of the modulus and phase of its order parameter $A = Re^{i\phi}$, in this case

$$\partial_t R = -\mu R + 2\partial_x R\partial_x\phi + R\partial_{xx}\phi + \gamma R\cos(2\phi), \quad (18)$$
$$R\partial_t\phi = -\nu R - R^3 - \partial_{xx}R + R(\partial_x\phi)^2 - \gamma R\sin(2\phi), \quad (19)$$

One simple solution of this equation is the uniform-phase soliton [24, 26]

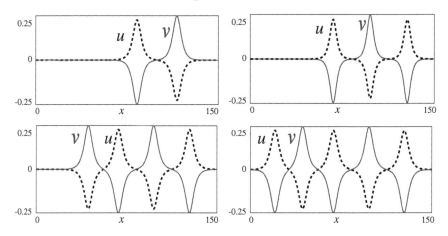

Fig. 5 Soliton-Antisoliton bound states. Snapshots of multiple localized solutions obtained from the LLG (5). Parameters are $g = -0.49781$, $h_a = -0.939895$, $\beta_x = 0.5$, $\beta_z = 1$, and $\alpha = 0.05$

$$\sin(2\phi_s) = \mu/\gamma,$$
$$R_s(x) = \sqrt{2\delta}\,\text{sech}(\sqrt{\delta}x), \tag{20}$$

where $\delta = -\nu + \sqrt{\gamma^2 - \mu^2}$. The right panel of Fig. 4 shows the typical profile of solitons given by the set of (20). Soliton amplitude and width are controlled by the detuning and dissipation, or equivalently by the external field and the spin-polarized current. Typical widths are about 30 nm. The uniform phase solitons $\partial_x \phi = \partial_{xx} \phi = 0$ become unstable when the size of the system surpasses a critical value, and a phase structure emerges [27, 33]. According to our simulations, nanostrips are usually small enough to ensure the stability of the uniform phase state. The magnetization components take the following form for the soliton solution

$$m_x = \frac{2\beta_x + \beta_z - R_s^2(x)}{2\beta_x + \beta_z + R_s^2(x)},$$
$$\begin{pmatrix} m_y \\ m_z \end{pmatrix} = \frac{2R_s(x)\sqrt{2\beta_x + \beta_z}}{2\beta_x + \beta_z + R_s^2(x)} \begin{pmatrix} \cos\varphi_s \\ \sin\varphi_s \end{pmatrix}, \tag{21}$$

One experimentally accessible quantity in nanopillars is the magentoresistance. The magnetoresistance depends on the magnetic configuration, and it can be approximated in terms of the average magnetization m_x^{av} along the polarization of the current, $\delta r = (1 - m_x^{av})/2$. Hence, the soliton fingerprint is a magnetoresistance similar to the one of the parallel state which increases with the square of the amplitude, it is $\delta r \sim \delta = h_a + \beta_x + \beta_z/2 + \sqrt{(\beta_z/2)^2 - g^2}$. This is a prediction of the parametric equivalence [6].

3.3.3 Soliton-Antisoliton Bound States

In parametrically driven systems, multiple coexisting solitons can emerge and interact among them [29, 42–46] and with walls [45–47]. The force between two remote solitons has been studied experimentally and theoretically in [42]. This interaction is attractive for soliton-soliton pairs [42], it is, for states that can be approximated by $R(x) \approx R_s(x_1) + R_s(x_2)$ where x_1 and x_2 are the positions of soliton cores. As a result of the attraction, solitons collapse [42, 43]. On the other hand, for solitons-antisoliton pairs, where $R(x) \approx R_s(x_1) - R_s(x_2)$, the interaction is repulsive [42, 44]. At short ranges, a third possibility exists, namely the bound states [29]. In the case of magnetic nanostripes, multiple soliton-antisoliton solutions can be found. Figure 5 illustrates these states, obtained from direct integration of (5). The multiple bumps are observed in a wide region of the parameter space, while we have not observed the soliton-soliton pair for this system. It is worth noting that this multi-stability of solitary structures makes the nanostripe a magnetoresistive memory of multiple levels.

4 Conclusions

Magnetic media forced by spin-polarized currents are equivalent to parametrically driven systems. This equivalence explains the existence of a wide variety of states in nanopillars, such as patterns and solitons. In this chapter we reviewed the parametric equivalence of [6], and we applied it to nanostripes. This configuration admits a one-dimensional description, moreover it reduces significantly the critical currents at which self-organization emerges. We found that nanostripe geometry permits the existence of soliton-antisoliton bound states. The creation and manipulation of solitary structures could be important for further developments of memory technologies and information carrying. Work in this direction is in progress.

Acknowledgments I thank Marcel G. Clerc, and Ignacio Bordeu for fruitful discussion. I gratefully acknowledge financial support from Becas Conicyt 2012, Contract No. 21120878.

References

1. I.D. Mayergoyz, G. Bertotti, C. Serpico, *Nonlinear Magnetization Dynamics in Nanosystems* (Elsevier, Oxford, 2009)
2. A. Slavin, V. Tiberkevich, Nonlinear auto-oscillator theory of microwave generation by spin-polarized current. IEEE Trans. Magn. **45**, 1875 (2009)
3. L.F. Álvarez, O. Pla, O. Chubykalo, Quasiperiodicity, bistability, and chaos in the Landau-Lifshitz equation. Phys. Rev. B **61**, 11613 (2000)
4. J. Bragard, H. Pleiner, O.J. Suarez, P. Vargas, J.A.C. Gallas, D. Laroze, Chaotic dynamics of a magnetic nanoparticle. Phys. Rev. E **84**, 037202 (2011)

5. A.O. León, M.G. Clerc, S. Coulibaly, Dissipative structures induced by spin-transfer torques in nanopillars. Phys. Rev. E **89**, 022908 (2014)

6. A.O. León, M.G. Clerc, Spin-transfer-driven nano-oscillators are equivalent to parametric resonators. Phys. Rev. B **91**, 014411 (2015)

7. P.-B. He, W.M. Liu, Nonlinear magnetization dynamics in a ferromagnetic nanowire with spin current. Phys. Rev. B **72**, 064410 (2005)

8. R.E. Troncoso, A.S. Núñez, Thermally assisted current-driven skyrmion motion. Phys. Rev. B **89**, 224403 (2014)

9. F. Zhao, Z.D. Li, Q.Y. Li, L. Wen, G. Fu, W.M. Liu, Magnetic rogue wave in a perpendicular anisotropic ferromagnetic nanowire with spin-transfer torque. Ann. Phys. **327**, 2085 (2012)

10. J.C. Slonczewski, Emission of spin waves by a magnetic multilayer traversed by a current. J. Magn. Mat. Mag. **159**, L1 (1996)

11. L. Berger, Emission of spin waves by a magnetic multilayer traversed by a current. Phys. Rev. B **54**, 9353 (1996)

12. M.D. Stiles, J. Miltat, in *Spin Dynamics in Confined Magnetic Structures*, edited by B. Hillebrands, A. Thiaville (Springer, Berlin, 2006) vol. 3, Chap. 7

13. S.I. Kiselev, J.C. Sankey, I.N. Krivorotov, N.C. Emley, R.J. Schoelkopf, R.A. Buhrman, D.C. Ralph, Microwave oscillations of a nanomagnet driven by a spin-polarized current. Nature (London) **425**, 380 (2003)

14. D. Berkov, N. Gorn, Transition from the macrospin to chaotic behavior by a spin-torque driven magnetization precession of a square nanoelement. Phys. Rev. B **71**, 052403 (2005)

15. L.D. Landau, E.M. Lifshiftz , *Mechanics*, vol. 1 (Course of Theoretical Physics) (Pergamon Press 1976)

16. M. Faraday, On a peculiar class of acoustical figures; and on certain forms assumed by groups of particles upon vibrating elastic surfaces. Philos. Trans. R. Soc. Lond. **121**, 299 (1831)

17. J.W. Miles, Parametrically excited solitary waves. J. Fluid Mech. **148**, 451 (1984)

18. I.V. Barashenkov, E.V. Zemlyanaya, Traveling solitons in the damped-driven nonlinear Schrödinger equation. SIAM J. Appl. Math. **64**, 800 (2004)

19. I.V. Barashenkov, E.V. Zemlyanaya, T.C. van Heerden, Time-periodic solitons in a damped-driven nonlinear Schrödinger equation. Phys. Rev. E **83**, 056609 (2011)

20. I.V. Barashenkov, E.V. Zemlyanaya, Soliton complexity in the damped-driven nonlinear Schrödinger equation: stationary to periodic to quasiperiodic complexes. Phys. Rev. E **83**, 056610 (2011)

21. B. Denardo, B. Galvin, A. Greenfield, A. Larraza, S. Putterman, W. Wright, Observations of localized structures in nonlinear lattices: domain walls and kinks. Phys. Rev. Lett. **68**, 1730 (1992)

22. J.N. Kutz, W.L. Kath, R.-D. Li, P. Kumar, Long-distance pulse propagation in nonlinear optical fibers by using periodically spaced parametric amplifiers. Opt. Lett. **18**, 802 (1993)

23. S. Longhi, Stable multipulse states in a nonlinear dispersive cavity with parametric gain. Phys. Rev. E **53**, 5520 (1996)

24. I.V. Barashenkov, M.M. Bogdan, V.I. Korobov, Stability diagram of the phase-locked solitons in the parametrically driven, damped nonlinear Schrödinger equation. Europhys. Lett. **15**, 113 (1991)

25. S.R. Woodford, I.V. Barashenkov, Stability of the Bloch wall via the Bogomolnyi decomposition in elliptic coordinates. J. Phys. A: Math. Theory **41**, 185203 (2008)

26. M.G. Clerc, S. Coulibaly, D. Laroze, Localized states beyond the asymptotic parametrically driven amplitude equation Phys. Rev. E **77**, 056209 (2008); Parametrically driven instability in quasi-reversible systems. Int. J. Bifurc. Chaos **19**, 3525 (2009); Localized states and non-variational IsingBloch transition of a parametrically driven easy-plane ferromagnetic wire. Physica D **239**, 72 (2010)

27. M.G. Clerc, S. Coulibaly, M.A. Garcia-Nustes, Y. Zárate, Dissipative localized states with Shieldlike phase structure. Phys. Rev. Lett. **107**, 254102 (2011)

28. P. Coullet, T. Frisch, G. Sonnino, Dispersion-induced patterns. Phys. Rev. E **49**, 2087 (1994)

29. D. Urzagasti, D. Laroze, M.G. Clerc, S. Coulibaly, H. Pleiner, Two-soliton precession state in a parametrically driven magnetic wire. J. Appl. Phys. **111**, 07D111 (2012)
30. D. Urzagasti, D. Laroze, M.G. Clerc, H. Pleiner, Breather soliton solutions in a parametrically driven magnetic wire. Europhys. Lett. **104**, 40001 (2013)
31. M.G. Clerc, S. Coulibaly, D. Laroze, Localized waves in a parametrically driven magnetic nanowire. Europhys. Lett. **97**, 30006 (2012)
32. A.O. León, M.G. Clerc, S. Coulibaly, Traveling pulse on a periodic background in parametrically driven systems. Phys. Rev. E. **91**, 050901 (2015)
33. M.G. Clerc, M.A. Garcia-Ñustes, Y. Zárate, S. Coulibaly, Phase shielding soliton in parametrically driven systems. Phys. Rev. E. **87**, 052915 (2013)
34. D.V. Berkov, J. Miltat, Spin-torque driven magnetization dynamics: micromagnetic modeling. J. Magn. Mag. Mater. **320**, 1238 (2008)
35. J.C. Slonczewski, Currents and torques in metallic magnetic multilayers. J. Magn. Magn. Mag. **247**, 324 (2002)
36. J. Xiao, A. Zangwill, M.D. Stiles, Boltzmann test of Slonczewskis theory of spin-transfer torque. Phys. Rev. B **70**, 172405 (2004)
37. J. Xiao, A. Zangwill, M.D. Stiles, Macrospin models of spin transfer dynamics. Phys. Rev. B **72**, 014446 (2005)
38. J. Barnas, A. Fert, M. Gmitra, I. Weymann, V.K. Dugaev, From giant magnetoresistance to current-induced switching by spin transfer. Phys. Rev. B **72**, 024426 (2005)
39. S.-W. Lee, K.-J. Lee, Effect of angular dependence of spin-transfer torque on zero-field microwave oscillation in symmetric spin-valves. IEEE Trans. Magn. **46**, 2349 (2010)
40. W. Kim, S-W. Lee, K-J. Lee, J. Micromagnetic modelling on magnetization dynamics in nanopillars driven by spin-transfer torque. Phys. D **44**, 384001 (2011)
41. M. Lakshmanan, The fascinating world of the LandauLifshitzGilbert equation: an overview. Philos. Trans. R. Soc. A **369**, 1280 (2011)
42. M.G. Clerc, S. Coulibaly, N. Mujica, R. Navarro, T. Sauma, Soliton pair interaction law in parametrically driven Newtonian fluid. Philos. Trans. R. Soc. A **367**, 3213 (2009)
43. M.G. Clerc, S. Coulibaly, L. Gordillo, N. Mujica, R. Navarro, Coalescence cascade of dissipative solitons in parametrically driven systems. Phys. Rev. E **84**, 036205 (2001)
44. D. Urzagasti, A. Aramayo, D. Laroze, Solitonantisoliton interaction in a parametrically driven easy-plane magnetic wire. Phys. Lett. A **378**, 2614 (2014)
45. X. Wang, R. Wei, Observations of collision behavior of parametrically excited standing solitons. Phys. Lett. A **192**, 1 (1994)
46. W. Wang, X. Wang, J. Wang, R. Wei, Dynamical behavior of parametrically excited solitary waves in Faraday s water trough experiment. Phys. Lett. A **219**, 74 (1996)
47. L. Gordillo, M.A. Garcia-Nustes, Dissipation-Driven Behavior of Nonpropagating Hydrodynamic Solitons Under Confinement. Phys. Rev. Lett. **112**, 164101 (2014)

Hyper-Chaotic and Chaotic Synchronisation of Two Interacting Dipoles

D. Urzagasti, D. Becerra-Alonso, L.M. Pérez, H.L. Mancini
and D. Laroze

Abstract In the present work we study numerically the deterministic spin dynamics of two interacting anisotropic magnetic particles in the presence of a time dependent external magnetic field. The particles are coupled through their dipole-dipole interaction. The applied magnetic field is composed of a constant amplitude longitudinal component and other transversal with time dependent amplitude. The system is modelled by the dissipative Landau-Lifshitz equation. The different types of synchronisation have been studied finding that the system presents chaotic anti-synchronisation of the canonical component, for a wide range of parameters. Finally, we also found that the system exhibits phase hyper-chaotic synchronisation.

1 Introduction

Theoretical studies on the dynamics of magnetisation reversal are normally based either on the Landau-Lifshitz [1] equation or on the Landau-Lifshitz-Gilbert equation [2]. These models have been used in both, discrete [3–11] and continuous magnetic systems [12–15]. Nonlinear time-dependent problems in magnetism have been studied in numerous cases and recent accounts of the state of the art can be found in [16–19]. In particular, phase diagrams of the chaotic regions for magnetic particles

D. Urzagasti
Instituto de Investigaciones Físicas, UMSA, P.O. Box 8635, La Paz, Bolivia

D. Becerra-Alonso
Universidad Loyola Andalucía, 14004 Córdoba, Spain

L.M. Pérez · H.L. Mancini
Departamento de Física y Matemática Aplicada,Universidad de Navarra,
31080 Pamplona, Spain

D. Laroze (✉)
Instituto de Alta de Investigación, Universidad de Tarapacá, Casilla 7D, Arica, Chile
e-mail: dlarozen@uta.cl

D. Laroze
SUPA School of Physics and Astronomy, University of Glasgow, Glasgow G12 8QQ, UK

© Springer International Publishing Switzerland 2016   261
M. Tlidi and M.G. Clerc (eds.), *Nonlinear Dynamics: Materials,
Theory and Experiments*, Springer Proceedings in Physics 173,
DOI 10.1007/978-3-319-24871-4_20

in the presence of time dependent magnetic fields have been given and explored [8–11]. Some of these models show new possible roots and ranges of physical parameters in chaotic domains that could motivate new experiments in this area.

Applications on two-particle systems under constant magnetic fields can be found in [20–23]. For conservative systems, when the interaction between the particles is based on energy exchange, it is possible to obtain analytical expressions for the magnetisations [20]. However, if the anisotropy energy is included, the system becomes non-linear and the analytical solutions are non-tractable. Furthermore, the system can exhibit chaotic states in the conservative case [23]. Two interacting dipoles in the presence of an external homogeneous magnetic field were studied in [24]. The authors found that the total magnetisation is not conserved. The non-dissipative case is a fluctuating function of time with a strong dependence on the strength of the dipolar term. In the dissipative case there is a transient time before the total magnetisation reaches its constant value. However, no permanent chaotic states were found. Recently, we have studied the effect of a time dependent magnetic field on the magnetisation processes of two anisotropic particles [25]. The main dynamical indicators were the Lyapunov exponents [26, 27]. The dynamics of this system was represented in the parameter space, displaying regions with different spatio-temporal behaviour that in this case could be regular or chaotic. Different combinations of parameter values bring the system to solutions that could be periodic, multi-periodic or more complex oscillations with chaotic or hyper-chaotic dynamics (when two or more Lyapunov exponents are positive).

On the other hand, the interaction between the two particles or two coherent dynamical systems can produce different kinds of synchronisation (complete, phase, lag, etc.) [28]. Synchronisation between chaotic systems has been studied in temporal systems and in spatio-temporal systems with symmetries beyond the classical master-slave coupling [29, 30] and it has been widely reviewed in [31–33]. Also, the effects of anisotropies in the coupling have been studied [34]. It is now well known that synchronisation in the presence of anisotropies could enhance synchronisation [35] or suppress chaotic behaviour in 3D systems [36]. In hyper-chaotic systems, synchronisation is achieved without chaos suppression, but the complexity of the system could be modified [37, 38]. Other works regarding synchronism in hyper-chaotic models can be found in [39–43].

The aim of this chapter is to analyse the influence of a time dependent external magnetic field on the synchronisation in a system of two interacting anisotropic magnetic particles. We consider that particles are coupled through a dipole-dipole interaction. The applied external time-dependent field produces a periodic driving in the direction perpendicular to the main anisotropy direction. Phase diagrams indicating different oscillation states, and details about chaotic and hyper-chaotic synchronisation regimes are presented. The manuscript is organised as follows, in Sect. 2, the theoretical model is briefly described. In Sect. 3, the numerical results are provided and discussed. Finally, some conclusions are presented in Sect. 4.

2 Theoretical Model

Let us consider two anisotropic magnetic particles in the presence of an external magnetic field, $\mathbf{H}_{ext}$. We assume that each particle can be represented by a magnetic mono-domain of magnetisation $\mathbf{M}_i$ with $i = (1, 2)$. This approximation is called *macrospin* approximation. The temporal evolution of the system can be modelled by the Landau-Lifshitz equation [1]

$$\frac{d\mathbf{M}_i}{dt} = -|\gamma|\mathbf{M}_i \times \mathbf{H}_i - \frac{\eta|\gamma|}{M_{iS}}\mathbf{M}_i \times (\mathbf{M}_i \times \mathbf{H}_i), \qquad (1)$$

where, γ is the gyromagnetic factor, which is associated with the electron spin and is approximately given by $|\gamma| = |\gamma_e|\mu_0 \approx 2.21 \times 10^5\,\mathrm{m\,A^{-1}\,s^{-1}}$. In the above equation, η denotes the dimensionless phenomenological damping coefficient that is characteristic of the material and has a typical value ranging from 10^{-4} to 10^{-3} in garnets and 10^{-2} or larger in cobalt or permalloy [44]. We assume that the coupling between the particles is the dipole-dipole interaction, hence the effective magnetic field for each particle, $\mathbf{H}_i$, is given by

$$\mathbf{H}_i = \mathbf{H}_{ext} + \beta_i \left(\mathbf{M}_i \cdot \hat{\mathbf{n}}_i\right) \hat{\mathbf{n}}_i + d^{-3} \left[3(\mathbf{M}_k \cdot \hat{\mathbf{r}})\hat{\mathbf{r}} - \mathbf{M}_k\right], \qquad (2)$$

being $(i, k) = 1, 2$ such that $i \neq k$. Here β_i measures the anisotropy along the $\mathbf{n}_i$ axis, d is the fixed distance between the two magnetic moments, and $\hat{\mathbf{r}}$ is a unit vector along the direction between the two particles. Notice that this special type of anisotropy is called uniaxial anisotropy and the constants β_i can be positive or negative depending on the specific substance and sample shape [44] in use. Let us assume that the particles have the same magnitude $M_{s,1} = M_{s,2} = M_s$ and the same anisotropy $\beta_1 = \beta_2 = \beta$ and $\mathbf{n}_1 = \mathbf{n}_2 = \mathbf{n}$. We apply an external magnetic field $\mathbf{H}_{ext}$ that comprises both, a constant longitudinal and a periodic transverse part with a fixed amplitude and frequency

$$\mathbf{H}_{ext} = \mathbf{H}_0 + \mathbf{H}_T \sin(\omega t), \qquad (3)$$

where $\mathbf{H}_0$ ($\|\hat{\mathbf{z}}$), $\mathbf{H}_T$ ($\perp\hat{\mathbf{z}}$) and ω are time independent. The axis $\hat{\mathbf{r}}$ is chosen perpendicular to the anisotropy axis, in particular $\hat{\mathbf{r}} = \hat{\mathbf{x}}$. We note that for zero damping, i.e. $\eta = 0$, and without parametric forcing, i.e. $\mathbf{H}_T = \mathbf{0}$, (1) is conservative. Hence, the dissipation and the oscillatory injection of energy move the magnetic particles in and out of the equilibrium. In such a circumstance the magnetisation of the particles may exhibit complex behaviour such as quasi-periodicity, and chaos [3, 7–9, 11].

Finally, let us remark that for each particle, the LL equation has a constant of motion, which is the magnetic modulus $|\mathbf{M}_i|$ [1]. This feature allows the description of the system with two effective variables. Hence, using spherical coordinates, $\mathbf{M}_i = M_{s,i} (\cos\phi_i \sin\theta_i, \sin\phi_i \sin\theta_i, \cos\theta_i)^T$, one can find that the canonical variables are $\{\phi_i, \cos\theta_i\} = \{\phi_i, m_{z,i}\}$ [18]. Therefore, these variables will be used for the synchronisation analysis.

3 Simulations

This section is divided in three parts. The first part briefly discusses the quantities used to characterise the different types of synchronisation. The second one, shortly comments the numerical protocols and the equation to be integrated. The last part covers some numerical results and the corresponding analysis.

3.1 Synchronisation and Dynamical Indicators

We study different types of synchronisations measuring the amplitudes $\{m_{z,i}\}$ and the angles $\{\phi_i\}$ with $i = (1, 2)$. The *common-signal based synchronisation* proposed in [29] can be extended to this case, where the interacting dipoles are enough to find different behaviours and degrees of synchronisation. In particular, two measurements are used in the present work to determine synchronisation.

The first one is the Pearson's coefficient. It has been previously used in chaotic systems [35], that in our case can be written as:

$$\rho_z = \frac{\langle (m_{z1} - \langle m_{z1} \rangle)(m_{z2} - \langle m_{z2} \rangle) \rangle}{\sqrt{\langle (m_{z1} - \langle m_{z1} \rangle)^2 \rangle} \sqrt{\langle (m_{z2} - \langle m_{z2} \rangle)^2 \rangle}}. \tag{4}$$

This coefficient can vary from -1 to 1. These values represent the complete anti-synchronised and synchronised regime, respectively. In the case of $\rho_z = 0$ the system is not synchronised.

In addition, the *phase synchronisation* will be used for interpretation [30, 31]. The similarity function needed in our case is given by:

$$\Delta\Omega = |\Omega_1 - \Omega_2|, \tag{5}$$

where $\Omega_i = \langle \dot{\phi}_i \rangle$ represents the averaged rate of change of ϕ_i. Note that when $\Delta\Omega = 0$, the system has a phase-synchronised state. Also, this is equivalent to $\Omega_1/\Omega_2 = 1$.

On the other hand, in order to characterise the dynamics of (1) the Lyapunov exponents (LEs) are evaluated. This method consists on quantifying the divergence between two initially close trajectories of a vector field [26, 27]. In general, for an effective N-dimensional dynamical system described by a set of equations, $dX^i/d\tau = F^i(\mathbf{X}, \tau)$, the ith-Lyapunov exponent is given by

$$\lambda_i = \lim_{\tau \to \infty} \left(\frac{1}{\tau} \ln \left(\frac{\|\delta X^i_\tau\|}{\|\delta X^i_0\|} \right) \right), \tag{6}$$

where $\|\delta X_{\xi}^{i}\|$ is the distance between the trajectories of the ith-component of the vector field at time ξ. They can be ordered by decreasing amplitude: $\lambda_1 \geq \lambda_2 \geq \cdots \geq \lambda_N$. The first two exponents are the largest Lyapunov exponent (LLE) and the second largest Lyapunov exponent (SLLE). Due to the fact that the LLG equation conserves the modulus of each particle $|\mathbf{m}_i|$ and that the applied magnetic field is time dependent, the effective dimension of the phase space is five. From a dynamical system point of view, more than one exponent may become positive for a system of dimension five. Therefore, by exploring the dependence of the LLE on the different parameters of the system, one can identify areas in the control parameter space, where the dynamics is chaotic (LLE positive), and those showing non-chaotic dynamics (LLE vanishing or negative). In addition, when both the LLE and the SLLE have positive values the system is at a *hyper-chaotic* regime. Nevertheless, since this is a one-frequency forced system, at least one of its Lyapunov exponents will always be zero; hence the simplest attractor is a periodic orbit. Another possibility is to have two or three Lyapunov exponents equal to zero. In these cases the system exhibits a two or three-frequency quasi-periodic behaviour. The Lyapunov exponents are presented in the form of 2-D maps as a function of the relevant parameters of the system [27].

The detailed analysis of the parameter space of the Lyapunov exponents for this system was previously presented in a recent work [25]. Hence, here we focus on the synchronised behaviours. We combined both indicators to quantify the synchronised dynamical regimes. For instance, for a fixed set of parameters if $\rho_z \approx -1$ and $\lambda_1 > 0$ the dipoles are in an anti-synchronised chaotic state; or if $\Delta\Omega = 0$ and $\lambda_1 > 0$ the system is in a chaotic phase-synchronised state.

3.2 Numerical Protocols

In order to simplify and speed-up the integration of the equations of motion, dimensionless units are used. This recasts (1) in terms of the magnetisation $\mathbf{m}_i = \mathbf{M}_i/M_s$ and time $\tau = t|\gamma|M_s$ [18]. This normalisation leads to $|\mathbf{m}_i| = 1$. In order to get a better physical insight into the problem, let us evaluate the scales introduced here. Typical values are, e.g. for cobalt materials, $M_s \approx 1.42 \times 10^6$ A/m [18], hence the time scale ($\tau = 1$) is in the picosecond range, $t_s = 1/(|\gamma|M_s) \approx 3.2$ ps. The dimensionless field is $\tilde{\mathbf{h}} = \mathbf{H}/M_s$ with $\tilde{h}_{x,y} \equiv h_{x,y}\sin(\varphi\tau)$, and $\tilde{h}_z \equiv h_z$ where $\varphi = \omega/(|H|M_s)$ is the dimensionless frequency. In order to avoid numerical artefacts [3], it is suitable to solve (1) in the Cartesian representation, namely,

$$\frac{dm_{x,i}}{d\tau} = -(m_{y,i} + \eta\, m_{x,i}m_{z,i})h_z + \left\{(m_{z,i} - \eta\, m_{x,i}m_{y,i})h_y + \eta\,(m_{y,i}^2 + m_{z,i}^2)h_x\right\}\sin(\varphi\tau)$$

$$+ d^{-3}(m_{z,k} + \eta\, m_{y,k}m_{x,i})m_{y,i} - \left\{\beta\, m_{y,i} + d^{-3}(m_{y,k} - \eta\, m_{z,k}m_{x,i})\right\}m_{z,i}$$

$$+ 2\eta\, d^{-3}m_{x,k}m_{y,i}^2 - \eta\,(\beta\, m_{x,i} - 2d^{-3}m_{x,k})m_{z,i}^2 , \tag{7}$$

$$\frac{dm_{y,i}}{d\tau} = (m_{x,i} - \eta\, m_{y,i} m_{z,i}) h_z - \left\{ (m_{z,i} + \eta\, m_{x,i} m_{y,i}) h_x - \eta\,(m_{z,i}^2 + m_{x,i}^2) h_y \right\} \sin(\varphi\tau)$$

$$- \left\{ \beta\, m_{y,i} + d^{-3} m_{y,k} \right\} \eta\, m_{z,i}^2 - d^{-3}(m_{z,k} + 2\eta\, m_{x,k} m_{y,i}) m_{x,i} - \eta\, d^{-3} m_{y,k} m_{x,i}^2$$

$$+ \left\{ \beta\, m_{x,i} - d^{-3}(2 m_{x,k} - \eta\, m_{z,k} m_{y,i}) \right\} m_{z,i}\,, \tag{8}$$

$$\frac{dm_{z,i}}{d\tau} = \left\{ (m_{y,i} - \eta\, m_{x,i} m_{z,i}) h_x - (m_{x,i} + \eta\, m_{y,i} m_{z,i}) h_y \right\} \sin(\varphi\tau)$$

$$+ d^{-3} \left\{ (m_{x,i} m_{y,k} + 2 m_{x,k} m_{y,i}) + \eta\,(m_{y,k} m_{y,i} - 2 m_{x,k} m_{x,i}) m_{z,i} \right\}$$

$$+ \eta\,(m_{x,i}^2 + m_{y,i}^2) \left\{ h_z + \beta\, m_{z,i} - d^{-3} m_{z,k} \right\}. \tag{9}$$

Equations (7)–(9) have been integrated by using a fifth order Runge-Kutta integration scheme with a variable step [45] that ensures a relative error of 10^{-7} on the magnetisation fields.

To take the time average in ρ_z and in $\Delta\Omega$ we evolve the system until its initial transient time has been finished and the system is in the persistent state. For both indicators we establish a tolerance lower than 3 % in order to consider the system in a synchronised state. Also, the LEs are calculated for a time span of $\tau = 80,000$ after an initial transient time of $\tau = 4000$ has been discarded. The Gram-Schmidt orthogonalisation process is performed after every $\delta\tau = 3.91$. The error has been estimated to be $E = 0.1\,\%$, which is sufficiently small for the purpose of the present analysis. Due to the large number of parameters involved in the system, $\beta = 1$, $\eta = 0.05$, $h_y = 1$ and $\varphi = 0.5$ will be fixed throughout the calculations. Parameters $\{h_x, h_z, d\}$ will vary depending on the simulation presented.

3.3 Numerical Results

In a system like the one presented in this chapter, synchronised periodic regimes are expected both when the external field or when the dipolar interaction dominates the dynamics. Whenever the external field is strong with respect to the dipolar interaction, dipoles can be expected to be influenced by the field in a synchronised manner. On the other hand, when dipolar interaction dominates, the field can do nothing but slightly perturb the tight orbits of the dipoles, and again, do so in synchronisation. This subsection will however show a broader variety of synchronised states and transitions.

The main results are given in Figs. 1, 2, 3 and 4. In particular, we concentrate the discussion on the influence of the driving field and the distante between particles. Figure 1a shows a color code phase diagram of the Pearson's coefficient ρ_{m_z} as a function of the field amplitudes h_x and h_z. The distance $d = 0.8$ has been chosen for being on the interphase between field prevalence and dipole interaction prevalence. The $h_x \lesssim 1.5$ periodic region shows a very diverse collection of interlaced periodic

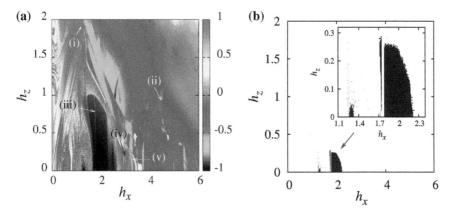

Fig. 1 **a** Phase diagram displaying ρ_z color coded as a function of the field amplitudes h_x and h_z. **b** Region of chaotic anti-synchronisation in frame (**a**) with $|1 + \rho_{m_z}| < 3\%$. The fixed parameters are: $\varphi = 0.5$, $\eta = 0.05$, $\beta = 1$, $d = 0.8$ and $h_y = 1$ (Color online)

synchronised and unsynchronised states. Higher h_z in this region mostly have ρ_{m_z} equal to 1 or close to it. Anti-synchronisation occurs in a smaller subregion for $\{0 \lesssim h_z \lesssim 0.2, 1 \lesssim h_x \lesssim 1.5\}$. The chaotic tongue-shaped region within $\{0 \lesssim h_z \lesssim 1, 1.5 \lesssim h_x \lesssim 2.5\}$, is anti-synchronised. This means that entering the tongue, from, for instance, lower values of h_x implies a transition from periodic unsynchronised regimes, to chaotic anti-synchronised ones. The detail of anti-synchronised chaos can be seen in Fig. 1b. However, there are values at the top h_z end of the tongue, and at the top h_x end of it where the dipoles go from synchronised periodicity to anti-synchronized chaos. This interesting transition to anti-synchronisation, precisely when the system earns a degree of instability due to chaos, is at this point unintuitive to explain in terms of analysis. The remaining chaotic region with $h_z \gtrsim 0.8(3.5 - h_x)$ is predominantly unsynchronised in terms of the Pearson's coefficient.

Several periodic-state parameters were indicated in Fig. 1a. Their trajectories of the first particle and the corresponding Fourier spectrum of $m_{z,1}$ are detailed in Fig. 2. We note that all trajectories presented have a high number of Fourier peaks. The simplest one is row (i), still found in the $h_z \lesssim 0.8(3.5 - h_x)$ region, where simple dynamics are more common (except for the chaotic tongue). Of the five cases presented, row (ii) has the most Fourier peaks. The figures show an intricate periodic orbit where the two dipoles are in antisymmetry. The other cases are all symmetric periodic orbits where one of the m_x poles prevails over the other, depending on the initial conditions. In addition, we can observe that there are periodic synchronisation in cases (i), (iv) and (v).

The first three frames of Fig. 3 show color code phase diagrams of the Pearson's coefficient as a function of the field amplitude h_x and the distance d. The Pearson's coefficient exhibits a complex topology in the parameters space. Multiple transitions

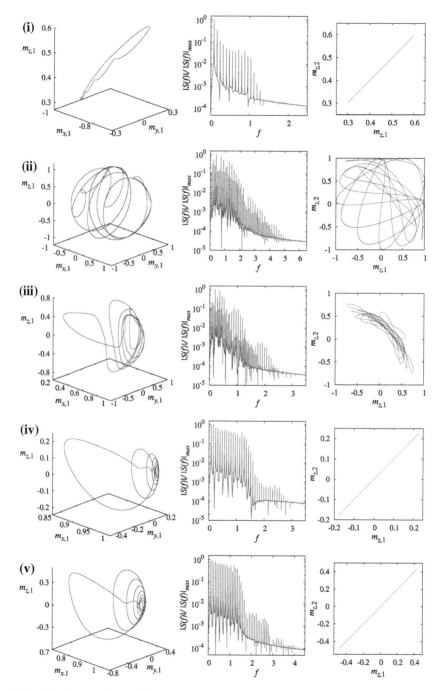

Fig. 2 3D trajectory of $\mathbf{m}_1$ (column 1), normalised FTT of $m_{z,1}$ (column 2) and parametric plot of $(m_{z,1}, m_{z,2})$ (column 3) of the instances indicated in Fig. 1a

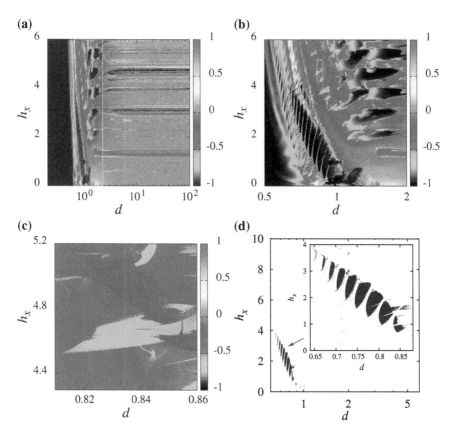

Fig. 3 **a** Phase diagram displaying ρ_{m_z} coded as a function of the field amplitude h_x and the distance d. **b** Magnification of the *white* box in (**a**). **c** Magnification of the white box in (**b**). **d** Zoom on the region (**b**) with chaotic anti-synchronisation with $|1 + \rho_{m_z}| < 3\%$. The fixed parameters are: $\varphi = 0.5$, $\eta = 0.05$, $\beta = 1$, $h_y = 1$ and $h_z = 0.1$ (Color online)

between synchronised to non-synchronised regimes can be observed. In fact, it outlines the similar features found in the LLE results of [25]. Frames (b) and (c) are successive magnifications of frame (a) for intermediate values of d. The characteristic shapes within $h_x \lesssim 2.0(1 - d)$ (as seen in frame (b)) are predominantly anti-synchronised, although sparse regimes of synchronised chaos are also found as it is shown in frame (d). The zoom presented in Fig. 3 frame (c) shows a granulated parametric space tending to anti-synchronisation, surrounded by non-synchronised regimes. Hyper-chaotic synchronisation was not found for these set of parameters.

Finally, Fig. 4 shows the phase synchronisation as a function of field amplitude h_x and the distance d. Here we focus on intermediate values of the distance. Phase synchronisation is prevalent for most of the range studied, with the exception of islands past $d \gtrsim 1$ (frames (d) and (e)). There is almost no phase synchronisation

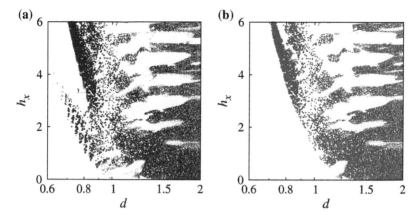

Fig. 4 Region of chaotic (**a**) and hyper-chaotic syncronisation (**b**) as a function of the field amplitude h_x and the distance d with $|1 - \Omega_1/\Omega_2| < 3\%$. The fixed parameters are: $\varphi = 0.5$, $\eta = 0.05$, $\beta = 1$, $h_y = 1$ and $h_z = 0.1$ (Color online)

for $1 \lesssim d \lesssim 1.5$ and $0 \lesssim h_x \lesssim 0.1$. We can observe that there are hyper chaotic phase-synchronised states as it shown in frame (b). These regions are subsets of the chaotic ones.

4 Final Remarks

The magnetisation dynamics of two anisotropic magnetic particles interacting via dipolar interaction in the presence of a constant longitudinal external magnetic field and a periodic transversal field has been studied using the dissipative Landau-Lifshitz equation. The system presents a variety of regular and (hyper-)chaotic regimes, such that the regular regions consist on either periodic or quasi-periodic solutions [25]. We have focused on the synchronisation phenomena. The different types of synchronisation have been measured using two main indicators: the Pearson's coefficient associated to the z-components of the magnetic moments and the difference between the phases. The latter is common for most of the parametric regimes presented, which is expectable whenever dipole interaction is predominant. However, the measurement of the Pearson's coefficient has provided with the analysis of four possible states: chaotic synchronisation, hyper-chaotic synchronisation, chaotic anti-synchronisation, and hyper-chaotic anti-synchronisation. The last one has not been found. Regarding the other three states: hyper-chaos is of course a subset of chaos in the parametric space. Synchronised hyper-chaos is never found in the vicinity of anti-synchronised chaos. Chaos-periodicity transitions, in terms of synchronisation, are also diverse. Periodic synchronisation is borderline with chaotic anti-synchronised states, as the field is increased both in the h_x and h_z directions.

Acknowledgments DU acknowledges the PhD fellowship from the Performance Agreement Project UTA/Mineduc (Universidad de Tarapacá). DBA was supported in part by the Spanish Inter-Ministerial Commission of Science and Technology under Project TIN2014-54583-C2-1-R, the European Regional Development fund, and the "Junta de Andalucía" (Spain), under Project P2011-TIC-7508. LMP and HLM acknowledge partial financial support from the Spanish Ministry of Science and Technology under Contract No. FIS2011-24642 and FIS2014-54101. DL acknowledges partial financial support from Basal Program Center for Development of Nanoscience and Nanotechnology (CEDENNA), UTA-project 8750-12 and Engineering and Physical Sciences Research Council Grant No. EP/L002922/1.

References

1. L. Landau, *Collected Papers of Landau* (Pergamon, New York, 1965)
2. T.L. Gilbert, A phenomenological theory of damping in ferromagnetic materials. IEEE Trans. Mag. **40**, 11613 (2004)
3. L.F. Alvarez, O. Pla, O. Chubykalo, Quasiperiodicity, bistability, andchaosinthe Landau-Lifshitz equation. Phys. Rev. B **61**, 11613 (2000)
4. D. Laroze, L.M. Perez, Classical spin dynamics of four interacting magnetic particles on a ring. Phys. B **403**, 473 (2008)
5. P. Diaz, D. Laroze, Configurational temperature for interacting anisotropic magnetic particles. Int. J. Bifurcat. Chaos **19**, 3485 (2009)
6. D.V. Vagin, P. Polyakov, Control of chaotic and deterministic magnetization dynamics regimes by means of sample shape varying. J. Appl. Phys. **105**, 033914 (2009)
7. R.K. Smith, M. Grabowski, R.E. Camley, Period doubling toward chaos in a driven magnetic macrospin. J. Magn. Magn. Mater. **322**, 2127 (2010)
8. J. Bragard, H. Pleiner, O.J. Suarez, P. Vargas, J.A.C. Gallas, D. Laroze, Chaotic dynamics of a magnetic nanoparticle. Phys. Rev. E **84**, 037202 (2011)
9. D. Laroze, J. Bragard, O.J. Suarez, H. Pleiner, Characterization of the chaotic magnetic particle dynamics. IEEE Trans. Mag. **47**, 10 (2011)
10. D. Laroze, D. Becerra-Alonso, J.A.C. Gallas, H. Pleiner, magnetization dynamics under a quasiperiodic magnetic field. IEEE Trans. Magn. **48**, 3567 (2012)
11. L.M. Pérez, J. Bragard, H.L. Mancini, J.A.C. Gallas, A.M. Cabanas, O.J. Suarez, D. Laroze, Netw. Heterog. Media **10**, 209 (2015)
12. A.M. Kosevich, B.A. Ivanov, A.S. Kovalev, Magnetic solitons. Phys. Rep. **194**, 117 (1990)
13. I.V. Barashenkov, M.M. Bogdan, V.I. Korobov, Stability diagram of the phase-locked solitons in the parametrically driven, damped nonlinear Schrödinger equation. Europhys. Lett. **15**, 113 (1991)
14. M.G. Clerc, S. Coulibaly, D. Laroze, Localized states and non-variational Ising-Bloch transition of a parametrically driven easy-plane ferromagnetic wire. Phys. D **239**, 72 (2010)
15. A.O. León, M.G. Clerc, S. Coulibaly, Traveling pulse on a periodic background in parametrically driven systems. Phys. Rev. E. **91**, 050901 (2015)
16. P.E. Wigen (ed.), *Nonlinear Phenomena and Chaos in Magnetic Materials* (World Scientific, Singapore, 1994)
17. B. Guo, S. Ding, *Landau Lifshitz Equations* (World Scientific, Singapore, 2008)
18. I.D. Mayergoyz, G. Bertotti, C. Serpico, *Nonlinear Magnetization Dynamics in Nanosystems* (Elsevier, Dordrech, 2009). and references therein
19. M. Lakshmanan, The fascinating world of the Landau-Lifshitz-Gilbert equation: an overview. Phil. Trans. R. Soc. A **369**, 1280 (2011)
20. D. Mentrup, J. Schnack, M. Luban, Spin dynamics of quantum and classical Heisenberg dimers. Phys. A **272**, 153 (1999)

21. D.V. Efremov, R.A. Klemm, Heisenberg dimer single molecule magnets in a strong magnetic field. Phys. Rev. B **66**, 174427 (2002)
22. D. Laroze, P. Vargas, Dynamical behavior of two interacting magnetic nanoparticles. Phys. B **372**, 332 (2006)
23. L.M. Pérez, O.J. Suarez, D. Laroze, H.L. Mancini, Classical spin dynamics of anisotropic Heisenberg dimers Cent. Eur. J. Phys. **11**, 1629 (2013)
24. D. Laroze, P. Vargas, C. Cortes, G. Gutierrez, Dynamics of two interacting dipoles. J. Magn. Magn. Mater. **320**, 1440 (2008)
25. D. Urzagasti, D. Becerra-Alonso, L.M. Pérez, H.L. Mancini, D. Laroze, *Hyper-chaotic magnetisation dynamics of two interacting dipoles* (Submitted)
26. A. Wolf, J.B. Swift, H.L. Swinney, J.A. Vastano, Determining Lyapunov exponents from a time series. Phys. D **16**, 285 (1985)
27. J.A.C. Gallas, The structure of infinite periodic and chaotic hub cascades in phase diagrams of simple autonomous flows. Int. J. Bifur. Chaos **20**, 197 (2010). and references therein
28. A. Pikovsky, M. Rosenblum, J. Kurths, *Synchronization* (A Universal Concept in Nonlinear Sciences). (Cambridge University Press, Cambridge, 2001)
29. L.M. Pecora, T.L. Carroll, Synchronization in chaotic systems. Phys. Rev. Lett. **64**, 821 (1990)
30. A.S. Pikovsky, M.G. Rosenblum, J. Kurths, Synchronization in a population of globally coupled chaotic oscillators. Europhys. Lett. **34**, 165 (1996)
31. M.G. Rosenblum, A.S. Pikovsky, J. Kurths, From phase to lag synchronization in coupled chaotic oscillators. Phys. Rev. Lett. **78**, 4193 (1997)
32. S. Boccaletti, J. Kurths, G. Osipov, D. Valladares, C. Zhou, The synchronization of chaotic systems. Phys. Rep. **336**, 1 (2002)
33. S. Boccaletti, *The Synchronized Dynamics of Complex Systems* (Elsevier, Dordrech, 2009)
34. J. Bragard, S. Boccaletti, H. Mancini, Asymmetric coupling effects in the synchronization of spatially extended chaotic systems. Phys. Rev. Lett. **91**, 064103 (2003)
35. J. Bragard, S. Boccaletti, C. Mendoza, H.G.E. Hentschel, H. Mancini, Synchronization of spatially extended chaotic systems in the presence of asymmetric coupling. Phys. Rev. E **70**, 036219 (2004)
36. J. Bragard, G. Vidal, H. Mancini, C. Mendoza, S. Boccaletti, Chaos suppression through asymmetric coupling. Chaos **17**, 043107 (2007)
37. G. Vidal, H. Mancini, Hyperchaotic synchronization under square symmetry. Int. J. Bifurcat. Chaos **19**, 719 (2009)
38. G. Vidal, H. Mancini, Hyperchaotic synchronization. Int. J. Bifurcat. Chaos **20**, 885 (2010)
39. Y-C. Lai, Synchronism in symmetric hyperchaotic systems, Phys. Rev. E **55**, R4861 (1997)
40. L. Yaowen, G. Guangming, Z. Hong, W. Yinghai, G. Liang, Synchronization of hyperchaotic harmonics in time-delay systems and its application to secure communication. Phys. Rev. E **62**, 7898 (2000)
41. A.A. Budini, Langevin approach to synchronization of hyperchaotic time-delay dynamics. J. Phys. A: Math. Theor. **41**, 445001 (2008)
42. C. Li, Y. Tong, H. Li, K. Su, Adaptive impulsive synchronization of a class of chaotic and hyperchaotic systems. Phys. Scr. **86**, 055003 (2012)
43. H.-M. Li, C.-L. Li, Switched generalized function projective synchronization of two identical/different hyperchaotic systems with uncertain parameters. Phys. Scr. **86**, 045008 (2012)
44. R.C. O'Handley, *Modern Magnetic Materials: Principles and Applications* (Wiley-Interscience, USA, 1999)
45. W.H. Press, S.A. Teukolsky, W.T. Vetterling, B.P. Flannery, *Numerical Recipes in FORTRAN* (Cambridge University Press, UK, 1992)

Part III
Robust Phenomena

Finger Dynamics in Pattern Forming Systems

Ignacio Bordeu, Marcel G. Clerc, René Lefever and Mustapha Tlidi

Abstract Macroscopic systems subjected to external forcing exhibit complex spatiotemporal behaviors as result of dissipative self-organization. We consider a paradigmatic Swift-Hohenberg equation in both variational and nonvariational forms in two-dimensions. We investigate in both equations the occurrence of a curvature instability generating spot multiplication, self-replication or fingering instability that affects the circular shape of two-dimensional localized structures. We show that when increasing the radius of localized structures, the first angular index m to become unstable is $m = 2$. This mode corresponds to an elliptical deformation of the circular shape of localized structures. We show also that for a fixed value of the radius of localized structures, the mode $m = 2$ becomes unstable for small values of the diffusion coefficient and higher modes become unstable for large angular index m. These result hold for both variational and nonvariational models. In addition, we analyze the stability of a single stripe localized structure and extended pattern for both variational and nonvariational Swift-Hohenberg equations.

1 Introduction

Localized structures (LSs) are nonlinear bright or dark peaks in spatially extended systems. They belong to the general class of dissipative structures found far from equilibrium [1, 2]. The conditions under which LSs and periodic patterns appear are

I. Bordeu (✉)
Departamento de Física, Facultad de Ciencias, Universidad de Chile, Santiago, Chile
e-mail: ibordeu@gmail.com

M.G. Clerc
Departamento de Física, Facultad de Ciencias Físicas y Matemáticas, Universidad de Chile, Casilla 487-3, Santiago, Chile
e-mail: mclerc@dfi.uchile.cl

R. Lefever · M. Tlidi
Faculté des Sciences, Université Libre de Bruxelles (U.L.B.), CP 231, Campus Plaine, 1050 Bruxelles, Belgium

© Springer International Publishing Switzerland 2016
M. Tlidi and M.G. Clerc (eds.), *Nonlinear Dynamics: Materials,*
Theory and Experiments, Springer Proceedings in Physics 173,
DOI 10.1007/978-3-319-24871-4_21

closely related. Typically, when the spatial symmetry breaking or Turing instability [3] becomes subcritical, there exists a pinning domain where localized structures are stable [4–7]. They occur in various fields of nonlinear science, such as chemistry [8], plant ecology [9, 10, 13], fluid dynamics [11, 12], magnetic systems [14], and optics [15–17]. Such LSs have been observed in the transverse section of coherently driven optical cavities and are often called cavity solitons. Currently they are attracting growing interest in the field of optics due to potential applications for all-optical control of light, optical storage, and information processing.

In two-dimensions, the circular shape of a single LS may exhibits a deformation and splitting leading to transition towards extended patterns that occupy the whole space available in (x, y) plane. This intriguing phenomenon is referred to as spot multiplication, self-replication or fingering instability depending on the authors. The question of stability of a LS is then central and the origin of curvature instabilities must be carefully examined. In the early nineties, it was reported numerically that spot multiplication phenomenon is possible in chemical systems [18]. Soon after and thanks to the development of open chemical reactors, spot multiplication was observed in various experiments such as ferrocyanide-iodate-sulphite reaction [19], Belousov-Zhabotinsky reaction [20–23], chloride dioxide-malonic-acid reaction [24]. This phenomenon is common to many nonequilibrium systems and occurs in other systems such as biology [28], material science [29, 30], and nonlinear optics [31]. In the limit of a large diffusivity ratio, spot like solutions in the two-dimensional Belousov-Zhabotinski reaction has been studied. It is shown analytically that for spots of small radius, the instability leads to a spot splitting into precisely two spots.

The patterns produced by curvature instability—spot multiplication, self-replication or fingering instability—are characterized by exhibiting elaborate textures. In the early nineties, fingering instability leading to the formation of labyrinthine pattern formation was experimentally observed in magnetic fluids [32] and cholesteric liquid crystals [33, 34]. The dynamic of these systems is governed by the minimization of a free energy. These are usually denominated variational or gradient systems. Theoretical studies were performed by use of non-variational models, that is, systems where its dynamics are not governed by the minimization of a functional. In this type of systems one expects to observe permanent dynamics such as propagative, oscillatory and chaotic state.

In this chapter, we investigate the curvature instability phenomenon in the framework of a Swift-Hohenberg equation, considering a variational and non-variational generalization.

2 Finger Instability of Localized Structure

In this section we focus on the the variational Swift-Hohenberg model. The Swift-Hohenberg equation is a well-known paradigm in the study of spatial periodic or localized patterns. It has been considered for that purpose in various field on nonlinear science such as hydrodynamics [35] , chemistry [36], plant ecology [37], and nonlinear optics [38]. This real order parameter equation generically applies to systems that undergo a symmetry breaking instability (often called Turing instability, [3]) close to a second-order critical point marking the onset of a hysteresis loop. The Swift-Hohenberg equation reads

$$\partial_t u = \eta + \varepsilon u - u^3 - \nu \nabla^2 u - \nabla^4 u \tag{1}$$

where $u = u(x, y, t)$ is a real scalar field, x and y are spatial coordinates and t is time. Depending on the context in which this equation has been derived, the physical meaning of the field variable $u(x, y, t)$ could be the electric field, phytomass density, or chemical concentration. The control or the bifurcation parameter η measures the input field amplitude, the aridity parameter, or chemical concentration. This parameter breaks the symmetry $u \to -u$ thus it accounts for the asymmetry between homogeneous states. The parameters ν and ε, are the diffusion coefficient and the cooperativity or bifurcation parameter, respectively. The local part of the equation— first three terms of the right side—account for the nascent bistability. Note that the model (1) is a variational equation so that

$$\partial_t u = -\frac{\delta f}{\delta u}, \tag{2}$$

where

$$f = \int\limits_{\infty}^{+\infty} dxdy \left[-\eta u - \frac{\varepsilon}{2} u^2 + \frac{1}{4} u^4 - \frac{\nu}{2} (\nabla u)^2 + \frac{1}{2} (\nabla^2 u)^2 \right]. \tag{3}$$

This implies that the equilibria of the Swift-Hohenberg equation are time independent, that is, the dynamics of this equation is characterized by minimizing f and the equilibria do not exhibit permanent dynamic. Equation (1) is well known for exhibiting localized solution and patterns [5]. For small ε, the model only exhibits one homogeneous equilibrium. The bifurcation diagram of the above model in $\{\varepsilon, \eta\}$ parameter space is shown in Fig. 1, where curves Γ_1 and Γ_2 represent the imperfect pitchfork bifurcation and the spatial instability of the homogeneous solution, respectively [6]. Hence, for positive ε the system presents coexistence between homogeneous and pattern state. In this parameter region one expects to observe localized pattern (or localized structure) [39]. Left and top panel of Fig. 2 show a typical localized structure of model (1), there is no analytical expression known for this solution.

Γ_3 curve, in Fig. 2, represent the bifurcation curve where a localized structure becomes unstable and give rise to a disordered periodic structure with a well defined wave number called labyrinthine pattern [40]. Figure 3 shows the emergence of labyrinthine pattern from a localized structure. This wavenumber is in the order of the characteristic size of the LS. We term this transition as finger instability, it happens when the parameters are modified from region II to region I, cf. Fig. 2.

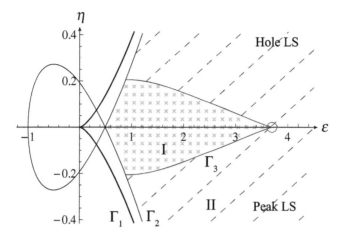

Fig. 1 Bifurcation diagram of Swift-Hohenberg model, (1), in (ε, η) space. In zone I generation of labyrinthine structures are observed from LS. In *dashed black* zone (II), stable localized structures are observed. Γ_1 and Γ_2 *curve* represent the imperfect pitchfork bifurcation and the spatial instability of the homogeneous solution. Γ_3 *curve* stands for the bifurcation where a localized structure becomes unstable and give rise to a disordered periodic structure with a well defined wave number called labyrinthine pattern

Fig. 2 Finger instability of Model (1) by which a LS generates a labyrinthine structure. The *left panel* represents the initial condition, and the *right one*, the final state. Parameters $\eta = -0.065$, $\varepsilon = 2.45$, $\nu = 2$, and a discretization of 1024×1024 points with $dx = 0.5$ and $dt = 0.03$

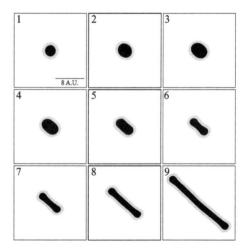

2.1 Numerical Observations

To study the complex finger dynamics exhibited by the Swift-Hohenberg (1), we have conducted numerical simulations using a pseudo-spectral method with a discretization of 1024×1024 points applying periodic boundary conditions, and verified with a finite-difference method. As we have mention, through simulations of (1) with periodic boundary conditions, one observes the destabilization of a localized structure, as shown in Fig. 3. This initial state first grows radially (cf. Fig. 3(1)). At some critical radius—which depend on the values of the control parameters $\{\eta, \varepsilon, \nu\}$—the LS destabilizes turning into a elliptically shaped state that leads to the formation of a single elongated structure, called Finger or Finger-like structure (cf. Fig. 3(2) or Fig. 2). The finger-like structure originally of a short length, grows larger and thinner up to a characteristic thickness (cf. Fig. 2).

Transversal instability occurs as a Finger-like structure becomes longer (at least an order of magnitude bigger than the LS typical size), at first, small oscillations are seen along the mid section of the structure (see Fig. 3(3)). As the finger continuously elongates, oscillation increase (Fig. 3(4)) in both size and number. These oscillations show a characteristic wavelength. The growth of the oscillations create a localized pattern structure which by a pattern instability generates disordered oscillations (Fig. 3(5)). The dynamic of the system does not saturate (i.e. finally the structure will invade all the system). Oscillations generate a disordered final state, Fig. 3(6). This labyrinthine structure is a stationary equilibrium state of the system, despite the complexity of the structure, it minimizes the free energy f, cf. expression (3). To understand the intriguing dynamic observed in Fig. 3, a step by step analysis is proposed. First, the transition from LS to finger structure will be studied. Secondly, the emergence of transversal oscillations on the finger-like structure and finally the extended pattern and non-saturating dynamics.

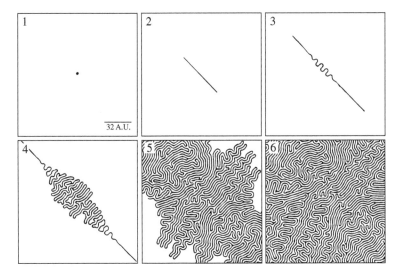

Fig. 3 Finger dynamic. Simulation of Swift-Hohenberg (1) with parameters: $\eta = -0.065$, $\varepsilon = 2.45$, $\nu = 2$, $dx = 0.5$, and $dt = 0.03$ and periodic boundary conditions. From the panel 1–6, it illustrated the evolution from a localized structure to converge in a labyrinthine pattern

2.2 Stability Analysis of Localized Structures

To understand the transition from localized structure to finger structure, a weakly non-linear analysis [41] is made on the Swift-Hohenberg model, (1) consider the limit of large radius for localized structures.

2.2.1 Weakly Non-linear Analysis

Starting from a stationary solution with azimuthal symmetry, i.e., a circular localized structure $u = u_s(r_s)$ where r_s characterizes the localized structure radius. Figure 4 depicts the typical circular localized structure obtained form the Swift-Hohenberg equation. Then, we perturb the localized structure

$$u(X, t) = u_s(X) + W(r_s, r_0),$$ (4)

where, X is the relative position $X = (r_s - r_0(\theta, t))$, r_0 is the perturbed radius position, θ stands for the angular coordinate, and $W \ll 1$ are the small non-linear corrections of the solution. Parameters are also perturbed: $\eta \to \eta - \delta\eta$ and $\varepsilon \to \varepsilon - \delta\varepsilon$, with $\delta\eta \ll 1$ and $\delta\varepsilon \ll 1$. Using polar representation of (1) and considering the perturbed solution (4) and parameters, at linear order in W one obtains

$$\mathscr{L}W = \partial_t r_0 \partial_X u_s + \eta - \delta\eta + \varepsilon u_s - \delta\varepsilon u_s - u_s^3 - \nu\nabla^2 u_s - \nabla^4 u_s,$$ (5)

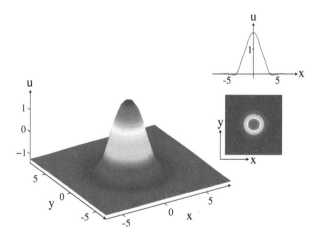

Fig. 4 Circular localized structure and its profile obtained from the simulation of (1) with periodic boundary conditions and parameters: $\eta = -0.065$; $\varepsilon = 2.45$; $\nu = 2$; $dx = 0.5$; $dt = 0.03$

where the lineal operator

$$\mathcal{L} \equiv -(\eta + \varepsilon + 3u_s^2 - \nu\nabla^2 - \nabla^4). \tag{6}$$

Then one can make use of a solvability condition (Fredholm alternative, [41]) by means of the inner product

$$\langle f|g \rangle = \lim_{L \to +\infty} \frac{1}{L} \int\limits_{-L}^{+L} f^* g \, dX, \tag{7}$$

here, L is the size of the system and f^* accounts for the complex conjugate of f. Under which $\mathcal{L} = \mathcal{L}^\dagger$ (i.e. $\mathcal{L}$ is a self-adjoint operator). Assuming that the radius of the localized structure is sufficiently large, the operator $\mathcal{L}$ which is explicitly dependent on the radial coordinate can be approximated by a homogeneous operator in this radial coordinate. This approach allows us to perform analytical calculations, which are not accessible when the operator is inhomogeneous. Using this approach the solvability condition yields

$$\frac{\partial r_0}{\partial t} = -\Delta \frac{\partial^2 r_0}{\partial^2 \theta} + 6\beta \frac{\partial^2 r_0}{\partial^2 \theta} \left(\frac{\partial r_0}{\partial \theta}\right)^2 - \frac{1}{r_s^4} \frac{\partial^4 r_0}{\partial^4 \theta} + \frac{2\beta}{r_s^3} \left(\frac{\partial r_0}{\partial \theta}\right)^2, \tag{8}$$

where

$$\beta \equiv \frac{\langle \partial_{XX} u_s | \partial_{XX} u_s \rangle}{\langle \partial_X u_s | \partial_X u_s \rangle r_s^4}, \tag{9}$$

and

$$\Delta \equiv (\nu - 2\beta) \frac{1}{r_s^2}. \tag{10}$$

Then the field $r_0(\theta, t)$ satisfies a nonlinear diffusion equation, which corresponds to the generalized Kuramoto-Sivashinsky equation [42, 43]. It gives the dynamical behavior of the interface in the vicinity of the bifurcation. It has been derived by Sivashinsky to describe the diffusive instabilities of planar flame fronts [43]. It has been concurrently established by Kuramoto [42] within the framework of phase dynamics in reaction-diffusion systems. Notice that (8) is of a non-variational type. The Kuramoto-Sivashinsky equation exhibits spatio-temporal chaotic behaviors and supplies an accurate description for phase turbulence in chemical systems and for turbulent flame fronts. The first, second, third and last term on the right side of (8) account for the linear diffusion, the non-linear diffusion, the hyper-diffusion and nonlinear advection, respectively. When $\Delta < 0$ the localized structure is stable, and for $\Delta > 0$, the localized state is unstable as result of the curvature instability. For a better understanding of (8) one can study the stability of different perturbation modes as follows.

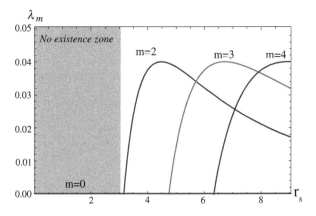

Fig. 5 λ_m as function of r_s for $m = 0, 2, 3, 4$. With $\nu = 2.0$. The *insets* show the different perturbation modes of a LS

2.2.2 Mode Analysis

For the stability analysis one considers the linear terms of nonlinear diffusion (25) which correspond to only the diffusion and hyper-diffusion terms. Taking now the Fourier mode expansion of r_0 in the form $r_0(\theta, t) = Re[\exp(im\theta + \Lambda_m t)]$, where, m is integer number, $\theta \in [0, 2\pi]$. By substituting in (25), a relation for the growth mode rate $(Re[\Lambda_m] = \lambda_m)$ is obtained

$$\lambda_m(r_s) = (\nu - 2\beta)\left(\frac{m}{r_s}\right)^2 - \left(\frac{m}{r_s}\right)^4 \tag{11}$$

where $\beta = \beta(r_s)$.

When λ_m is positive, the perturbation is unstable (see Fig. 5). Keeping the parameters fixed, the stability of the LS changes with its radius, which is a function of the system parameters. For small radius, mode $m = 0$ causes that the localized structure increases de radius without any change of its shape. In this prescription, mode $m = 1$ corresponds to an asymmetric deformation, which is not observed in the system under study. Therefore, this mode will be not analyze on the present study. After a determined point mode $m = 2$ becomes unstable generating an elliptically shaped structures, see Fig. 3(2). It has been shown that for fixed parameter values, different modes emerge by means of the LS radius increment. Nevertheless, if the radius of the LS is fixed, one expects that by the variation of a control parameter, lets say ν, one could control the emergence of certain unstable modes (cf. Fig. 6).

Note that for every fixed radius the first mode to become unstable is Mode 2 (Fig. 3(2)), increasing ν provokes the appearance of higher order modes. However, as ones considers larger structures, the unstable modes appear at lower range of values for ν. Therefore, the localized structure will be deformed, to a finger-like structure.

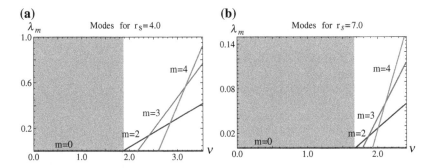

Fig. 6 Emergence of instability modes for fixed radii as function of the value of the diffusion parameter v, for (1) $r_s = 2$, and (2) $r_s = 6$

2.3 Stability Analysis of and Infinite Stripe

2.3.1 A Single Infinite Stripe

In Sect. 2.2.1 was shown that mode 2 is an instability which generates an elongation of the circular localized structure into a single finger-like structure. The stability of the central section of the finger structure is studied to see how transversal oscillations appear. For the sake of simplicity we will focus on an infinitely long structure. Figure 7 shows a stripe state obtained from numerical simulation of the Swift-Hohenberg (1). Under this approach a similar stability analysis as the previous is done, where

$$u = u_f(\xi) + W(x, X_0) \tag{12}$$

where u_f is the infinite stripe (or finger) solution, $\xi = x - X_0(y, t)$ the relative position and X_0 stands for the deformation of the infinite finger. Applying the anzats

Fig. 7 Infinite finger solution and its profile obtained from the simulation of he Swift-Hohenberg (1) with periodic boundary conditions and parameters: $\eta = -0.065; \varepsilon = 2.45; v = 2.0; dx = 0.5; dt = 0.03$

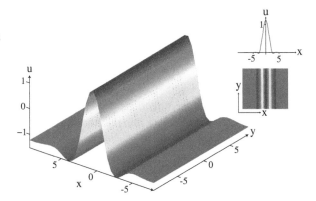

(12) in (1) at first order in W and applying the solvability condition, the following equation is obtained for the dynamic of X_0

$$\partial_t X_0 = -(\nu - 2\beta)\partial_{yy} X_0 + 6\beta\partial_{yy} X_0(\partial_y X_0)^2 - \partial_{yyyy} X_0, \tag{13}$$

where diffusion coefficient is defined by $\mathscr{D}' = (\nu - 2\beta)$, and

$$\beta' = \frac{\langle \partial_{\xi\xi} u_f | \partial_{\xi\xi} u_f \rangle}{\langle \partial_\xi u_f | \partial_\xi u_f \rangle}. \tag{14}$$

Notice that (13) is variational type. This equation is well known for exhibiting a zig-zag instability [44], however in our case we have a dynamic with no saturation thus, no zig-zag is generated. To account for this apparent contradiction, one must take into account that our weakly non-linear analysis is only valid for small perturbations, when the perturbation increase, the system enters in the regime of non-linear instabilities that have been studied in this sense [45].

If the limit $r_s \to +\infty$ is taken in (8) considering the change of coordinate system, one recovers (13) and so the adequate diffusive coefficient $\mathscr{D}'$. The growth of the oscillations created by the transversal instability, locally a roll-like pattern is formed in the middle section of the structure, see Fig. 3(4). Now it will be shown how through a pattern instability this structure destabilizes generating its own oscillations.

2.4 Stability Analysis of a Uniform Pattern

To study the stability of the pattern, a periodic solution is considered, $u_p = A \exp[iq_0 x]$, where A is the amplitude and q_0 the wave number of the stable pattern in the x direction (see Fig. 8). Analogously to the previous sections, we consider a linear perturbation

$$u(\boldsymbol{r}, t) = u_p(\boldsymbol{r}) + \delta u(\boldsymbol{r}, t), \tag{15}$$

with $\delta u \ll 1$ such that

$$\delta u = \delta A_0 e^{\sigma_{q_0} t} [e^{(iq_0 x + \boldsymbol{K}\boldsymbol{r})} + e^{-(iq_0 x + \boldsymbol{K}\boldsymbol{r})}], \tag{16}$$

where, $\boldsymbol{K} = k_x \hat{x} + k_y \hat{y}$ is the perturbation wave vector, and $\boldsymbol{r} = x\hat{x} + y\hat{y}$ the position vector. Introducing the perturbed ansatz (15) in (1), and considering only linear terms in δA_0, one obtains the growth rate relation

$$\sigma_{q_0}(\boldsymbol{K}) = \varepsilon - 3A^2 + \nu[(q_0 + k_x)^2 + k_y^2] - [(q_0 + k_x)^4 + (q_0 + k_x)^2 k_y^2 + k_y^4] \tag{17}$$

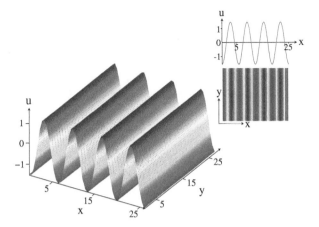

Fig. 8 Uniform pattern and its profile obtained from the simulation of (1) with periodic boundary conditions and parameters: $\eta = -0.065$; $\varepsilon = 2.45$; $\nu = 2.0$; $dx = 0.5$; $dt = 0.03$

Thus, when $\sigma_{q_0}(K) > 0$, the regular pattern structure is unstable, generating oscillations out of the x-axis. shows the region of instability of stripe pattern in the wave number space. Inside this region, the striped pattern becomes unstable generating the emergence oscillatory stripes, as illustrated in Fig. 9.

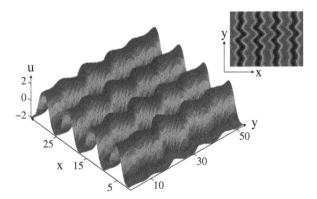

Fig. 9 Spatial instability of a stripe pattern of the Swift-Hohenberg (1) with periodic boundary conditions and parameters: $\eta = -0.065$; $\varepsilon = 2.45$; $\nu = 2.0$; $dx = 0.5$; $dt = 0.03$. The image shown in a given time, the appearance of undulations on the stripes pattern. The *inset* shows a colormap plot of the undulated stripe pattern

3 Non-variational Model

Near the second order critical point and close to long-wave pattern forming regime, the space-time dynamics of many natural out of equilibrium systems can be described by a single real order parameter equation that includes non-variational effects. This equation is refereed to as non-variational generalized Swift-Hohemberg model [46, 47].

$$\partial_t u = \eta + \varepsilon u - u^3 - \nu\nabla^2 u - \nabla^4 u + c(\nabla u)^2 + bu\nabla^2 u \qquad (18)$$

where the order parameter $u(x, y, t)$ is a real scalar field $u = u(x, y, t)$, with x and y are spatial coordinates and t is time. Depending on the context in which this equation has been derived, the physical meaning of the field variable $u(x, y, t)$ could be the electric field or chemical concentration or average molecular orientation. The model (18) has been derived for many far from equilibrium systems such as reaction-diffusion, biology and optics [46, 47]. The control parameter η measures the input field amplitude or chemical concentration. ε is the bifurcation parameter, η brakes the symmetry $u \to -u$ thus it accounts for the asymmetry between homogeneous states, ν is the diffusion coefficient, c and b coefficients of the last two terms account for the non-variational dynamic. The terms proportional to c and b, respectively, account for the nonlinear advection and diffusion. Notice that only for the case where $b = 2c$, (18) has a Lyapunov functional (or free energy)

$$\partial_t u = -\frac{\delta F[u, \nabla u, \nabla^2 u]}{\delta u}, \qquad (19)$$

where

$$F = \iint_{\infty}^{+\infty} dx dy \left[-\eta u - \frac{\varepsilon}{2}u^2 + \frac{1}{4}u^4 - \frac{\nu}{2}(\nabla u)^2 + \frac{1}{2}(\nabla^2 u)^2 + cu(\nabla u)^2 \right].$$

$$(20)$$

When $b \neq 2c$ the (18) is non-variational, i.e., there is no Lyapunov functional associated with this equation [48].

Localized structures are persistent in the presence of non-variational terms. As a result of these non variational terms LSs may be exhibit permanent dynamics as oscillation or chaotic dynamical behavior, among others [46, 49]. The dynamics of spots deformation, accompanied by the emergence of unstable fingers with the final appearance of labyrinthine patterns also we observe in the non variational Swift-Hohenberg model, (18). Figure 10 illustrates the process of emergence of labyrinthine patterns from a localized structure.

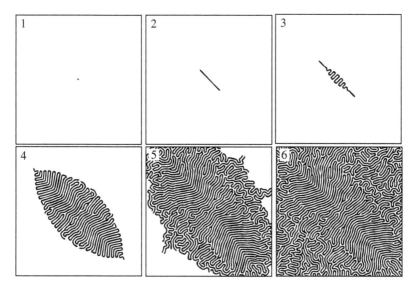

Fig. 10 Transition from a single peak LS to labyrinthine pattern in the nonvariational Swift-Hohenberg (18). Different images sequentially illustrate the temporal evolution of the Swift-Hohenberg (18) with parameters: $\eta = -0.225$; $\varepsilon = 1.0$; $\nu = 2.0$; $dx = 0.5$; $dt = 0.03$ $c = 0.1$ and $b = -0.1$. *1* Localized structure, *2* finger-like structure, *3* undulations in the finger structure, *4* and *5* localized transient patterns, and *6* stationary labyrinthine pattern

3.1 Spot Deformation

To study the stability of two dimensional localized structures solutions of (20), we apply a weakly non-linear analysis [41] for the non-variational Swift-Hohenberg model, (18). Let first consider a single localized structure peak. It is stationary radially symmetric solutions $u = u_s(r_s)$ where r_s is LS radius. Then we consider the following perturbation around this solution

$$u(X, t) = u_s(X) + W(r_s, r_0(\theta, t)) \tag{21}$$

with X is the relative position $X = r_s - r_0(\theta, t)$ with respect to the r_s LS radius. The perturbed radius position is r_0 and the angular coordinate is denoted θ. The nonlinear corrections to the stationary radially symmetric solutions $u = u_s(r_s)$ is $W \ll 1$. We consider also the following perturbation of the parameters: $\eta \to \eta - \delta\eta$ and $\varepsilon \to \varepsilon - \delta\varepsilon$, with $\delta\eta \ll 1$ and $\delta\varepsilon \ll 1$. Using polar representation of (18) and considering the perturbed solution (21) and the above parameter perturbation, we obtain analogously to previous section at the linear order in W, the following equation

$$
\mathcal{L}W = \partial_t r_0 \partial_X u_s + \eta - \delta\eta + \varepsilon u_s - \delta\varepsilon u_s - u_s^3
$$
$$
- \nu\nabla^2 u_s - \nabla^4 u_s + c(\nabla u_s)^2 + b u_s \nabla^2 u_s, \tag{22}
$$

where the lineal operator

$$
\mathcal{L} \equiv -\eta - \varepsilon - 3u_s^2 - 2c\nabla u_s \nabla + (\nu + b u_s)\nabla^2 - \nabla^2 u_s + \nabla^4. \tag{23}
$$

In order to apply the the solvability condition [41], we consider the following inner product

$$
\langle f(X) | g(X) \rangle = \lim_{L \to +\infty} \frac{1}{L} \int_{-L}^{+L} f^* g \, dX. \tag{24}
$$

Here, L is the size of the system. The operator $\mathcal{L} = \mathcal{L}^\dagger$ (i.e. $\mathcal{L}$ is a self-adjoint operator), admits a Goldstone mode given by $\partial_X u_s$. By applying the solvability condition and assuming that the radius of the localized structure is large enough, after straightforward calculations we obtain the equation

$$
\partial_t r_0 = -\Delta \partial_\theta^2 r_0 + \frac{6\beta}{r_s^4} \partial_\theta^2 r_0 (\partial_\theta r_0)^2 - \frac{1}{r_s^4} \partial_\theta^4 r_0 + \frac{2\beta}{r_s^3} (\partial_\theta r_0)^2 \tag{25}
$$

with,

$$
\Delta = (\nu - 2\beta(r_s) - b\gamma)\frac{1}{r_s^2},
$$
$$
\beta = \frac{\langle \partial_{XX} u_s | \partial_{XX} u_s \rangle}{\langle \partial_X u_s | \partial_X u_s \rangle},
$$
$$
\gamma = \frac{\langle \partial_X u_s | u_s \partial_X u_s \rangle}{\langle \partial_X u_s | \partial_X u_s \rangle}. \tag{26}
$$

Then we obtain again the generalized Kuramoto-Sivashinsky (8), where the only coefficient renormalized by non-variational terms is the diffusion. Therefore, dynamics exhibited by LSs in variational or non-variational case are similar. Thus the growth rate relation ($r_0(\theta, t) = Re[R_0 \exp(im\theta + \Lambda_m t)]$), has the form

$$
\lambda_m(r_s) = (\nu - 2\beta - b\gamma)\left(\frac{m}{r_s}\right)^2 - \left(\frac{m}{r_s}\right)^4, \tag{27}
$$

where $\beta = \beta(r_s)$.

The eigenvalue λ_m as a function of r_s is plotted for different values of the angular index m shown in Fig. 11. For small radius, mode $m = 0$ causes an increase of the radius of LS without any change of its shape. The mode $m = 0$ is marginally stable since for $\lambda_0(r_s) = 0$. As in the case of the variational Swift-Hohenberg (1), the first

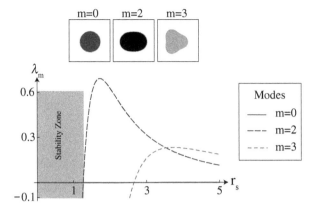

Fig. 11 (Color online) Growth rate as function of the localized structure radius for the generalized Swift-Hohenberg model, (18). Theoretical λ_m as function of r_s for $m = 0, 2, 3$, using the formula (27) with $\nu = 2.0$ and $b = 0.015$. The *insets* show the different perturbation modes of a LS

mode that generates an elliptically deformation of the circular shape of localized structures is $m = 2$. Similarly, to the previous section, one can perform similar studies of stability for a single stripe and stripe patterns, the results are similar.

4 Conclusions and Remarks

We have described the stability of circular shape localized structures in two-dimensional variational and non-variational Swift-Hohenberg equations. We first draw a bifurcation diagram showing different solutions that appear in different regime of parameters. Then, we have shown that the angular index $m = 2$ become first unstable as consequence of curvature instability. This instability leads to an elliptical deformation of the circular shape of localized structure. Other type of deformations corresponding to $m = 3, 4$ and 5 occur also in the Swift-Hohenberg equation. We performed this study analytically and numerically. We have shown also that for a fixed values of the localized structures radius, the angular index $m = 2$ becomes unstable for small diffusion (anti diffusion) coefficient while higher order modes $m > 2$ become unstable for large values of a diffusion coefficient. We have performed also the stability of a single stripe localized structure and an extended pattern. Finally, we have study the spot deformation instability that affects the circular shape of two-dimensional localized structures in a non-variational Swift-Hohenberg model.

We have characterized the emergence of labyrinthine patterns from the instability of localized structures where the main mechanism behind this intricate dynamic is the curvature instability. Therefore, this type of dynamical behavior is common to a wide class phenomenon of out equilibrium systems. This phenomenon has been observed in magnetic fluids and cholesteric liquid crystals. Despite the complexity of the dynamics of this may simply be piloted by the free energy minimization.

Acknowledgments M.G.C. thanks the financial support of FONDECYT project 1120320. I.B. is supported by CONICYT, Beca de Magister Nacional.

References

1. P. Glansdorff, I. Prigogine, *Thermodynamic Theory of Structures. Stability and Fluctuations* (Wiley, New York, 1971)
2. G. Nicolis, I. Prigogine, Self-Organization in Non Equilibrium Systems (Wiley, New York, 1977)
3. A.M. Turing, Philos. Trans. R. Soc. Lond. 237, 37 (1952); I. Prigogine, R. Lefever, J. Chem. Phys. 48, 1695 (1968)
4. Y. Pomeau, Front motion, metastability and subcritical bifurcations in hydrodynamics. Physica D **23**, 3 (1986)
5. M. Tlidi, P. Mandel, R. Lefever, Phys. Rev. Lett. **73**, 640 (1994)
6. P. Coullet, C. Riera, C. Tresser, Prog. Theor. Phys. Supp. 139 (2000)
7. M.G. Clerc, C. Falcon, Physica A 356, 48 (2005); U. Bortolozzo, M.G. Clerc, C. Falcon, S. Residori, R. Rojas, Phys. Rev. Lett. 96, 214501 (2006); M.G. Clerc, E. Tirapegui, M. Trejo, Phys. Rev. Lett. 97, 176102 (2006)
8. V.K. Vanag, L. Yang, M. Dolnik, A.M. Zhabotinsky, I.R. Epstein, Nature 406, 389 (2000); I. Lengyel, I.R. Epstein, Proc. Natl. Sci. U.S.A. 89, 128301 (2004); V.K. Vanag, I.R. Epstein, Phys. Rev. Lett. 92, 128301 (2004); V.S. Zykov, K. Showalter. Phys. Rev. Lett. 94, 068302 (2005)
9. O. Lejeune, M. Tlidi, P. Couteron, Phys. Rev. E 66, 010901(R) (2002); E. Meron, E. Gilad, J. von Hardenberg, M. Shachak, Y. Zarmi, Chaos, Solitons Fractals 19, 367 (2004); M. Rietkerk, S.C. Dekker, P.C. de Ruiter, J. van de Koppel, Science 305, 1926 (2004); M. Tlidi, R. Lefever, A. Vladimirov, On vegetation clustering, localized bare soil spots and fairy circles. Lect. Notes Phys. 751, 381 (2008); E. Sheffer, H. Yizhaq, M. Shachak, E. Meron. J. Theor. Biol. 273, 138 (2011)
10. P. Couteron, F. Anthelme, M.G. Clerc, D. Escaff, C. Fernandez-Oto, M. Tlidi, Plant clonal morphologies and spatial patterns as self-organized responses to resource-limited environments. Philos. Trans. R. Soc. A **372**, 20140102 (2014)
11. J. Wu, R. Keolian, I. Rudnick, Observation of a nonpropagating hydrodynamic soliton. Phys. Rev. Lett. **52**, 1421–1424 (1984)
12. M.G. Clerc, S. Coulibaly, N. Mujica, R. Navarro, T. Sauma, Soliton pair interaction law in parametrically driven Newtonian fluid. Philos. Trans. R. Soc. A **367**, 3213–3226 (2009)
13. E. Meron, Nonlinear Physics of Ecosystems (CRC Press, Boca Raton and Taylor & Francis group, Boca Raton, 2015)
14. M.G. Clerc, S. Coulibaly, D. Laroze, Localized states beyond asymptotic parametrically driven amplitude equation. Phys. Rev. E 77, 056209 (2008); Localized states of parametrically driven easy-plane ferromagnetic wire. Physica D 239, 72 (2010); Interaction law of 2D localized precession states. EPL 90, 38005 (2010); Localized waves in a parametrically driven magnetic nanowire. EPL 97, 3000 (2012); A.O. Leon, M.G. Clerc, S. Coulibaly, Traveling pulse on a periodic background in parametrically driven systems. Phys. Rev. E 91, 050901(R) (2015); M.G. Clerc, S. Coulibaly, D. Laroze, A.O. Leon, A.S. Nunez, Alternating spin-polarized current induces parametric resonance in spin valves. Phys. Rev. B 91, 224426 (2015)
15. A.J. Scroggie, W.J. Firth, G.S. McDonald, M. Tlidi, R. Lefever, L.A. Lugiato, Chaos, Solitons Fractals 4, 1323 (1994); V.B. Taranenko, K. Staliunas, C.O. Weiss, Phys. Rev. A 56, 1582

(1997); V.B. Taranenko, K. Staliunas, C.O. Weiss, Phys. Rev. Lett. 81, 2236 (1998); S. Barland, J.R. Tredicce, M. Brambilla, L.A. Lugiato, S. Balle et al., Nature 419, 699 (2002); X. Hachair, L. Furfaro, J. Javaloyes, M. Giudici, S. Balle et al., Phys. Rev. A 72, 013815 (2005)
16. F. Haudin, R.G. Rojas, U. Bortolozzo, S. Residori, M.G. Clerc, Homoclinic snaking of localized patterns in a spatially forced system. Phys. Rev. Lett. 107, 264101 (2011)
17. E. Averlant, M. Tlidi, H. Thienpont, T. Ackemann, K. Panajotov, Opt. Express 22, 762 (2014); A.B. Kozyreff, I.V. Shadrivov, Y.S. Kivshar, Appl. Phys. Lett. 104, 084105 (2014)
18. J.E. Pearson, Science 261, 189 (1993)
19. K. Lee, W.D. McCormick, J.E. Pearson, H.L. Swinney, Nature (London) 369, 215 (1994)
20. A.P. Munuzuri, V. Percz-Villar, M. Markus, Phys. Rev. Lett. 79, 1941 (1997)
21. A. Kaminaga, V.K. Vanag, I.R. Epstein, Angew. Chem. 45, 3087 (2006)
22. A. Kaminaga, V.K. Vanag, I.R. Epstein, J. Chem. Phys. 122, 174706 (2005)
23. T. Kolokolnikov, M. Tlidi, Phys. Rev. Lett. 98, 188303 (2007)
24. P.W. Davis, P. Blanchedeau, E. Dulos, P. De Kepper, J. Phys. Chem. A 102, 8236 (1998)
25. C. Muratov, V.V. Osipov, Phys. Rev. E 53(3101), 1996; 54(4860), (1996); C. Muratov. Phys. Rev. E 66, 066108 (2002)
26. Y. Hayase, T. Ohta, Phys. Rev. Lett. 81, 1726 (1998)
27. Y. Hayase, Phys. Rev. E 62, 5998 (2000)
28. E. Meron, E. Gilad, J. von Hardenberg, M. Shachak, Chaos Solitons Fractals 19, 367 (2004)
29. X. Ren, J. Wei, SIAM. J. Math. Anal. 35, 1 (2003)
30. Y. Nishiura, H. Suzuki, SIAM. J. Math. Anal. 36, 916 (2004)
31. M. Tlidi, A.G. Vladimirov, P. Mandel, Phys. Rev. Lett. 89, 233901 (2002)
32. A.J. Dickstein, S. Erramilli, R.E. Goldstein, D.P. Jackson, S.A. Langer, Labyrinthine pattern formation in magnetic fluids. Science 261, 1012 (1993)
33. P. Ribiere, P. Oswald, Nucleation and growth of cholesteric fingers under electric field. J. De Physique 51, 1703 (1990)
34. P. Oswald, J. Baudry, S. Pirkl, Static and dynamic properties of cholesteric fingers in electric field. Phys. Rep. 337, 67 (2000)
35. J. Swift, P.C. Hohenberg, Phys. Rev. A 15, 319 (1977)
36. MF. Hilali, G. Dewel, P. Borckmans, Phys. Lett. A 217, 263 (1996)
37. R. Lefever, N. Barbier, P. Couteron, O. Lejeune, J. Theor. Biol. 261, 194 (2009)
38. M. Tlidi, M. Georgiou, P. Mandel, Phys. Rev. A 48, 4605 (1993)
39. P. Coullet, Int. J. Bifurcation Chaos 12, 2445 (2002)
40. M. Le Berre, E. Ressayre, A. Tallet, Y. Pomeau, L. Di Menza, Phys. Rev. E. 66, 026203 (2002)
41. L.M. Pismen, Patterns and Interfaces in Dissipative Dynamics (Springer Series in Synergetics, Berlin, 2006)
42. Y. Kuramoto, T. Tsuzuki, Persistent propagation of concentration waves in dissipative media far from thermal equilibrium. Prog. Theor. Phys. 55, 356 (1976)
43. G.I. Sivashinsky, Nonlinear analysis of hydrodynamic instability in laminar flames I: derivation of basic equations. Acta Astronautica 4, 1177 (1977)
44. C. Chevallard, M.G. Clerc, P. Coullet, J.M. Gilli, Interface dynamics in liquid crystals. Eur. Phys. J. E 1, 179 (2000); Zig-zag instability of an Ising wall in liquid crystals. Europhys. Lett. 58, 686-692 (2002)
45. A. Hagberg, A. Yochelis, H. Yizhaq, C. Elphick, L. Pismen, E. Meron, Linear and nonlinear front instabilities in bistable systems. Physica D 217, 186–192 (2006)
46. M.G. Clerc, A. Petrossian, S. Residori, Bouncing localized structures in a liquid-crystal light-valve experiment. Phys. Rev. E 71, 015205(R) (2005)
47. G. Kozyreff, M. Tlidi, Nonvariational real Swift-Hohenberg equation for biological, chemical, and optical systems Chaos: an interdisciplinary. J. Nonlinear Sci. 17, 037103 (2207)
48. S. Wiggins, Introduction to Applied Nonlinear Dynamical Systems and Chaos, 2nd edn. (Springer, Berlin, 2003)
49. N. Verschueren, U. Bortolozzo, M.G. Clerc, S. Residori, Spatiotemporal chaotic localized state in liquid crystal light valve experiments with optical feedback. Phys. Rev. Lett. 110, 104101 (2013); Chaoticon: localized pattern with permanent dynamics. Philos. Trans. R. Soc. A. 372, 20140011 (2014)

Random Walk Model for Kink-Antikink Annihilation in a Fluctuating Environment

Daniel Escaff

Abstract In this report, the kink-antikink interaction, in one spatial dimension, is revised in the framework of a paradigmatic model for bistability (real Ginzburg-Landau equation). In particular, it is pointed out that, when it is taking into account the fluctuations, drastically changes the main features of the interaction. To wit, since the long distance interaction is exponentially weak, the kink-antikink movement is ruled by the fluctuations. A simple random walk model, that incorporates the pair self-annihilation, is proposed. We discussed the implications that, consider the fluctuations, has in the coarsening dynamics. That is, the coarsening law, for the growing of the domains in each stable state, changes from being logarithmic to becoming in the power law $\sqrt{t}$.

1 Introduction

When considering spatially extended bistable systems, where there are two stable equilibrium states associated with some symmetry, the slow or ulterior dynamics consists in a domain interaction or coarsening. The origin of these states can be a spontaneous breaking of symmetry, like a ferromagnetic transition, where the symmetrical states are spin up and down [1]. The fast dynamics in this transition is characterized by the establishment of local magnetization and the slow dynamics consists in a coarsening of magnetic domains. Hence, a bistable system that initially is arranged in the unstable state, as consequence of the fluctuations the system first goes locally to one of these equilibrium states, given rise to domains. The typical size of these domains are small at the beginning, but start to grow. This last process corresponds to the slow dynamics. In one spatial dimension, two different domain are separated by a defect, often called kinks and antikinks depending on the sense of connection. In the context of ferromagnetic transition, this defect corresponds to a

---

D. Escaff (✉)
Complex Systems Group, Facultad de Ingeniería y Ciencias Aplicadas,
Universidad de Los Andes, Mon. Alvaro Del Portillo, 12.455 Las Condes, Santiago, Chile
e-mail: escaffnetmail@yahoo.com

© Springer International Publishing Switzerland 2016   293
M. Tlidi and M.G. Clerc (eds.), *Nonlinear Dynamics: Materials,*
Theory and Experiments, Springer Proceedings in Physics 173,
DOI 10.1007/978-3-319-24871-4_22

point with zero magnetization. In two spatial dimensions, two different domains are separated by an interface, which is a trivial extension of the one-dimension defect. We can imagine this interface as a curve composed by defect points.

Here, we will focus in one spatial dimension, where the key to understand the slow dynamics is the kink-antikink interaction. In particular, we are interested in how the presence of fluctuation affects this interaction. In fact, it is shown that the presence noise changes radically the pair interaction regardless of the noise intensity. It is worth to emphasize that, the order parameter approach is based on the adiabatic elimination of many degrees of freedom of the underlying microscopic dynamics. However, these ignored degrees of freedom play a fundamental role in the macroscopic description, the fluctuation of the order parameter.

We base our analysis in a paradigmatic model for bistability, the real Ginzburg-Landau equation. This model does has any conserved quantity, such as the Cahn-Hilliard model that conserve mass, where the coarsening dynamics has been widely studied [2–7]. Or the Van der Waals normal form, which, beside mass, also conserve momentum, and exhibits an interesting spinodal decomposition that leads to a a coarsening of domains [8]. Furthermore, the model does not exhibit any pattern forming instability, as the Swift-Hohenberg model [9, 10], where also can be observed a coarsening-types of dynamics [11]. The real Ginzburg-Landau equation is, therefore, a simple model that allow us to analyze the main effects of fluctuations in one spatial dimensions.

The manuscript is organized as follow: In Sect. 2 we present the model a summarize the main feature of its dynamics. In Sect. 3 we present our simple random walk approach for kink-antikink annihilation. And, in Sect. 4, we present our concluding remarks.

2 Non-conservative Model for Kink-Antikink Interaction

Our starting point is the paradigmatic model for bistability

$$\partial_t u = -\frac{\delta \mathscr{F}[u]}{\delta u}, \tag{1}$$

where

$$\mathscr{F}[u] = \int \left\{ \frac{1}{2} (\partial_x u)^2 + V(u) \right\} dx, \tag{2}$$

with the double-well potential

$$V(u) = \frac{1}{4} \left(u^2 - 1 \right)^2. \tag{3}$$

Therefore, we are in the presence of a relaxation dynamics that leads to the minimization of the functional (2). Clearly, the global minima are $u = \pm 1$, with the unstable fix point $u = 0$.

Notice that, the above dynamics do not have any conservative quantity, in contrast with, for instance, the CahnHilliard model that conserves the mass. Choosing a random initial condition, the typical dynamic behavior exhibited by system (1) can be depicted in two steps: a fast time scale, where the system sets local domains where it is in one of its two equilibria states $u = -1$ or 1, separated by defects; and a slow time scale, where the system displays domain dynamics (or coarsening dynamics) aimed at establishing a global equilibrium that consists in a completely uniform state (with $u = -1$ or 1).

The coarsening dynamics (slow time scale) is ruled by the defects interaction. A single defect has the form

$$u_{\pm}(x) = \pm \tanh \left(\frac{x - X^{\pm}}{\sqrt{2}} \right), \tag{4}$$

where $X^{\pm}$ are the positions of the defect, that is $u_{\pm}(X^{\pm}) = 0$. Note that this parameter is an arbitrary number because of the spatial translation invariance of the system. The kink solution corresponds to u_{+}, which links -1 from $-\infty$ to 1 at ∞. The opposite connection (u_{-}) is called antikink. They can be treated as inertialess particles, characterized by their positions $X^{\pm}$.

2.1 Single Kink-Antikink Pair Interaction

The simplest setup to analyze the defects interaction is a single isolated kink-antikink pair, as shown Fig. 1. Since the system has the symmetry $u \longrightarrow -u$, without lose of generality we will assume that the kink is rightward to the anti-kink ($X^{+} < X^{-}$), as shown Fig. 1. The long distance interaction is exponentially weak, in fact [12]

$$\frac{dX}{dt} = -12\sqrt{2}e^{-\sqrt{2}X} \quad \text{and} \quad \frac{dY}{dt} = 0, \tag{5}$$

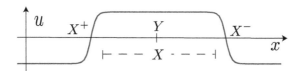

Fig. 1 Localized domain composes by a pair kink-antikink, where X^{+} and X^{-} are the positions of the kink and antikink, respectively

where $Y = \left(X^+ + X^-\right)/2$ corresponds to the *center of mass coordinate* (cf. Fig. 1), while $X = \left(X^- - X^+\right)$ corresponds to the *relative coordinate*, which decrease logarithmically with time.

2.2 Kink-Antikink Gas

In a more general scenario, we can consider a gas of kink anti-kink pairs. It is important to remark that a gas of kink and anti-kink is a frequent situation, because the system arrives to the global minima by means of defects annihilation. Then, if the ith kink occupies the position X_i^+, while the ith anti-kink the position X_i^- and it is placed between the ith and the $(i+1)$th kink, the gas will be described by the dynamical system [12]

$$\dot{X}_i^+ = 6\sqrt{2}\left(e^{-\sqrt{2}(X_i^- - X_i^+)} - e^{-\sqrt{2}(X_i^+ - X_{i-1}^-)}\right),$$
$$\dot{X}_i^- = 6\sqrt{2}\left(e^{-\sqrt{2}(X_{i+1}^+ - X_i^-)} - e^{-\sqrt{2}(X_i^- - X_i^+)}\right). \tag{6}$$

Hence, due to the attractive interaction and the annihilation process, the number of particles or defects, $n(t)$, decreases with time, and the system exhibits coarsening dynamics. The number of particles is related with the average size of the domains, $\langle L(t)\rangle$. In fact, if we consider periodic boundary condition then the average size satisfies

$$\langle L\rangle = \frac{\sum_{i=1}^{n}\left[\left(X_i^- - X_i^+\right) + \left(X_i^+ - X_{i-1}^-\right)\right]}{n} = \frac{L_0}{n},$$

where L_0 is the total size of the system. On the other hand, since the dynamical system (6) has the symmetry

$$\left(X_i^- - X_i^+\right) \longrightarrow \left(X_i^- - X_i^+\right) + \alpha, \tag{7}$$
$$\left(X_i^+ - X_{i-1}^-\right) \longrightarrow \left(X_i^+ - X_{i-1}^-\right) + \alpha, \tag{8}$$
$$t \longrightarrow e^{\sqrt{2}\alpha}t, \tag{9}$$

the average size must satisfy then the self similarity law

$$\langle L(t)\rangle + \alpha = \left\langle L\left(e^{\sqrt{2}\alpha}t\right)\right\rangle,$$

therefore

$$\langle L(t)\rangle = \log(t)/\sqrt{2},$$

and the number of particles obey the law

$$n(t) = \sqrt{2}L_0/\log(t), \tag{10}$$

i.e. this number decreases until the system goes to the homogeneous equilibrium state [the global minima of the potential (2)].

2.3 Including the Presence of Fluctuation

The other ingredient that we will include in our analysis, is the presence of fluctuation. That is,

$$\partial_t u = -\frac{\delta \mathcal{F}[u]}{\delta u} + \sqrt{\gamma}\zeta(x, t), \tag{11}$$

where $\mathcal{F}[u]$ comes from (2). γ is the parameter that gives account of the noise intensity (for equilibrium systems this parameter is proportional to the temperature) and $\zeta(x, t)$ is a white gaussian noise with zero mean value, $\langle \zeta(x, t) \rangle = 0$, and correlation

$$\langle \zeta(x, t) \zeta(x', t') \rangle = \delta(t' - t)\delta(x' - x).$$

The single kink (4) moves as a Brownian particle (see [13])

$$\frac{dX^{\pm}}{dt} = \sqrt{\eta}\xi(t) \tag{12}$$

where $\eta \equiv 3\gamma/\sqrt{8}$, and $\xi(t)$ is a white noise with zero mean and correlation $\langle \xi(t)\xi(t') \rangle = \delta(t' - t)$.

Now, in a system of interacting kink-antikink pairs, since the interaction exponentially decreases with the distance, the main driven force will be the fluctuations. That is, one might figure out such system as a gas of Brownian particles that annihilate when collide. The effect of the attractive deterministic interaction may be also included by writing an appropriated Fokker-Planck equation [14]. However, for a dilute enough kink-antikink gas, the main effects should be: (i) the Brownian motion of the defect; and (ii) the annihilation process. Since average path size of each defect grows as $\sqrt{t}$, then we can conjecture that the average size of the domains satisfies a power law of the form $\langle L(t) \rangle \sim \sqrt{t}$ (i.e. faster than the logarithmic deterministic growing). In Fig. 2, is shown the spatiotemporal evolution of the both systems from the same initial condition and clearly the domain dynamics in the stochastic system is more efficient, that is, the number of defect decrease faster than in the deterministic system.

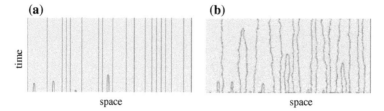

Fig. 2 Spatiotemporal density plot of $|u|$ from numerical simulations of model (11), where *red* denotes near to zero, while *yellow* near to one. The same initial condition: **a** without noise $\eta = 0.0$; **b** with noise $\eta = 0.5555$

A detailed analytical characterization of the annihilation process might be complicated, because it required a characterization of the near kink-antikink interaction. In the next section it is presented a simple random walk model to deal with this process.

3 The Simple Random Walk Approach for Kink-Antikink Annihilation

Let us call S the set of all possible states of a system composes by a single pair kink-antikink. Since the process is Markovian, it is entirely characterized by the conditional probability $P(\psi, t | \acute{\psi}, \acute{t}) \equiv P(\psi, t)$ [15], with $\psi, \acute{\psi} \in S$, which satisfies the master equation

$$\partial_t P(\psi, t) = \sum_{\phi \in S} [w(\phi| \psi) P(\phi, t) - w(\psi| \phi) P(\psi, t)], \qquad (13)$$

where $w(\phi| \psi)$ is the transition probability from state ϕ to ψ.

We can separate $S = S_2 \cup S_0$, where S_2 contain all the states where the pair has not self-annihilated yet, and the kink and anti-kink are in the position $\{X^+, X^-\}$; while, S_0 contain the state where the pair is already annihilated, i.e. it has only one element. Then, we can write

$$P(\psi, t) = \begin{cases} P_2(X^+, X^-, t) & if \quad \psi \in S_2 \\ P_0(t) & if \quad \psi \in S_2 \end{cases} \qquad (14)$$

In order to write explicitly the master equation (13), we need the expression of the transition probability. Note that, since there is no probability of transition from S_0 to S_2, P_2 satisfies an autonomous equation. The transport in S_2 is of diffusive type. And P_0 only increases for the annihilation process that occur at S_2, that is, its dynamic is completely dependent of P_2. Then, the master equation (13) can be decomposed in the following hierarchy of equations

$$\partial_t P_2 = \left\{ \tfrac{\eta}{2} \left(\partial_{X^+ X^+} + \partial_{X^- X^-} \right) - W \left(X^+ - X^- \right) \right\} P_2, \tag{15}$$

$$\partial_t P_0 = \int_\Omega W \left(X^+ - X^- \right) P_2 \left(X^+, X^-, t \right) dX^+ dX^-, \tag{16}$$

where Ω is the domain under study, and $W \left(X^+ - X^- \right)$ is the transition probability from S_2 to S_0, it is positive and only depend on the distance between the kink and anti-kink pair. Let us introduce the relative coordinate $X = \left(X^+ - X^- \right)$, and the center of mass $Y = \left(X^+ + X^- \right) /2$. In these coordinates the above P_2-equation is separable and takes the form

$$\partial_t V = \frac{\eta}{4} \partial_{YY} V,$$

$$\partial_t U = (\eta \partial_{XX} - W(X)) U, \tag{17}$$

where $P_2 = U(X, t) V(Y, t)$. Therefore, the center of mass satisfies a diffusion equation, that is, it behaves as a simple Brownian particle. While, the relative coordinate satisfies a modified diffusion equation, where the extra term $W(X)$ plays the role of a probability sink, and it gives account of the annihilation process.

Therefore, to model the annihilation process, we must to model $W(X)$. If we approach the kink anti-kink interaction as a local process, that is, the defects are points that move independently and when they collide instantaneously these defects disappear (annihilate process). Then, $W(X)$ can be modeled as an infinite barrier

$$W(X) = \begin{cases} 0 & if \quad X > 0 \\ \infty & else \end{cases}. \tag{18}$$

In other work, we are solving U-equation (17), with the boundary condition $U(0, t) = 0$, in the region $X > 0$ (we are assuming that the kink is right-ward to the anti-kink). In addition, we consider the deterministic initial condition $U(X, 0) = \delta(X - X_0)$, then

$$U(X, t) = \frac{1}{\sqrt{4\pi \eta t}} \left(\exp \left[-\frac{(X - X_0)^2}{4\eta t} \right] - \exp \left[-\frac{(X + X_0)^2}{4\eta t} \right] \right). \tag{19}$$

Hence, we can interpret the number of particles as two time the probability that the kink and the anti-kink have not self-annihilated yet, that is

$$n(t) = 2 \int_0^\infty U(X, 0) \, dX, \tag{20}$$

therefore

$$n(t) = 2 erf \left(\frac{X_0}{2\sqrt{\eta t}} \right). \tag{21}$$

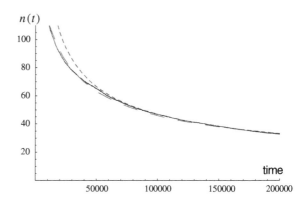

Fig. 3 The number of particles as function of time. The *continuous curve* is obtained from the average of 100 realization of (11) for $\eta = 4.0$. The *dashed line* is a fitting with the single pair formula (21), with $X_0 = 54.3$. The *dotted curve* is the power law $2X_0/\sqrt{\eta t}$

Note that asymptotically the particles number has the form

$$n\,(t \sim \infty) \simeq \frac{2X_0}{\sqrt{\eta t}} + \mathcal{O}\left(t^{-3/2}\right),$$

that is, for a large time, the number of particles decreases with the power law $\sqrt{t}$. And, therefore, average size of the domains grows with the power law $\langle L(t) \rangle \sim \sqrt{t}$.

In Fig. 3 is displayed a comparison, for $n(t)$, between direct numerical simulations from (11), for a kink-antikink gas, and the power law $n(t) \sim 1/\sqrt{t}$.

4 Discussion and Final Remarks

We have deeply revised the kink-antikink interaction in the framework of the paradigmatic model for bistability (11). Showing that this interaction is drastically affected by the presence of fluctuation, regardless of the noise intensity. To wit, since the long distance interaction is exponentially weak, the defects movement is ruled by the fluctuations. We discussed the implications that this fact has in the coarsening dynamics. Thai is, the coarsening law changes from being logarithmic, to becoming a power law ($\sqrt{t}$). Probably, for low noise intensity, there is a logarithmic crossover that takes more time, if the noise intensity is lower. However, for enough large time, the behavior will follow the power law.

To have a complete description, however, we must account for the possibility that the fluctuation creates a new kink-antikink pair. The creation process supplies new pairs to the system, competing with the annihilation process. In Fig. 4 is shown numerical result from model (11) for high noise intensity. Figure 4a displays a creation event for a given realization of the stochastic process. In Fig. 4b is shown the inverse of particle numbers a given time (t^*) as function of the noise intensity. Therefore, there is an optimal noise intensity where the coarsening process is more efficient. Namely, for too low intensity the defect movement is slow and the

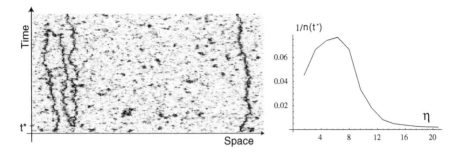

Fig. 4 **a** Spatio-temporal diagram of (11) with $\eta = 5.0$, showing a creation event at t^*. *Grey scale* is inversely proportional to absolute value of field u. **b** The inverse of particle numbers a given time (t^*) as function of the noise intensity

annihilation events less often. On the other hand, for too large noise intensity, even through the annihilation events are more frequent, the creation events become more probable, and the number of particles does not decrease. Probably, the steady state consists in a average density of pairs, per unit of lengh, not in the minima of the potential (2).

Acknowledgments I would like to thank Prof. M.G. Clerc for fruitful discussions, and FONDE-CYT (Project N 1140128) for financial support.

References

1. A. Hubert, R. Scafer, *Magnetic domains* (Springer, Berlin, 1998)
2. A.J. Bray, Domain-growth scaling in systems with long-range interactions. Phys. Rev. E **47**, 3191–3195 (1993)
3. C.L. Emmott, A.J. Bray, Coarsening dynamics of a one-dimensional driven Cahn-Hilliard system. Phys. Rev. E **54**, 4568–4575 (1996)
4. A.J. Bray, Defect relaxation and coarsening exponents. Phys. Rev. E **58**, 1508–1513 (1998)
5. S.J. Watson, F. Otto, B.Y. Rubinstein, S.H. Davis, Coarsening dynamics of the convective Cahn-Hilliard equation. Physica D **178**, 127–148 (2003)
6. H. Calisto, M. Clerc, R. Rojas, E. Tirapegui, Bubbles Interactions in the Cahn-Hilliard Equation. Phys. Rev. Lett. **85**, 3805–3808 (2000)
7. A.A. Golovin, A.A. Nepomnyashchy, S.H. Davis, M.A. Zaks, Convective Cahn-Hilliard models: from coarsening to roughening. Phys. Rev. Lett. **86**, 1550–1553 (2001)
8. M. Argentina, M.G. Clerc, R. Soto, van der Waals-like transition in fluidized granular matter. Phys. Rev. Lett. **89**, 044301 (2002)
9. J. Swift, P.C. Hohenberg, Hydrodynamic fluctuations at the convective instability. Phys. Rev. A **15**, 319 (1977)
10. M. Bestehorn, H. Haken, Transient patterns of the convection instability: a model-calculation. Z. Phys. B **57**, 329 (1984)
11. M. Tlidi, P. Mandel, R. Lefever, Kinetics of localized pattern formation in optical systems. Phys. Rev. Lett. **81**, 979–982 (1998)
12. K. Kawasaki, Defect-phase dynamics for dissipative media with potential. Progr. Theory Phys. **80**, 123–138 (1984)

13. T. Funaki, The scaling limit for a stochastic PDE and the separation of phases. Probab. Theory Relat. Fields. **102**, 221–288 (1995)
14. D. Contreras. M.G. Clerc, Internal noise and system size effects induce nondiffusive kink dynamics. Phys. Rev. E **91**, 032922 (2015)
15. N.G. van Kampen, *Stochastic processes in physics and chemistry* (Elsevier, North-Holland, 1981)

Time-Periodic Forcing of Spatially Localized Structures

Punit Gandhi, Cédric Beaume and Edgar Knobloch

Abstract We study localized states in the Swift–Hohenberg equation when time-periodic parametric forcing is introduced. The presence of a time-dependent forcing introduces a new characteristic time which creates a series of resonances with the depinning time of the fronts bounding the localized pattern. The organization of these resonances in parameter space can be understood using appropriate asymptotics. A number of distinct canard trajectories involved in the observed transitions is constructed.

1 Introduction

Fourth order reversible systems capture the behavior of a host of systems in physics, chemistry, and biology [9, 20] that exhibit localized structures in the form of a time-independent patch of pattern embedded in a homogeneous background. Examples of systems that support localized structures of this type include buckling of slender structures [18, 19], ferrofluids [30], shear flows [32], convection [2–5, 24, 26], nonlinear optical media [12], urban criminal behavior [23, 34] and desert vegetation [33, 36]. We consider the following Swift–Hohenberg equation (SHE):

$$u_t = ru - \left(1 + \partial_x^2\right)^2 u + bu^2 - u^3, \tag{1}$$

where $u(x, t)$ is a real scalar field, r is a forcing parameter, and $b > 0$ is a constant. The Swift–Hohenberg equation provides an excellent qualitative description of spatially

P. Gandhi (✉) · E. Knobloch
Department of Physics, University of California, Berkeley, CA 94720, USA
e-mail: punit_gandhi@berkeley.edu

E. Knobloch
e-mail: knobloch@berkeley.edu

C. Beaume
Department of Aeronautics, Imperial College London, London SW7 2AZ, UK
e-mail: ced.beaume@gmail.com

© Springer International Publishing Switzerland 2016
M. Tlidi and M.G. Clerc (eds.), *Nonlinear Dynamics: Materials,*
Theory and Experiments, Springer Proceedings in Physics 173,
DOI 10.1007/978-3-319-24871-4_23

localized structures in the systems mentioned above. These solutions live on a pair of branches that *snake* within a snaking or pinning parameter interval. Analysis reveals that within this interval a large number of such states can be simultaneously stable [8, 21].

We consider the SHE (1) on a sufficiently long spatially periodic domain, and set $b = 1.8$ so that, for constant forcing, there exists an interval $r_{sn} \leq r \leq 0$ of bistability between the trivial $u \equiv 0$ state and a spatially periodic state $u = u_P(x)$ with $r_{sn} \approx -0.3744$. We characterize localized states comprised of a patch of u_P embedded in a $u = 0$ background through the location $x = f$ of the front that defines their right edge relative to their center $x = 0$:

$$f = 2 \frac{\int_0^{\Gamma/2} x u^2 \, dx}{\int_0^{\Gamma/2} u^2 \, dx}. \tag{2}$$

For forcing between $r_- \approx -0.3390$ and $r_+ \approx -0.2593$, the dynamics are organized around a series of stable localized solutions in which the fronts remain pinned at locations an integer number of wavelengths of u_P apart. The fronts of a localized initial condition within the pinning interval will move either inward or outward until f reaches a value corresponding to a stable state. Outside the pinning interval but inside the bistability interval, the fronts are no longer pinned and the localized patterns either steadily expand ($r_+ < r < 0$) or shrink ($r_{sn} < r < r_-$) via repeated wavelength nucleation or annihilation. In both cases, the average speed of the front increases monotonically with the distance to the pinning interval.

Near the edge of the pinning interval, leading order asymptotic theory [8] predicts that the time to nucleate or annihilate a wavelength of the pattern on either side is given by $(T_{\pm}^{dpn})^{-1} \approx \Omega_{\pm} \sqrt{|r - r_{\pm}|}/\pi$ when $0 < \pm(r - r_{\pm}) \ll 1$. We have used the subscript $+$ (resp. $-$) for nucleation (resp. annihilation) events on the right (resp. left) of the pinning interval. For quantitatively accurate predictions of the depinning time outside of this limit, we employ the following numerical fit [13]:

$$\left(T_{\pm}^{dpn}\right)^{-1} = \sum_{j=1}^{5} \sigma_j^{\pm} |r_0 - r_{\pm}|^{\frac{j}{2}}, \tag{3}$$

where the coefficients $\sigma_j^{\pm}$ are obtained from SHE simulations with constant forcing (Table 1).

To the left of the bistability interval (i.e., for $r < r_{sn}$), the dynamics of localized states are better described by overall amplitude decay, and we use a numerical fit of the form (3) to quantify the amplitude collapse time T_{sn}^{col} of a periodic solution below r_{sn} in terms of $|r - r_{sn}|$.

In practice, the forcing can fluctuate in time and induce the creation of new states [6, 25, 38]. Systems can be noisy [1, 29, 31] leading to front propagation [10] or temporally periodic [7, 22, 37] providing control opportunities [35]. We focus here on the dynamics of pre-existing localized structures under the influence of time-

Table 1 Values of the coefficients σ_j determined from a least squares fit of the depinning/collapse time to simulations of spatially localized initial conditions with constant forcing

	Ω	σ_1	σ_2	σ_3	σ_4	σ_5
T_+^{dpn}	0.5285	0.1687	0.1141	0.7709	-0.4000	0.0803
T_-^{dpn}	0.7519	0.2381	-0.8445	33.37	-306.4	1067
T_{sn}^{col}	0.7705	0.2081	0.4431	2.962	-34.15	79.52

The frequency Ω is calculated from leading order asymptotic theory [8] for perturbations of localized states that are marginally stable at $r_\pm$ and periodic states that are marginally stable at r_{sn}

periodic forcing. The effect of such oscillations on the growth of vegetation patches near the transition to desertification is of particular interest: over the course of a year, seasonal variations may place the system alternately within conditions where only the bare soil state is stable and within conditions where bistability between bare soil and vegetation patterns is observed. Steady models of this process predict the presence of patchy patterns [28] and only limited results are available on their reaction to time-dependence in external conditions [14, 39].

We introduce the time-periodic forcing in the simplest way:

$$r(t) = r_0 + \rho \sin(2\pi t/T), \tag{4}$$

where $r(t)$ is hereafter referred to as the forcing parameter, and restrict attention to localized structures satisfying reflection symmetry: $u(x) \to u(-x)$. The oscillation amplitude ρ is chosen to straddle the snaking interval $r_- \leq r_0 \leq r_+$ with $\rho > (r_+ - r_-)/2$ and $T > 0$. The time-dependent forcing may cause the localized patterns to breathe, or grow for part of the cycle via nucleation of new wavelengths of the pattern followed by wavelength annihilation during the remainder of the cycle as we shall see. The period of the forcing, T, introduces a new characteristic time in the system that interacts with the depinning time to create resonances. The origin of these resonances as well as their impact on the way parameter space is structured is described in the next section. Section 3 discusses a class of peculiar periodic orbits called canard orbits that are involved in the transitions between resonances. The paper concludes in Sect. 4.

2 Temporal Resonances

To understand the series of resonances underpinning the partitioning of the parameter space described below, we begin by considering the effect of an asymptotically small forcing amplitude $\rho \ll 1$ when r_0 is located near one of the edges of the pinning interval ($|r_0 - r_\pm| \ll 1$). When $(r_0 - r_\pm)/\rho$ is finite the depinning time scales like $|r_0 - r_\pm|^{-1/2}$ and we therefore choose a forcing period such that $\sqrt{\rho}T \sim \mathcal{O}(1)$ in order to allow enough time for depinning to occur while the system is outside of the

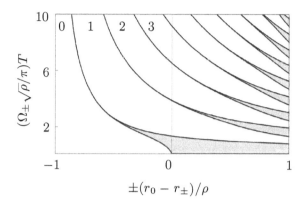

Fig. 1 Resonance bands (*white*) in the (r_0, T) plane obtained from the Mathieu description [13]. Localized states that are marginally stable at $r = r_\pm$ undergo a fixed number of depinning events per forcing cycle within each resonance band. This number is indicated in the most prominent bands. The transition zones (*gray*) indicate parameter values where the average number of depinning events per forcing cycle is non-integer

constant r pinning interval. It turns out that this limit is described, after appropriate transformation, by the Mathieu equation [13] which captures precisely the periodicity of the nucleation process in the frame of the front. This equation predicts a set of resonance bands within an $\mathcal{O}(\rho)$ vicinity of either edge of the pinning interval. The resonances occur when the system spends an integer number of nucleation times outside of the constant r pinning interval. The r-dependence of the depinning time (3) allows the resonance bands to persist even when the system remains outside the constant r pinning interval throughout the entire forcing cycle.

Figure 1 shows the predicted dynamics of a localized state that is marginally stable at $r = r_\pm$ in the presence of the periodic forcing (4) in terms of $\sqrt{\rho}T$ (in units of $\pi/\Omega_\pm$) and $\pm(r_0 - r_\pm)/\rho$. A series of resonance bands (shown in white) separated by transition zones (shown in gray) is observed. In each of these bands, the number of depinning events per forcing cycle is locked to an integer number, starting with 0 for the leftmost band, and increasing by 1 for each successive band. The transition zones indicate parameters with a non-integer average number of depinning events per forcing cycle, resulting either in periodic motion with period greater than T or nonperiodic motion [13].

2.1 Creation of Sweet Spots and Pinched Zones

We can analyze the interaction between the resonance bands occurring at either edge of the pinning interval in the following asymptotic limit. We tune the amplitude and average value of the forcing such that the extrema of the forcing remain within an

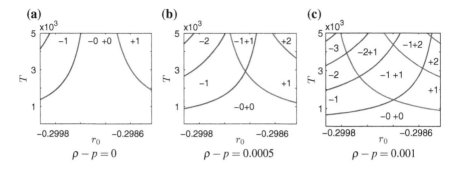

Fig. 2 Predictions from the asymptotic theory (5). The *blue* (resp. *red*) *lines* correspond to parameter values where n_- (resp. n_+) changes (color online)

asymptotically small vicinity of the edges of the pinning interval: $|\rho - p| \ll 1$, where $p = (r_+ - r_-)/2$, and assume $|r_0 - r_c| \ll 1$, where $r_c = (r_+ + r_-)/2$ is the center of the pinning interval, so that $|\rho - p|/|r_0 - r_c| \sim \mathscr{O}(1)$. Additionally, we choose the period of the forcing cycle such that $|\rho - p|T \sim \mathscr{O}(1)$ in order to obtain slow-fast dynamics involving slow drifts along the constant forcing localized state branch separated by fast depinning events.

This limit predicts the number of nucleation events n_+ and annihilation events $-n_-$ during one forcing period [13]:

$$
n_\pm = \begin{cases} \left[\frac{\Omega_\pm T}{2\pi\sqrt{2p}}(r_0 \pm \rho - r_\pm)\right] & \text{if } \pm(r_0 \pm \rho - r_\pm) > 0 \\ 0 & \text{if } \pm(r_0 \pm \rho - r_\pm) \leq 0 \end{cases}, \tag{5}
$$

where the brackets indicate rounding to the nearest integer and come from the settling of the state to a stable localized solution upon re-entry into the pinning interval.

For $\rho - p < 0$, the resonance bands associated with the left and right edges of the pinning interval are disjoint but asymptotically approach $r_0 = r_c$ as $T \to \infty$ for $\rho = p$. For $\rho - p > 0$, an asymptotically small sweet spot and pinching structure begins to form as a result of successive crossings between the resonance bands, as shown in Fig. 2.

2.2 Phase Space Partitioning

We ran simulations for $\rho = 0.1$ and $10 \leq T \leq 400$, initialized with a stable steady-state localized solution at $r(t = 0)$. The simulations revealed four different behaviors that we exemplify in Fig. 3. The localized structure expands (f increases) by the nucleation of new wavelengths on either side of the pattern when $r > r_+$ while it contracts (f decreases) by the decay of side wavelengths when $r < r_-$. In the case

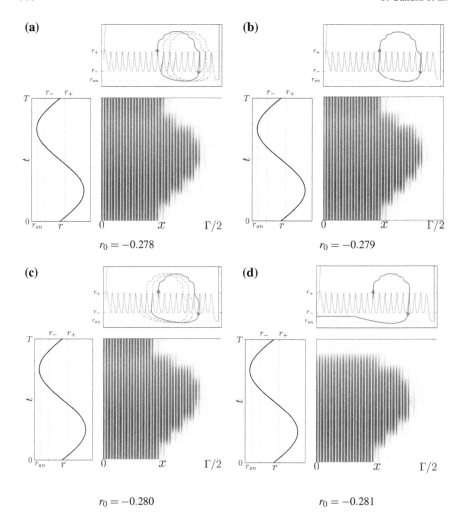

Fig. 3 Breathing localized structures observed in a periodic domain of size $\Gamma = 80\pi$ in the SHE (1) with the forcing (4) for $\rho = 0.1$, $T = 300$, $b = 1.8$ and different values of r_0. *Left panels* the forcing r represented as a function of time with thin *dashed lines* indicating the boundaries of the pinning and bistability regions. *Right panels* space-time diagrams over one forcing cycle for the right half of each state with positive (negative) values of the field u shown in *red* (*blue*). *Top panels* trajectories of the position $x = f$ of the right front of the localized state in (f, r) space superposed on the constant forcing snaking diagram

$r < r_{sn}$, an overall amplitude decay mode kicks in that may destroy the localized state within a single forcing period. We can therefore observe growing (Fig. 3a) or decaying (Fig. 3c) solutions, or collapse to the trivial state (Fig. 3d). When not enough time is spent below r_{sn}, the dynamics can be balanced by suitably choosing the parameter values and spatially localized periodic orbits can be obtained (Fig. 3b).

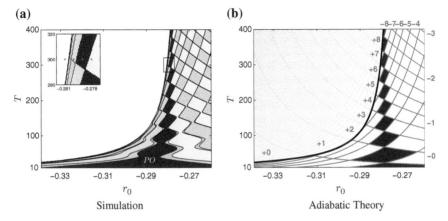

Fig. 4 **a** Color map of the different behaviors observed from simulations in the (r_0, T) plane for $\rho = 0.1$ and $b = 1.8$ [13]. Periodic orbits exist within region *PO*. The *yellow/orange* (*light blue/blue*) regions to the *right* (*left*) of *PO* correspond growing (decaying) solutions where the pattern experiences net growth (decay) by 1, 2, ... wavelengths on either side per cycle. All regions are defined by $-0.25 + N < \langle \Delta f \rangle / 2\pi < 0.25 + N$ with $\pm N \in \mathbb{N}$. We note that a more stringent definition would produce slightly narrower regions, particularly for shorter periods, but qualitatively similar. Transition zones are shown in *gray*. The *white region* indicates parameter values at which the amplitude of the localized pattern decays within one cycle independently of its original length. The *inset* shows the location of the simulations shows in Fig. 3. **b** The *red* (*blue*) lines show predictions from (7). The bands are labeled with *red* (*blue*) signed integers, and the *thick black line* marks the prediction of the cliff beyond which amplitude decay is expected. Both panels are plotted over the static pinning interval $r_- < r_0 < r_+$

We characterize the results in a (r_0, T) diagram through the average motion of the fronts per forcing cycle:

$$\langle \Delta f \rangle = \frac{f(t = t_0 + N_t T) - f(t = t_0)}{N_t}, \tag{6}$$

where t_0 is large enough to bypass initial transients and N_t is a large number of forcing cycles over which the dynamics is averaged. The numerical results are reported in Fig. 4a. Periodic orbits exist in the region *PO* that displays a series of contractions and expansions and progressively shrinks as T increases. Around this region, similarly shaped regions display states that expand or shrink over time. They are structured in a regular fashion: the region right next to *PO* displays states that grow or decay by one wavelength on each side of the pattern during each forcing cycle, and farther regions display successively faster growing or decaying states. The transition between each of these regions is not abrupt and occurs via transition zones [shown in gray in Fig. 4a] [13].

With constant forcing, one can approximate the signed number of depinning events that occur outside of the pinning interval by integrating the depinning rate over the time of interest:

$$n_\pm = \pm \int \frac{dt}{T_\pm^{\text{dpn}}(r)}. \tag{7}$$

In the limit $T \to \infty$, we can treat the parameter $r(t)$ quasi-statically and make use of (3) for $T_\pm^{\text{dpn}}$. We construct an adiabatic prediction by assuming the following series of events during each forcing cycle:

- $r > r_+$: the localized state begins to nucleate wavelengths of the pattern. We count the total number of depinning events using positive real numbers $n_+ > 0$ obtained from (7) using (3) and Table 1.
- $r_+ > r > r_-$: Upon entry into the pinning interval, the state converges to the closest stable localized solution, corresponding to rounding n_+ or n_- to the nearest integer $[n_+]$ or $[n_-]$.
- $r < r_-$: We count the number of wavelengths annihilated on either side using negative real numbers $n_- < 0$ also obtained from (7). If the time spent with $r < r_{sn}$ exceeds that required for $\int_{r<r_{sn}} \left(T^{\text{col}}(r(t)) \right)^{-1} dt = 1/2$, then the state decays irrevocably.

Figure 4b shows that this prediction bears a striking resemblance to the numerical results in Fig. 4a. In fact, most of the features obtained during simulations can be explained using the adiabatic theory even far away from the limit for which it is constructed [13, 14]. The series of contractions and expansions observed in the numerical simulation is therefore a reflection of the sweet spot and pinching structure created as in Fig. 2.

3 Canard Trajectories

We have, up to this point, described *stable* localized breathing states that we obtained by time-stepping a *stable* steady-state solution to (1) with constant forcing $r = r_0$. As the snaking structure indicates, each stable localized state is connected to the next one by a branch of unstable localized states. These unstable states also generate spatially localized periodic orbits under periodic forcing. These orbits are similar to those presented in the last section but instead of tracking the stable part of the snaking branches as the forcing varies, they track the unstable part and are therefore unstable as well.

Near the transition between neighboring resonance bands in Fig. 2, one can find trajectories that follow a stable snaking branch during one traversal of the pinning interval and an unstable one during the return trip. A small change in parameter values can cause the trajectory to jump to one of the two nearby stable solution branches during its passage along the unstable branch before completing the journey across the pinning interval. During such a jump the fronts bounding the localized

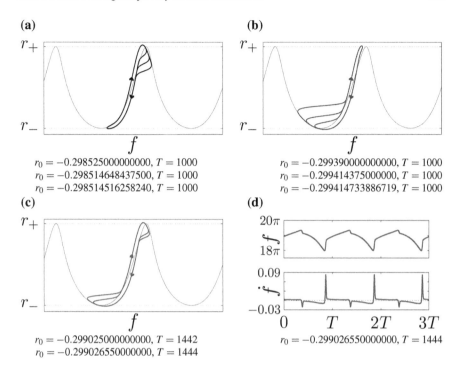

(a)

$r_0 = -0.298525000000000, T = 1000$
$r_0 = -0.298514648437500, T = 1000$
$r_0 = -0.298514516258240, T = 1000$

(b)

$r_0 = -0.299390000000000, T = 1000$
$r_0 = -0.299414375000000, T = 1000$
$r_0 = -0.299414733886719, T = 1000$

(c)

$r_0 = -0.299025000000000, T = 1442$
$r_0 = -0.299026550000000, T = 1444$

(d)

$r_0 = -0.299026550000000, T = 1444$

Fig. 5 **a** C^+ canards, **b** C_- canards and **c** C_-^+ canards represented through the front location $f(t)$ versus the forcing strength $r(t)$ for $\rho - p = 0.001$ (Fig. 2c). The *thin blue line* represents the stable ($\partial_f r > 0$) and unstable ($\partial_f r < 0$) parts of the branch of localized solutions for constant forcing ($\rho = 0$). In each case the parameters are listed in order of increasing time spent on the unstable branch. **d** Three periods of a C_-^+ canard from panel (**c**) shown using the front location $x = f$ and its speed $\dot{f}$ as functions of time (color online)

state move outward or inward depending on whether the stable state reached is longer or shorter than the unstable state. In the following we refer to trajectories that drift along unstable states for part of the forcing cycle as *canard* trajectories [11].

It is possible to control how far along the unstable solution branch the system reaches before jumping to a stable branch and thereby generate a family of canard trajectories. Figure 5 shows three families of periodic canard trajectories computed from (1) such that $||u(t) - u(t+T)||_{L^2} < 10^{-10}$ for some sufficiently large t. Solutions in the family of C^+ canards follow the unstable branch close to the saddle-node at $r = r_+$ but deviate before reaching the saddle-node at $r = r_-$ [panel (a)]. Solutions of this type are found near the transition between one growth band and the next. The C_- canards shown in panel (b) follow the unstable branch close to the r_- saddle-node but do not reach the r_+ saddle-node; these are found near transitions between adjacent decay bands. Both sets of transitions are approximated by (5). In regions where both bands intersect, it is possible to obtain C_-^+ canards [panel (c)] which temporarily follow two different unstable branches; the associated front location $x = f$ and its

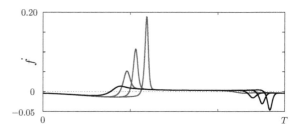

Fig. 6 One period of the C_- (*green*) and C^+ (*black*) canards from Fig. 5a, b represented through the speed $\dot{f}$ of the right front as a function of time. The larger amplitude peaks are associated with the larger canards in Fig. 5 (color online)

speed $\dot{f}$ is represented in panels (d). When the trajectory is drifting along the branch of steady states the fronts move slowly inward or outward; however, the jumps from the unstable state to the stable state manifest themselves in abrupt changes in the front location, or equivalently in dramatic peaks in the front speed $\dot{f}$. Figure 6 shows how a small change in r_0 (T remaining fixed) impacts the time evolution of canard trajectories. Decreasing r_0 delays the onset of the bursts and increases the front speed, a consequence of the fact that the trajectory now departs from an unstable state farther from the saddle-node and hence with a larger unstable eigenvalue. However, canards that manage to traverse almost the entire unstable part of the branch of steady states are expected to display once again slower dynamics.

The canards shown in Fig. 5 correspond to the simplest canard families, organized by a single stable portion of the branch of steady states with no depinning. However, a careful tuning of the parameters reveals the presence of canards displaying depinning. Figure 7 shows several examples of the corresponding trajectories.

The periodic orbits described by these canards are organized around *two* segments of stable steady states and the adjacent unstable steady states. The transitions between these segments are associated with the addition or loss of one wavelength on either side of the localized structure. A whole flock of canards can thus be obtained involving more segments of stable states and therefore displaying more depinning events per cycle.

In a similar fashion, we can obtain periodic orbits whose solution amplitude follows that of the lower branch spatially periodic state of the steady SHE. This gives rise to C_- canards characterized by a monotonic decrease in amplitude followed by a sudden jump to larger amplitude. Since the spatially periodic state u_p only displays one saddle-node no C^+ or C_-^+ canards can be obtained. These canard trajectories, represented in Fig. 8 (left panel), can be made to follow the unstable periodic state for a longer amount of time by choosing r_0 closer to the transition to amplitude collapse. Such trajectories spend more time in the depinning regime ($r < r_-$) as well as more time in the collapse regime ($r < r_{sn}$). As a result this regime is characterized by a competition between depinning and amplitude collapse as illustrated in Fig. 9 but the state ultimately collapses as exemplified by the spiraling trajectory in Fig. 8 (right panel).

(a)   **(b)**

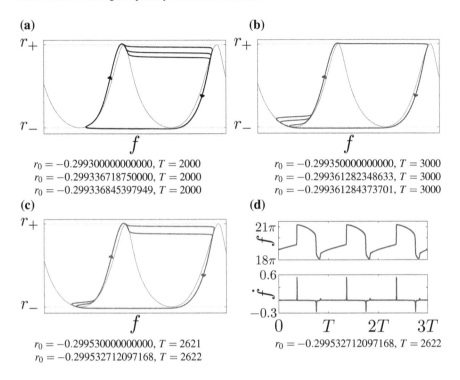

Fig. 7 "Larger" canards represented in the same fashion as in Fig. 5

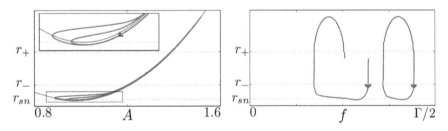

Fig. 8 Amplitude and front position of canard trajectories of spatially *localized* states that follow the unstable amplitude of the spatially *periodic* state u_p for some amount of time. Here $\rho = -0.1$, $r_0 = -0.276055$, -0.276220, $T = 1100$, and $\Gamma = 640\pi$. The fold on the u_p branch is at $r_{sn} \approx -0.374370$

4 Discussion

In this chapter, we have used the SHE with a sinusoidal time-periodic forcing to describe how steady localized states are impacted by temporal variations in a parameter that temporarily take them outside their existence range. Numerical simulations complemented with asymptotic predictions were used to determine the location in

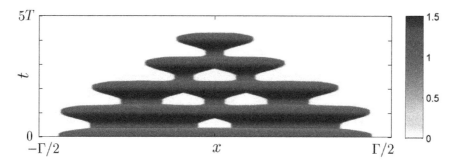

Fig. 9 Space-time plot of an amplitude canard trajectory similar to the shrinking canard seen in Fig. 8. Here, new fronts are generated in the interior in addition to the breathing dynamics at the edges. Parameters are $\rho = -0.1$, $r_0 = -0.276228387$, $T = 1100$, and $\Gamma = 640\pi$. Owing to the large extent of the domain, the pattern is not fully represented, only its local maxima are plotted against time

parameter space of time-periodic spatially localized states and to reveal an unexpected sweet spot–pinching structure in the (r_0, T) plane for a fixed amplitude ρ of the forcing. This structure is a consequence of a series of resonances between the forcing period and the nucleation time for new cells outside the pinning interval, and can be reproduced accurately using adiabatic theory as summarized in (7). Close to the resonance bands, a series of canard trajectories can be found. Two types of canards have been identified: phase canards, in which the spatial extent of the localized pattern changes abruptly as an additional wavelength is nucleated or annihilated on either side, and amplitude canards, in which the amplitude temporarily drops to the amplitude of the unstable lower branch of spatially periodic states before abruptly increasing to the amplitude of the stable upper states. Canards are normally considered to be a property of finite-dimensional systems, although there are indications that they should be observable in pattern-forming (i.e., spatially extended) systems and in particular in the Faraday system [16]. It is therefore of particular interest to present clear evidence for such orbits in a partial differential equation.

The Swift–Hohenberg equation provides a dependable framework for studies and control of the dynamics of spatially localized states as proved time and time again [17, 27]. For this reason the robustness of the resonance structure predicted by adiabatic theory for parameter values far from the adiabatic limit leads us to expect similar dynamics in related systems, and in particular in models of desert vegetation [39]. The reason for this expectation is that the phenomena described here are fundamentally low-dimensional. Indeed, a similar series of resonances is present in a simple ordinary differential equation, the periodically forced Adler equation, as described elsewhere [15].

Acknowledgments This work was supported by the National Science Foundation under grant CMMI–1233692.

References

1. D. Barkley, L.S. Tuckerman, Computational study of turbulent laminar patterns in Couette flow. Phys. Rev. Lett. **94**, 014502 (2005)
2. C. Beaume, A. Bergeon, H.-C. Kao, E. Knobloch, Convectons in a rotating fluid layer. J. Fluid Mech. **717**, 417–448 (2013)
3. C. Beaume, A. Bergeon, E. Knobloch, Convectons and secondary snaking in three-dimensional natural doubly diffusive convection. Phys. Fluids **25**, 024105 (2013)
4. C. Beaume, H.-C. Kao, E. Knobloch, A. Bergeon, Localized rotating convection with no-slip boundary conditions. Phys. Fluids **25**, 124105 (2013)
5. C. Beaume, E. Knobloch, A. Bergeon, Nonsnaking doubly diffusive convectons and the twist instability. Phys. Fluids **25**, 114102 (2013)
6. I. Belykh, V. Belykh, R. Jeter, M. Hasler, Multistable randomly switching oscillators: the odds of meeting a ghost. Euro. Phys. J. Spec. Top. **222**, 2497–2507 (2013)
7. P. Binder, D. Abraimov, A.V. Ustinov, S. Flach, Y. Zolotaryuk, Observation of breathers in Josephson ladders. Phys. Rev. Lett. **84**, 745–748 (2000)
8. J. Burke, E. Knobloch, Localized states in the generalized Swift–Hohenberg equation. Phys. Rev. E **73**, 056211 (2006)
9. A.R. Champneys, Homoclinic orbits in reversible systems and their applications in mechanics, fluids and optics. Phys. D **112**, 158–186 (1998)
10. M.G. Clerc, C. Falcon, E. Tirapegui, Additive noise induces front propagation. Phys. Rev. Lett. **94**, 148302 (2005)
11. W. Eckhaus, *Relaxation oscillations, including a standard chase on French ducks*. In: Asymptotic Analysis II, Lecture Notes in Mathematics, vol. 985 (Springer, New York, 1983), pp. 449–494
12. W.J. Firth, L. Columbo, A.J. Scroggie, Proposed resolution of theory-experiment discrepancy in homoclinic snaking. Phys. Rev. Lett. **99**, 104503 (2007)
13. P. Gandhi, C. Beaume, E. Knobloch, A new resonance mechanism in the Swift–Hohenberg equation with time-periodic forcing. SIAM J. Appl. Dyn. Sys. **14**, 860–892 (2015)
14. P. Gandhi, E. Knobloch, C. Beaume, Localized states in periodically forced systems. Phys. Rev. Lett. **114**, 034102 (2015)
15. P. Gandhi, E. Knobloch, C. Beaume, Periodic phase-locking and phase slips in active rotator systems. arXiv:1509.03582 (2015)
16. M. Higuera, E. Knobloch, J.M. Vega, Dynamics of nearly inviscid Faraday waves in almost circular containers. Phys. D **201**, 83–120 (2005)
17. S.M. Houghton, E. Knobloch, Swift–Hohenberg equation with broken cubic-quintic nonlinearity. Phys. Rev. E **84**, 016204 (2011)
18. G.W. Hunt, H.M. Bolt, J.M.T. Thompson, Structural localization phenomena and the dynamical phase-space analogy. Proc. R. Soc. London A **425**, 245–267 (1989)
19. G.W. Hunt, M.A. Peletier, A.R. Champneys, P.D. Woods, M.A. Wadee, C.J. Budd, G.J. Lord, Cellular buckling in long structures. Nonlinear Dyn. **21**, 3–29 (2000)
20. E. Knobloch, Spatially localized structures in dissipative systems: open problems. Nonlinearity **21**, T45–T60 (2008)
21. E. Knobloch, Spatial localization in dissipative systems. Annu. Rev. Condens. Matter Phys. **6**, 325–59 (2015)
22. O. Lioubashevski, Y. Hamiel, A. Agnon, Z. Reches, J. Fineberg, Oscillons and propagating solitary waves in a vertically vibrated colloidal suspension. Phys. Rev. Lett. **83**, 3190–3193 (1999)
23. D.J.B. Lloyd, H. O'Farrell, On localised hotspots of an urban crime model. Phys. D **253**, 23–39 (2013)
24. D. Lo Jacono, A. Bergeon, E. Knobloch, Three-dimensional binary fluid convection in a porous medium. J. Fluid Mech. **730**, R2 (2013)
25. B. Marts, A. Hagberg, E. Meron, A.L. Lin, Resonant and nonresonant patterns in forced oscillators. Chaos **16**, 037113 (2006)

26. I. Mercader, O. Batiste, A. Alonso, E. Knobloch, Convectons, anticonvectons and multiconvectons in binary fluid convection. J. Fluid Mech. **667**, 586–606 (2011)

27. I. Mercader, O. Batiste, A. Alonso, E. Knobloch, Travelling convectons in binary fluid convection. J. Fluid Mech. **772**, 240–266 (2013)

28. E. Meron, Pattern-formation approach to modelling spatially extended ecosystems. Ecol. Model. **234**, 70–82 (2012)

29. A. Prigent, G. Grégoire, H. Chaté, O. Dauchot, W. van Saarloos, Large-scale finite-wavelength modulation within turbulent shear flows. Phys. Rev. Lett. **89**, 014501 (2002)

30. R. Richter, I.V. Barashenkov, Two-dimensional solitons on the surface of magnetic fluids. Phys. Rev. Lett. **94**, 184503 (2005)

31. B. Schäpers, M. Feldmann, T. Ackemann, W. Lange, Interaction of localized structures in an optical pattern-forming system. Phys. Rev. Lett. **85**, 748–751 (2000)

32. T.M. Schneider, J. Gibson, J. Burke, Snakes and ladders: localized solutions of plane Couette flow. Phys. Rev. Lett. **104**, 104501 (2010)

33. J.A. Sherratt, An analysis of vegetation stripe formation in semi-arid landscapes. J. Math. Biol. **51**, 183–197 (2005)

34. M.B. Short, A.L. Bertozzi, Nonlinear patterns in urban crime: hotspots, bifurcations and suppression. SIAM J. Appl. Dyn. Sys. **9**, 462–483 (2010)

35. J.V.I. Timonen, M. Latikka, L. Leibler, R.H.A. Ras, O. Ikkala, Switchable static and dynamic self-assembly of magnetic droplets on superhydrophobic surfaces. Science **341**, 253–257 (2013)

36. M. Tlidi, R. Lefever, A. Vladimirov, *On vegetation clustering, localized bare soil spots and fairy circles.* In: Dissipative Solitons: From Optics to Biology and Medicine, Lecture Notes in Physics, vol. 751 (Springer, Berlin, 2008), pp. 1–22

37. P.B. Umbanhowar, F. Melo, H.L. Swinney, Localized excitations in a vertically vibrated granular layer. Nature **382**, 793–796 (1995)

38. A. Yochelis, C. Elphick, A. Hagberg, E. Meron, Frequency locking in extended systems: The impact of a Turing mode. Europhys. Lett. **69**, 170–176 (2005)

39. Y.R. Zelnik, S. Kinast, H. Yizhak, G. Bel, E. Meron, Regime shifts in models of dryland vegetation. Phil. Trans. R. Soc. A **371**, 20120358 (2013)

Formation and Interaction of Two-Kink Solitons

Mónica A. García-Ñustes

Abstract Two-kink soliton solutions have received attention due to its implications in string theory and particles physics. In the present work we analyze the necessary conditions for the appearance of two-kink solitons in the sine-Gordon model under the perturbation of a space-dependant force. Interactions of two-kink solitons are also investigated. We show that kink repulsive force eventually leads to a fragmentation of the solution before interaction and finally to dissolution of these bound states.

1 Introduction

The appearance of kink solitons in scalar field models such as ϕ^4 and sine-Gordon (SG) equation depends on the configuration of the energy potential $U(\phi)$. When the potential has, at least, two minima, the system exhibits a solution that connects both energetic states [1].

For example, in the case of ϕ^4 model the solution connects the potential minima $\bar{\phi} = \pm 1$. In the SG case, the potential supports infinite minima. Thus, if ϕ only varies from 0 to 2π, the kink soliton interpolates between the two adjacent minima $\bar{\phi} = 0, 2\pi$. In general, it has been proved that depending on the shape of the potential $U(\phi)$ various classes of localized solutions in the form of kinks, pulses, and semisolitons are possible [2].

An interesting variation of kink soliton is the two-kink soliton introduced by Bazeia et al. [3, 4]. These solutions are present in a family of models described by polynomial potentials $U_p(\phi)$ in (1, 1) dimensions which support minima at $\phi = 0$ and ± 1 for odd values of the control parameter p. The solution connects the minima

M.A. García-Ñustes (✉)
Instituto de Física, Pontificia Universidad Católica de Valparaíso,
Avenida Brasil, Valparaíso, Casilla 2950, Chile
e-mail: mgarcia16@gmail.com

© Springer International Publishing Switzerland 2016
M. Tlidi and M.G. Clerc (eds.), *Nonlinear Dynamics: Materials,
Theory and Experiments*, Springer Proceedings in Physics 173,
DOI 10.1007/978-3-319-24871-4_24

317

$\phi = \pm 1$ passing through the local minimum at $\phi = 0$. Therefore, the two-kink soliton profile consists of two standard kinks separated by a distance which is proportional to p, the parameter that specifies the potential.

The two-kink soliton has received attention by its implications in the brane world scenario [5–7] and in particle physics. This latter possesses a particular interest. Kink solitons can exhibit properties typically associated with particle-like states. In this manner, kink-antikink configurations can be regarded as particle-antiparticle interactions and may lead to a better understanding of fundamental properties of particles. Additionally, experimental realization of kink-antink configurations can be easily achieved in long Josephson junction (LJJ) and its discrete version called Josephson array. A regular kink in LJJ corresponds to a Josephson vortex or fluxon— a quantum of magnetic flux—which is well described by the SG kink solution. The fusion of two regular fluxons (kinks) into a bound state (two-kink state) has been already reported in an annular Josephson array subject to a constant bias current [8–10]. Similar solutions called multikinks have been reported in the discrete version of the SG model—the Frenkel-Kontorova model—with higher order dispersive terms [11, 12].

Further investigations [6, 7] revealed that the two-kink solution can be observed in models whose potentials are controlled by a parameter and not by the degree of the self interacting scalar field. In other words, the energy configuration of the potential can be changed by a control parameter. Indeed, in [13] we have shown the formation of a pair of two-kink solitons in a perturbed sine-Gordon model due to internal mode instabilities. A suitable space-dependant force parametrized by a control parameter acts as a forming force of two-kink solutions. For a certain value of the control parameter, an internal shape mode destabilizes, creating consecutive pairs of kinks. Concurrently, the external force generates an effective potential that permits that the consecutive kinks formed a bound state, a two-kink solution appears.

In the present work, we present a brief summary of the necessary conditions under which the formation of a two-kink soliton is possible in a perturbed SG model, following the general approach introduced in [14]. This procedure allows us to conduct an exact stability analysis of the internal modes when the system is perturbed by a suitable space-dependent force. Additionally, we analyze the interaction of two-kink solutions formed by internal mode instabilities. Due to the kink-antikink attractive interaction, the two-kink solitons are fragmented before their collision, resulting in a total breaking of the bound state afterwards.

This work is organized as follows. In Sect. 2, we review the exact stability analysis approach in the SG model perturbed by a external force. In Sect. 3 we obtain an expression of the energy that permits us to predict and control multi-kink formation. An analysis of the required conditions for the appearance of a pair of two-kink solitons is given in Sect. 4. Interactions of two-kink solitons formed by internal instabilities is studied in Sect. 5. Finally we present our concluding remarks in Sect. 6.

2 Stability Analysis

The sine-Gordon model has been applied in many branches of physics as particle physics and condensed matter theory. The model can describe many phenomena from domain walls in ferromagnets and fluxons in long Josephson junctions in solid state physics [15, 16] to DNA models in biology [17].

In the unperturbed version of the SG model, the solution does not have internal modes. However, following the stability analysis technique, it has been shown that when the system is perturbed by an inhomogeneous external force, SG soliton internal modes can exist. Furthermore, inhomogeneous perturbations of the soliton equations both in the form of external forces or parametric impurities, generate effective potentials for the motion of the kink, i.e., the zeros of $F(x)$ represent equilibrium positions for the kink (antikink). The activation of extra internal modes and the zeros of the force can lead to the appearance of different phenomena such as kink explosion, tunneling escape from potential wells, creation of a kink-antikink pair, among others [14, 18–20].

To understand the conditions for the appearance of a two-kink soliton solution, let us review the stability conditions for a kink in the presence of an inhomogenous external force $F(x)$, with at least one x_* such that $F(x_*) = 0$ [14].

The perturbed Klein-Gordon equation is

$$\phi_{tt} - \phi_{xx} + \frac{\partial U}{\partial \phi} = F(x), \tag{1}$$

where $U(\phi)$ is a potential that possesses at least two minima. The sine-Gordon and ϕ^4 equations are particular examples of (1) when $F(x) = 0$.

To solve exactly the stability problem, it is necessary to build a general function ϕ_k with the topological properties of the kink-soliton solution. Then, via an inverse problem, we are able to obtain a suitable $F(x)$ with the properties of the physical system under study. In particular, if $F(x)$ possesses only one zero ($F(x_*) = 0$) and $\frac{\partial F(x)}{\partial x}\big|_{x=x_*} > 0$, then the point $x = x_*$ is a stable (unstable) equilibrium position for the kink (antikink). Otherwise, if $\frac{\partial F(x)}{\partial x}\big|_{x=x_*} < 0$, the equilibrium position $x = x_*$ is unstable (stable) for the kink (antikink).

Considering $\phi(x, t) = \phi_k(x) + f(x)e^{\lambda t}$ and performing a linear stability analysis we get the following eigenvalue problem,

$$\hat{L} f = \Gamma f, \tag{2}$$

where $\hat{L} = -\partial_x^2 + \left\{ \frac{\partial^2 U}{\partial u^2}\big|_{\phi=\phi_k(x)} \right\}$ and $\Gamma = -\lambda^2$. The translational and internal shape modes associated to the kink can be obtained from (2). For further explanations read [14, 18–20] and references therein.

Let us see the following perturbed SG model,

$$\phi_{tt} - \phi_{xx} + \sin\phi = F(x) \tag{3}$$

where

$$F(x) = \frac{2(B^2 - 1)\sinh(Bx)}{\cosh^2(Bx)} \tag{4}$$

is an external force with a single zero (or equilibrium position for the kink soliton) in $x_* = 0$.

The exact stationary solution of (3) is $\phi_k(x) = 4\arctan(\exp(Bx))$. This solution represents a kink-soliton at the position $x = 0$. The stability problem (2) can be solved exactly. The eigenvalues of the discrete spectrum are given by the formula

$$\Gamma_n = B^2(\Lambda + 2\Lambda n - n^2) - 1, \tag{5}$$

where $\Lambda(\Lambda + 1) = \frac{2}{B^2}$.

The integer part of Λ ($[\Lambda]$) yields the number of eigenvalues in the discrete spectrum, which correspond to the soliton modes (including the translational mode Γ_0, and the internal shape modes $\Gamma_n(0 < n < [\Lambda])$. Everything concerning the stability of the soliton in this situation can be obtained from the equation for Γ_n.

The parameter B acts as a control parameter. For $B^2 > 1$, the translational mode is stable and there are no internal (shape) modes. In this regime, the kink stays at the equilibrium position ($x_* = 0$). If $1/3 < B^2 < 1$, then the translational mode is unstable but there are no internal modes. The kink moves from its unstable equilibrium position without shape deformations. When $1/6 < B^2 < 1/3$, the translational mode and one internal shape mode arise. This internal shape mode is stable. For $B^2 < 1/6$ many other internal modes can appear. For $B^2 < 2/[\Lambda_*(\Lambda_* + 1)]$, where $\Lambda_* = (5 + \sqrt{17})/2$, the first internal mode becomes unstable. This instability leads to a kink break-up by an internal shape instability and consequently to the emission of a pair kink-antikink [14, 18, 20].

As expected, numerical simulations performed close to the threshold $B^2 < 2/[\Lambda_*(\Lambda_* + 1)]$ have shown the emission of a kink-antikink pair. Below the threshold, a consecutive series of kink-antikink pairs are emitted leading to the formation of multi-kink bound states, among them, the two-kink soliton [13]. The number of emitted pairs depends of the control parameter B. To evaluate and therefore predict the number of consecutive emitted pairs is necessary some energy considerations.

Despite the particular construction of the external force, any physical model topologically equivalent will exhibit the same dynamical behavior around an equilibrium position. This scenario is also valid for any $F(x)$ with several zeros. Additionally, the stability conditions are stated in terms of the control parameter B. Therefore, we can easily control the kink dynamics by varying the external force parameters. This is specially important in experimental realizations of these physical systems.

3 Energy Considerations

As we have mentioned above, the appearance of the two-kink soliton is linked to the successive emission of kink-antikink pairs. Such emission requires an extra amount of energy to be able to afford consecutive pairs formation. This energy is given by the external force and consequently depends of the control parameter B. The energy carried by a kink, in the unperturbed SG model, described by $\phi_0 = 4\arctan(e^{\pm\gamma(x-vt)})$ is $E = 8\gamma$ where v stands for the kink velocity and $\gamma = 1/(1 - v^2)$.

The perturbed nonlinear Klein-Gordon equation (1) can be derived from the following Lagrangian density

$$\mathscr{L} = \frac{1}{2}\phi_t^2 - \frac{1}{2}\phi_x^2 - U(\phi) + F(x)\phi, \tag{6}$$

with the associated Hamiltonian density

$$\mathscr{H} = \frac{1}{2}\phi_t^2 + \frac{1}{2}\phi_x^2 + U(\phi) - F(x)\phi. \tag{7}$$

The energy conservation law can be written as $\frac{dH}{dt} = 0$, where $H = \int_{-\infty}^{\infty} \mathscr{H}\,dx$. If $\phi_k = 4\arctan(e^{Bx})$ and $U(\phi) = (1 - \cos\phi)$, then the total energy can be written as

$$H_T = H_k + H_{ext} \tag{8}$$

where $H_k = \frac{4}{B}(B^2 + 1)$ represents the energy associated with the solution ϕ_k. H_{ext} is the extra energy given by the external force. Replacing $F(x)$ by expression (4), we obtain that,

$$H_{ext} = -8(B^2 - 1) \int\limits_{-\infty}^{\infty} \frac{\sinh(Bx)}{\cosh^2(Bx)} \arctan(e^{Bx})\,dx. \tag{9}$$

After an integration by parts, H_{ext} reads,

$$\begin{aligned} H_{ext} &= -\frac{8(B^2 - 1)}{B} \left[\frac{2e^{Bx}\arctan(e^{Bx}) + 1}{e^{2Bx} + 1} \right] \Big|_{-\infty}^{\infty} \\ &= \frac{8(B^2 - 1)}{B}. \end{aligned} \tag{10}$$

Therefore, the expression for the total energy is,

$$H_T = \frac{4}{B}\left(3 - B^2\right). \tag{11}$$

As we can see, the total energy increases with decreasing values of B. From the stability conditions, the first emission of a pair kink-antikink takes place when $B^2 < 1/6$. Using (11) with $B^2 = 1/6$, we are able to estimate the energy minimum value for the first kink-antikink pair creation: $H_{I'} = 34\sqrt{6}/3 \approx 27.76$. By comparison with the energy related to ϕ_k, it is clear that the external force provides enough energy to produce an additional kink-antikink pair.

Using this criteria, we can get an estimate of the control parameter B at which the formation of the two-kink soliton pair takes place. The required energy to produce at least seven motionless solitons is $H_T = 56$. From (11), we obtain that the value of the control parameter that fulfills this condition is $B = 0.2111$.

4 Formation of a Pair of Two-Kink Soliton

Formerly, we have discussed about the relevance of the emission of successive kink-antikink pairs in the two-kink formation. However, once two consecutive emissions of pairs are attained, the formation of a bound state between two kinks (antikinks) requires an additional ingredient.

In a system with no external forces ($F(x) = 0$), two single solitons interact each other through a repulsive force which decays with the distance between their centres of mass [21]. Such an interaction will separate them, preventing a bonding process. On the other hand, the zeros of the external force (e.g. (4)) are equilibrium positions. When these positions are unstable, the kink solitons move away from them. This cause an opposite effect to the repulsive force. Making possible the formation of a two-kink soliton.

Numerical simulations considering the perturbed SG model (3) where $F(x)$ is given by (4) show good agreement with theoretical predictions. For $B = 0.2564$, the internal shape mode is unstable and generates a pair of two-kink solitons. Figure 1

Fig. 1 Numerical simulation of (3) with $F(x)$ given by (4) under the threshold of shape instability ($B = 0.2564$). We can observe the formation of a two-kink soliton pair

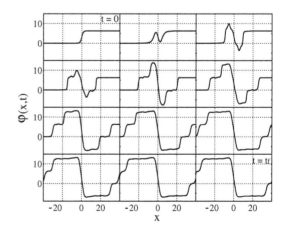

Fig. 2 Space-time diagram
of the two-kink soliton pair
formation under the
threshold of shape instability
($B = 0.2564$). *Inset*
Schematic plot that
illustrates the direction of
solitonic interaction forces
(*blue arrows*) and the
unstable (repeller)
equilibrium position (*dashed
red arrow*) (color online)

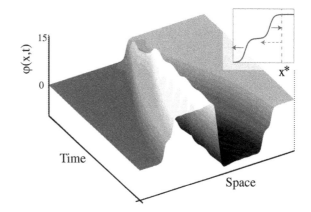

displays different stages of two-kink formation process. Two consecutive kink-antikink pairs are emitted. Meanwhile, the external force creates an unstable point for emitted kinks which move away from $x_* = 0$ in both directions. At the same time, a repulsive force between kinks takes place allowing the appearance of a pair of two-kink solitons moving in opposite directions. In contrast, a motionless antikink soliton appears interpolating between -4π and 2π. For an antinkink, the point $x^* = 0$ is a stable equilibrium. This attracting point compresses anti-kinks together such that there is no distance between them, forming a packing state. Figure 2 depicts the whole process. The inset shows a schematic representation of forces interplay.

Is noteworthy that two-kink soliton formation takes place for a larger value of B ($B = 0.2564$) in comparison with theoretical predictions ($B = 0.2111$). The origin for this discrepancy can be understood on the basis that the effective mass of propagative solitons differs from the motionless ones, resulting in a lower energy. In fact, numerical and experimental works report that the formation of bound states of kinks or multi-kink solitons can lead to a decrease of the associated energy due to a grow of their velocity [8–10]. Despite this, the energy criteria provide a very good estimation of the number of solitons produced after a kink internal mode instability.

To complete the view, we further decrease the value of B. At $B = 0.200$ a multi-kink soliton pair is emitted and a large antikink (interpolating between 8π and -7π) is formed at $x_* = 0$ (see Fig. 3). Proving the formation of multi-kinks solitons.

5 Interaction of Two-Kink Solitons Formed by Internal Modes Instabilities

Next, let us analyze the interaction of two-kink solutions formed by internal mode instabilities. Now, we consider a lattice from -180 to 180 in the x-direction. The values of the system parameters are chosen as $dt = 0.1$ for the time step, $\gamma = 0.01$ for the damping coefficient and $t_{\max} = 300$ for integration time. To create the interacting

Fig. 3 Numerical
simulations of (3) given by
(4) for a lower value of B
($B = 0.2000$). A multi-kink
soliton pair is emitted in
opposite directions

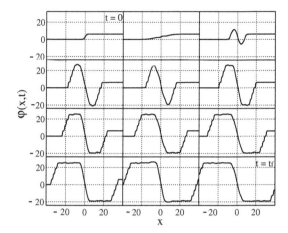

two-kink solitons, we promote formation processes at symmetric positions of the
lattice, i.e., at $x = \pm120$. Accordingly to previous numerical simulations, the control
parameter is setting at $B = 0.2564$ in order to create two-kinks solitons.

The initial conditions are

$$\phi(x, t = 0) = 4\arctan(e^{x-x_0}) + 4\arctan(e^{-x+x_0}) - 2\pi,$$
$$\phi_t(x, t = 0) = 0 \tag{12}$$

with $x_0 = 120$.

Figure 4 displays the interaction of a pair of two-kink solitons moving in opposite
directions. At first stages, both solitons move with constant speed remaining as
bounded states. We have to note that external forces, which enable the two-kink
formation, are still present. Both two-kink solitons are moving away from the unstable
points (located at $x_0^* = \pm120$) created by such forces. However, the external forces
are decaying functions with the distance from x_0^*. In such way that kink repulsive force
become relevant enough to produce a fragmentation of the two-kink soliton before the
interaction. This effect produces that the upper kinks attract each other interacting as
single solitons in the usual way. They annihilate at certain point to emerge undistorted
with twists in opposite directions in order to preserve the topological charge [22].
Afterwards, a cascade of single kink-antikink interactions occurs, finishing with a
dissolution of the two-kink soliton. For multi-kink solitons interaction with $B = 0.200$ the scenario is quite similar. Nevertheless, a decaying breather is formed at
the last stage of interaction (see Fig. 5). The formation of such decaying breather is
not well understood.

For two-kink solitons close enough, opposite forces (external and kink repulsive
forces) are still balanced avoiding the fragmentation of the two-kink soliton before
the interaction. In this case, the two-kink soliton emerge undistorted preserving the

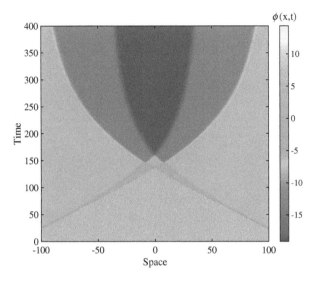

Fig. 4 Two-kink solitons approaching each other. The repelling force between kink leads to disintegration of the solution before interaction

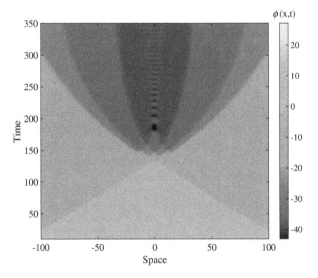

Fig. 5 Multi-kink solitons approaching each other. After a cascade of kink-antikink interactions, a decaying breather is formed

topological charge (see Fig. 6). This result is in contrast with Bazeia's two-kink solitons which interaction leads to oscillon states, bouncing dynamics or even the total destruction of the two-kink solitons [23, 24]. Further investigations are required.

Fig. 6 Two-kink solitons interacting. Both solutions emerge undistorted after the interaction

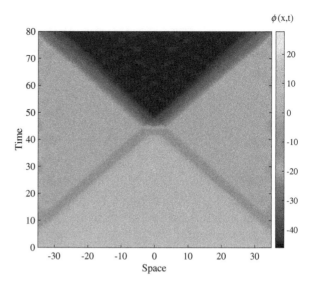

6 Conclusions

We show the formation of a two-kink soliton pair in the perturbed sine-Gordon model due to internal mode instabilities. The conditions for the formation of this kind of solutions are a combination of instabilities in the shape internal modes and an effective potential created by an external force. These results show that the external force plays an important role in the formation process of other types of solutions as multi-kinks. Interaction of two-kink solitons has been also considered. However, given that the external force decays with the distance, the repulsive force between kink become relevant enough to fragmented the two-kink soliton before the interaction. However, for two-kink solitons close enough, both solutions emerge undistorted after interaction. Further investigations are required.

Acknowledgments M.A.G-N. thanks FONDECYT project 11130450 for funding support.

References

1. T. Dauxois, M. Peyrard, *Physics of Solitons* (Cambridge University Press, 2006)
2. J.A. González, J.A. Holyst, Phys. Rev. **35**, 3643 (1987)
3. D. Bazeia, J. Menezes, R. Menezes, Phys. Rev. Lett. **91**, 241601 (2003)
4. D. Bazeia, L. Losano, J. Malbouisson, R. Menezes, Phys. D **237**, 937 (2008)
5. D. Bazeia, J. Furtado, A.R. Gomes, JCAP **02**, 002 (2004)
6. A. de Souza Dutra, Phys. D **D238**, 798 (2009)
7. A.E.R. Chumbes, M.B. Hott, Phys. Rev. D **81**, 045008 (2010)
8. A.V. Ustinov, B.A. Malomed, S. Sakai, Phys. Rev. B **57**, 11691 (1998)
9. J. Pfeiffer, M. Schuster, A. Abdumalikov Jr, A. Ustinov, Phys. Rev. Lett. **96**(3), 34103 (2006)

10. B.A. Malomed, Phys. Rev. B **47**(2), 1111 (1993)
11. A.R. Champneys, Y.S. Kivshar, Phys. Rev. E **61**, 2551 (2000)
12. A.A. Aigner, A.R. Champneys, V.M. Rothos, Phys. D **186**, 148 (2003)
13. M.A. García-Nustes, J.A. González, Phys. Rev. E **86**, 066602 (2012)
14. J.A. González, J.A. Holyst, Phys. Rev. B **45**, 10338 (1992)
15. Y.S. Kivshar, B.A. Malomed, Rev. Mod. Phys. **61**, 763 (1989)
16. A. Barone, G. Paternó, *Physics and Applications of the Josephson Effect* (Interscience, 1982)
17. C.B. Tabi, A. Mohamodau, T.C. Kofané, Phys. Scr. **77**, 045002 (2008)
18. J.A. González, A. Bellorín, L.E. Guerrero, Phys. Rev. E **65**, 065601 (2002)
19. J.A. González, A. Bellorín, L. Guerrero, Phys. Rev. E **60**, R37 (1999)
20. J.A. González, M.A. García-Nustes, A. Sánchez, P.V.E. McClintock, New J. Phys. **10**, 113015 (2008)
21. B.A. Mello, J.A. González, L.E. Guerrero, E. Lopez-Atencio, Phys. Lett. A **244**, 277 (1998)
22. J.V. José, E.J. Saletan, *Classical Dynamics: A Contemporary Approach* (Cambridge University Press, 1998)
23. T.S. Mendonca, H.P. de Oliveira, J. High Energy Phys. **133**, 1 (2015)
24. T.S. Mendonca, H.P. de Oliveira, Arxiv. **1502.03870**, 1 (2015)

Around the Ising Model

Fernando Mora, Felipe Urbina, Vasco Cortez and Sergio Rica

Abstract This chapter discuss several features and connections arising in a class of Ising-based models, namely the Glauber-Ising time dependent model, the Q2R cellular automata, the Schelling model for social segregation, the decision-choice model for social sciences and economics and finally the bootstrap percolation model for diseases dissemination. Although all these models share common elements, like discrete networks and boolean variables, and more important the existence of an Ising-like transition; there is also an important difference given by their particular evolution rules. As a result, the above implies the fact that macroscopic variables like energy and magnetization will show a dependence on the particular model chosen. To summarize, we will discuss and compare the time dynamics for these variables, exploring whether they are conserved, strictly decreasing (or increasing) or fluctuating around a macroscopic equilibrium regime.

1 Introduction

The Ising model, introduced in the early 1920s, by Lenz [1] and Ising [2] as a thermodynamical model for describing ferromagnetic transitions has evolved as one of the most prolific theories in the twenty century opening a huge number of new areas of knowledge (for an historical review see [3]). The importance of the Ising model raises in its universality and robustness, indeed despite its simplicity, this

F. Mora · F. Urbina · V. Cortez · S. Rica (✉)
Facultad de Ingeniería y Ciencias, Universidad Adolfo Ibáñez,
Avda. Diagonal las Torres 2640, Peñalolén, Santiago, Chile
e-mail: sergio.rica@uai.cl

F. Mora
e-mail: femora@alumnos.uai.cl

F. Urbina
e-mail: felipe.urbina@uai.cl

V. Cortez
e-mail: vcortez@alumnos.uai.cl

© Springer International Publishing Switzerland 2016
M. Tlidi and M.G. Clerc (eds.), *Nonlinear Dynamics: Materials,
Theory and Experiments*, Springer Proceedings in Physics 173,
DOI 10.1007/978-3-319-24871-4_25

model has been the starting point for the emergence of various subfields in physical (and social) sciences, namely, phase transitions, renormalisation group theory, spin-glasses, lattice field theories, etc.

In the current contribution, we shall discuss four distinct applications of Ising-based models with applications to both statistical mechanics as social sciences. The first one is devoted to the Glauber-Ising time dependent model with applications to decision-choice theory in economics and social sciences. In the 60s Glauber [4], introduced an stochastic time dependent rule to mimic the statistical properties of the original Ising problem. Glauber's dynamics has been considered in the context of social sciences by Brock and Durlauf [5, 6], and, more recently, by Bouchaud [7].

The second topic is Q2R automata model introduced in the 80s by Vichniac [8]. The Q2R[1] possess time reversal symmetry, which is at the core of any fundamental theory in physics. Moreover, the temporal evolution of this automata conserves a quantity which is closely related to the energy of the Ising model [9]. We are interested in this model because is a natural starting point for studying the statistical and typical irreversible behavior of reversible systems. As shown in [10], this system evolves in an irreversible manner in time towards an "statistical attractor", moreover the macroscopic observable, the so called global magnetization, depends on the value of the initial energy following a law which is exactly the one obtained theoretically by Onsager [11] and Yang [12], more than 60 years ago. Moreover, in [13] it is shown how this model exhibits the same features of Hamiltonian systems with many degrees of freedom, that is, a sensibility to initial conditions, positive Lyapunov exponents, among others.

The third model that we shall discuss in this article concerns the Schelling model of social segregation, introduced in the early 70s by Thomas C. Schelling [14–16]. This model became one of the paradigm of an individual-based model in social science. Schelling's main contribution is that shows on the formation of a large scale pattern of segregation as a consequence of purely microscopic rules. More recently, it has been shown that the Ising energy, which is a good measure of segregation, acts as a Lyapunov potential of the system is driven, under particular conditions, by a strictly decreasing energy principle [17].

Finally, we shall discuss a model for dissemination's disease known as Bootstrap percolation, first introduced in the late 70s by Chalupa et al. [18]. In this model a healthy individual may be infected if the majority of its neighbors are infected. On the other hand an infected individual never recovers, so it remains infected forever. This model has been used as a model for disease's propagation. One of the most important questions arising is the determination of the critical number of infected individuals to contaminate the whole population.

The paper is organized as follows, in Sect. 2, some common features, as well as, the precise rules for each particular model are explicitly described. Next, in Sect. 3 the main dynamical behavior, the salient properties and the phase transitions are shown and explained, for each of them. Finally, we conclude.

---

[1]Q by four, *quatre*, in french, 2 by two steps automata rule as explicitly written below, and R by reversible.

2 Ising-Based Models

2.1 Generalities

2.1.1 The Lattice and the Neighborhood

All models discussed below, display similar features, the system consisting of a lattice with $N \gg 1$ nodes, in which each node, k, may take a binary value $S_k(t) = \pm 1$ at a given time. Each node k on the lattice interacts, in general, with all other individuals, with an interaction coefficient J_{ik} (i denotes an arbitrary node). But in particular, a node, k, may interact only with a finite neighborhood denoted by V_k. The number of neighbors for site k, $|V_k|$, is the total number of non zero J_{ik} for each node. In Fig. 1 we show, as an example, four possible lattice configurations.

2.1.2 The "Energy" and the "Magnetization"

We define the macroscopic observables of the system, by analogy with the original Ising model of ferromagnetism, as follows:

$$E[\{S\}] = -\frac{1}{2} \sum_{i,k} J_{ik} S_i(t) S_k(t) , \tag{1}$$

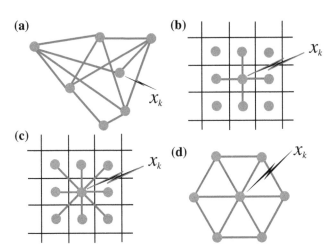

Fig. 1 Examples of lattices and neighborhoods. We illustrate explicitly: **a** an arbitrary network with a random number of neighborhoods; and three periodic regular lattices in two space dimensions: **b** a square lattice with a von-Neuman neighborhood of 4 individuals (the original lattice of the Ising model with the nearest neighborhood); **c** a square lattice with a Moore neighborhood of 8 individuals, and **d** a hexagonal lattice with 6 neighborhoods

$$M[\{S\}] = \sum_{k=1}^{N} S_k(t). \tag{2}$$

These quantities will be the pertinent observables, and we shall use them to classify the distinct cases that we will be described in the next sections.

2.2 The Time-Dependent Glauber-Ising Model

Glauber [4], in the 60s, introduced a dynamical model for the study of the Ising model. The rule governing Glauber's model is the following:

Let, the local magnetization at the site k and at a time t, be:

$$U_k(t) = B + \sum_{i} J_{ik} S_i(t), \tag{3}$$

with B being an external magnetic field. Then, the spin's value at the next time step, $t+1$, will be

$$S_k(t+1) = \text{sgn}(U_k(t)), \tag{4}$$

that is $S_k(t+1) = +1$ if $U_k(t) \geq 0$ and $S_k(t+1) = -1$ if $U_k(t) < 0$. We call (4) the deterministic rule. In probability language, if $U_k(t) \geq 0$, then $S_k(t+1)$ would be $+1$ with probability 1, and it would be -1 with probability 0. This rule is updated in parallel fashion.

Next, this deterministic rule may be modified by a probabilistic rule, in the following way:

$$S_k(t+1) = \begin{cases} +1 & \text{with probability } p = \frac{1}{1+e^{-\beta U_k(t)}} \\ -1 & \text{with probability } p = \frac{1}{1+e^{\beta U_k(t)}} \end{cases} \tag{5}$$

Notice that in the limit $\beta \to \infty$ one recovers the deterministic behavior (4), while in the limit $\beta \to 0$ one reaches a completely random (binomial) dynamics regardless of the value of U, that is $S_k(t+1)$ would be $+1$ with probability 1/2.

The Glauber rule is indeed a Markov chain which manifests, in a perfect way, the statistical properties of the Ising phase transition for the case of Von-Neuman neighbourhoods, and it also agrees with the mean field approximation for the case of a large number of neighbours. Finally, nowadays the Glauber dynamics is the starting point for numerical simulations of spin glasses systems with random values for the J_{ik} coefficients.

2.2.1 Random Decision-Choice Model

Let us consider now a random choice model [5–7] in the context of social sciences. An individual takes a choice based on a combination of decision quantities, namely an individual "decision parameter" f_k, a "global decision" or "public information" parameter $F(t)$ (which could be included in the previous individual decision parameter) and a "social pressure" $\sum_i J_{ik} S_i(t)$.

Next take,[2] $U_k(t) = f_k + F(t) + \sum_i J_{ik} S_i(t)$, and follow the Glauber deterministic dynamics (4) or more generally the Glauber random dynamics (5).

Due to both, the Ising-like feature as the Glauber Dynamics evolution rule, a phase transition is known to appear. This transition favors the decision into one or another of the two options of the binary variable.

2.3 The Q2R Automata

The Q2R rule considers the following two-step rule which is updated in parallel [8][3]:

$$
S_k(t+1) = S_k(t-1) \times
\begin{cases}
+1 & \text{if } \sum_i J_{ik} S_i(t) \neq 0 \\
-1 & \text{if } \sum_i J_{ik} S_i(t) = 0
\end{cases}
\tag{6}
$$

Naturally, it is possible to add, without any difficulty, an external magnetic field B. However, some caution should be taken into account: the model works if $U_k(t) = B + \sum_i J_{ik} S_i(t)$, may vanish, therefore, B and the J_{ik} factors should be integers. For instance in the case of a finite neighborhood, $B + |V_k|$ should be an even number.

The rule (6) is explicitly invariant under a time reversal transformation $t + 1 \leftrightarrow t - 1$. Moreover, as shown by Pomeau [9], the following quantity, that we may call an energy, despite not being exactly the energy (1)

$$
E[\{S(t), S(t-1)\}] = -\frac{1}{2} \sum_{i,k} J_{ik} S_k(t) S_i(t-1),
\tag{7}
$$

is preserved under the dynamics defined by the Q2R rule (6). Moreover, the energy is bounded by $-2N \leq E \leq 2N$.

The rule (6) is complemented with an initial condition $S_k(t = 0)$ and $S_k(t = 1)$ that will be described more precisely in the next section.

---

[2]The so called "perceive overall incentive agent function", by Bouchaud [7].

[3]This two-step rule may be naturally re-written as a one-step rule with the aid of an auxiliary dynamical variable [9].

2.4 Schelling Model for Social Segregation

Schelling model, is also characterized by a binary variable S_k which may take values
$+1$ and -1. We shall say that an individual S_k at the node k "happy" at his site,
if and only if, there are less than θ_k neighbors at an opposite state. θ_k is a tolerance
parameter that depends in principle on the node and, it may take all possibles integer
values, such that $0 < \theta_k < |V_k|$ (we exclude the cases $\theta_k = 0$ and $\theta_k = |V_k|$ from
our analysis). The satisfaction criterion reads[4]

An individual S_k is unhappy at the node k *if and only if*:

$$\sum_{i\in V_k} S_i = \begin{cases} |V_k| - 2n_k(-1) \leq |V_k| - 2\theta_k, \text{ if } S_k = +1 \\ 2n_k(-1) - |V_k| \geq 2\theta_k - |V_k|, \text{ if } S_k = -1. \end{cases} \tag{8}$$

Here $n_k(+1)$ is the number of neighbors of S_k that are in the state $+1$; and, $n_k(-1)$
the number of neighbors of S_k in the state -1, naturally $n_k(+1) + n_k(-1) = |V_k|$.

Having labeled all different un-happy individuals, one takes randomly two of
them in opposite states (one $+1$, and one -1) and exchanges them. Even when this is
not exactly the original Schelling's rule, the present *Schelling's protocol* is a simpler
one. In any case, it can be modified in a straightforward way to include for example
vacancies [19, 20], different probabilities of exchange [19], multiple states variables
[21], etc.

If k and l are these random nodes, then the evolution rules:

$$S_k(t) \rightarrow S_k(t+1) = -S_k(t), \quad S_l(t) \rightarrow S_l(t+1) = -S_l(t)$$

and for all other nodes $i \neq k \& l$ remain unchanged $S_i(t) \rightarrow S_i(t+1) = S_i(t)$.

The protocol is iterated in time forever or until the instant when one state does
not have any unhappy individuals to be exchanged.

Notice, that Schelling criteria (8) is deterministic, however the exchange is a random process, therefore two initial configurations will not display the same behavior
in detail, but they will evolve to the same statistical attractor [22].

---

[4]The criteria (8) may be unified in a single criteria [17] (multiplying both sides of the two inequalities
by S_k): an individual S_k is unhappy at the node k if , and only if, $S_k \sum_{i\in V_k} S_i \leq |V_k| - 2\theta_k$, which
is a kind of energy density instead of the threshold criteria found in Glauber dynamics (4).

Schelling's protocol, defined above, has a remarkable property: if $\theta_k > \frac{|V_k|}{2}$ then any exchange $k \leftrightarrow l$, will always decrease the energy

$$E[\{S\}] = -\frac{1}{2}\sum_k \sum_{i \in V_k} S_i(t)S_k(t). \qquad (9)$$

The energy (9) follows from (1), whenever $J_{ik} = 1$ for neighbors and $J_{ik} = 0$ otherwise.

For a proof, we refer to [17]. We shall only add the following remark: if $\theta_k > \frac{|V_k|}{2}$, then the evolution necessarily stops in finite time. This is because the energy (9) is bounded from below by $E_0 = -\frac{1}{2}\sum_{k=1}^N |V_k|$ and because the energy (1) decreases strictly. On the other hand, for $\theta_k < \frac{|V_k|}{2}$, the energy may increase or decrease after an exchange indistinctly.

2.5 Bootstrap Percolation

We shall consider the problem of bootstrap percolation for a given lattice [18]. As in the previous models, each node k interacts with $|V_k|$ neighbors, the neighborhood defined by the set V_k. As before the state, S_k may take values $+1$ and -1 depending on if it is "infected" or not. At a given "time" the state $S_k(t)$ evolves into $S_k(t + 1)$ under the following parallel rule: if a site is not infected, and if the *majority* of its neighbors are infected, then the site becomes infected [23]. On the other hand, if the site is already infected it keeps its infected state.

Summarizing, the evolution rule, which is updated in parallel, may be written in the following general way:

$$\text{if } S_k(t) = -1 \text{ and } \sum_k S_k(t) > 0, \text{ then } S_k(t+1) = +1, \qquad (10)$$

otherwise, if $S_k(t) = 1$ then $S_k(t+1) = 1$.

From the dynamics it follows directly that the energy (9) decreases in time, $E(t + 1) \leq E(t)$, as well as the magnetization increases in time: $M(t + 1) \geq M(t)$. As in the case of the Schelling model, because the energy is a strictly decreasing functional, and because it is bounded from below in a finite network, then the dynamics always stops in finite time.

Finally, let us comment that a problem that has increased in interest in recent times deals with the question of how the total infection depends on the initial configuration which is randomly distributed and such that a site will be at the state $S_k = +1$ with a probability p [24].

Table 1 Recapitulation of the four above mentioned models, and its main conservation properties

Dynamics	Evolution criteria	Energy	Magnetisation
Glauber	$\text{sgn}(B + \sum_i J_{ik} S_i(t))$	Not conserved	Not conserved
Q2R	$\sum_i J_{ik} S_i(t) = 0$	Conserved	Not conserved
Schelling	$\text{sgn}(S_k(t)) \sum_{i \in V_k} S_i(t) \le |V_k| - 2\theta_k$	Not conserved[a]	Conserved
Bootstrap	$\sum_{i \in V_k} S_i(t) > 0$	$\Delta E < 0$	$\Delta M > 0$

[a] If $\theta_k > |V_k|/2$ then $\Delta E < 0$

Naturally, if initially $p \approx 1/2$, then every site has in average the same number of $S_k = +1$ states and $S_k = -1$ in its neighborhood, then the system would percolate almost in one step. However, as p decreases, one can define a probability, $P(p)$, which is the probability that the system would percolate at the end of the evolution process. At the end this probability can be numerically determined.

2.6 Recapitulation

The afore mentioned models have in common a threshold criteria (4), (6), (8), and (10) the subsequent dynamics follows different rules. Therefore one should expect distinct properties.

The Glauber Dynamics does not preserve neither the energy or magnetization, however the Q2R dynamics (Sect. 2.3) does preserve only the energy but does not preserve the magnetization. The Schelling model (Sect. 2.4) does preserve only the magnetization, but if $\theta_k > |V_k|/2$ the system's energy is strictly a decreasing function. Finally, in the infection model of Sect. 2.5, the energy strictly decreases whereas the magnetization is an increasing function of time (Table 1).

3 Ising Patterns, Transitions, and Dynamical Behavior

In this section, we shall roughly describe the essential phenomenology of the Ising-like models and rules described in the previous section, whether they are governed (or not) by the rules of conservation of magnetization energy.

3.1 Glauber and Decision-Choice Model Dynamics

The time dependent Glauber-Ising model shows a very rich phenomenology. As such, the model's behavior has been explored using mean field approximation (the Curie-Weiss law) as well as by direct simulations of the rule (5). Here our macroscopic

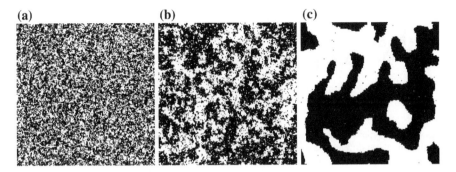

Fig. 2 Snapshots of the patterns for the Glauber-Ising model. The simulation is for a $N = 256 \times 256$ periodic lattice with von Neuman neighborhood. Moreover we take $f_k = 0$ and $F = 0$. The parameter of "irrationality" and the magnetization averages are, respectively: **a** corresponds to a paramagnetic phase for $\beta = 0.53$ and $\langle M \rangle / N = 0.0006$; **b** a critical phase for $\beta = 0.82$ and $\langle M \rangle / N = 0.02$; and **c** corresponds to a ferromagnetic phase $\beta = 1.8$, and $\langle M \rangle / N = 0.39$

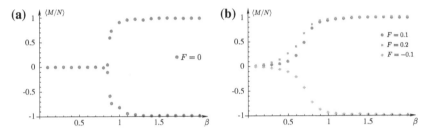

Fig. 3 Average magnetization $\langle M \rangle$ versus β. The average are taken from long time simulations of approximately 20000 time steps. In both cases the random external field is settled to zero $f_k = 0$. **a** Case of $F = 0$; and **b** Cases of $F = \pm 0.1$ and $F = 0.2$

observable is the total magnetization per site, namely $M(t)/N$ and were $M(t)$ is defined in equation (2). In what it follows, we will only show results for the direct simulation of the Glauber-Ising model (4) and we shall use the terminology of social sciences [7]. In Fig. 2 we show three distinct states characterized by different values of the parameter of "irrationality" β,[5] and a null value for the public information parameter $F(t)$.

In Fig. 3 we show two different figures for the mean magnetization $\langle M \rangle / N$ versus the irrationality parameter β, divided into two groups depending on the non-zero or null value for the public information parameter $F(t)$. Each point, was calculated for a total of approximately 2×10^4 time steps. We can readily observe the appearance of a bifurcation for the case $F = 0$ and β greater than $\beta_c = 0.8$.

Therefore, the time dependent Glauber-Ising model displays a transition from a paramagnetic to a ferromagnetic phase for $\beta_c \approx 0.8$ which is in agreement with the critical threshold value of the Ising model, $\beta_c = \log(1 + \sqrt{2}) \approx 0.881 \ldots$

---

[5]In statistical physics, β is the inverse of the thermodynamical temperature, $\beta \sim 1/T$.

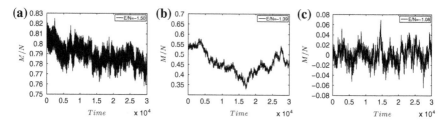

Fig. 4 Three types of magnetization dynamics for a running time of $T = 3 \times 10^4$ time steps, and considering three different values of energy. **a** Corresponds to an initial energy $E/N = -1.50$, **b** corresponds to an initial energy $E/N = -1.39$ and **c** corresponds to an initial energy $E/N = -1.08$. The three figures show the fluctuations in the macroscopic observable $M(t)$

3.2 Q2R Dynamics

We shall now present the dynamics of the Q2R model for the case of von Neuman vicinity (the coupling interaction $J_{ik} = 1$ for the four closest neighbors), which is the original Q2R cellular automata [8].

The time evolution of magnetization, given an initial energy value E/N, provides a direct observation of the spin's dynamics and fluctuations. In what it follows, we will base our results and analysis taking a periodic grid of size $N = 256 \times 256$.

When the initial energy value is $E/N = -1.50$, which refers to Fig. 4a, it can be seen that the system's dynamics fluctuates without significative changes in the magnetization's value. This means that the overall set of spins are oriented in a preferred direction. This is known as a ferromagnetic state. If we raise the initial energy value and take $E/N = -1.39$, which corresponds to Fig. 4b, the dynamics abruptly fluctuates because of the closeness to the critical energy value: E_c/N [10]. Finally, if the initial value of the energy is greater than in the previous cases, e.g. $E/N = -1.08$, Fig. 4c shows how the dynamics of magnetization decays reaching a zero mean value $\langle M \rangle \approx 0$.

Similarly, Fig. 5 shows some characteristic snapshots of the spin field patterns at a given time for the same energy per site. When the energy value is $E/N = -1.50$ (see Fig. 5a), it can be seen how the spins are organized with a well defined magnetization, namely $S_k = +1$ or $S_k = -1$. This is a ferromagnetic phase. However, when the initial energy value is $E/N = -1.39$ (close to the critical energy), as shown in Fig. 5b, the system generates patterns characterized by well defined clusters of states. Finally, for an energy $E/N = -1.08$ (see Fig. 5c) the system shows an homogeneous state with the spin distributed more or less randomly, which characterizes a paramagnetic phase.

Also it can be shown that the average magnetization $\langle M \rangle$ depends critically on the initial energy, E/N, of the system (Fig. 6).[6]

---

[6]Q2R is a micro canonical description of the Ising transition, therefore we use the energy in absence of any temperature. In [10] it is shown the excellent agreement among the Q2R bifurcation diagram with the Ising thermodynamical transition.

(a)   **(b)**   **(c)**

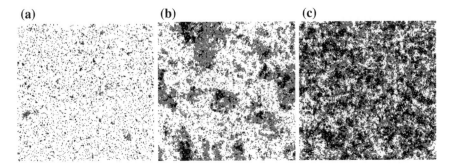

Fig. 5 Snapshots of spin structure at $T = 3.5 \times 10^4$ considering three initial values. **a** Corresponds to an initial energy $E/N = -1.50$ and a magnetization $M/N = 0.79$, **b** corresponds to an initial energy $E/N = -1.39$ (which is close to the transition energy $E/N = -\sqrt{2}$) and $M/N = 0.455$; and **c** corresponds to an initial energy $E/N = -1.08$ and $M/N = 0.012$

Fig. 6 Phase transition diagram for the average Magnetization $\langle M \rangle$ versus initial energy E/N, for a grid size $N = 256 \times 256$

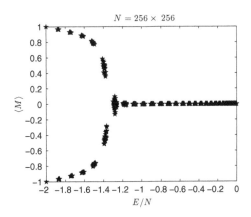

Finally, we can state three fundamental features from the above phase diagram. First, there exists a zone in which the system stays into a ferromagnetic state when the value of the energy is lower than the critical energy $E < E_c$. Secondly, there is a second order phase transition at $E_c/N = -\sqrt{2}$ and it is formally equivalent to the Ising critical temperature [10]. Third, when the initial energy value is greater than the critical energy $E > E_c$, the system presents a paramagnetic phase, with a magnetization value $\langle M \rangle = 0$.

3.3 Schelling Dynamics

We shall characterize the dynamics of Schelling model for the particular case in which the system is a two dimensional periodic lattice, and each site possess the same neighborhood consisting in the $|V|$ closest individuals. We shall consider also that the parameter θ_k is uniform, that is, $\theta_k = \theta$.

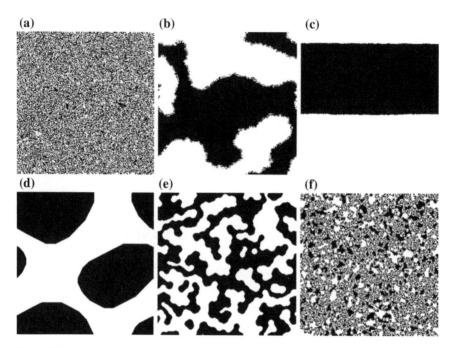

Fig. 7 Schelling's patterns for various satisfaction parameter θ in a square periodic lattice of $N = 256$ nodes. The vicinity is uniform and contains $|V| = 20$ elements. **a** $\theta = 5$; **b** $\theta = 6$; **c** $\theta = 9$; **d** $\theta = 10$ (eventually this case the two spots observed merges into a single one, this coalescence dynamics, however, it happens after a longtime); **e** $\theta = 11$ and **f** $\theta = 15$, are two cases whenever the energy is a strictly decreasing function so the dynamics stops in finite time, in the former case this happens after a time so segregation is possible, however in the later case the dynamics stops shortly after the Schelling algorithm started. For $\theta = 15$ we say that this is a frustrated dynamics, because the system cannot reach the ground state energy because the dynamics stops after one of the population is completely happy

Figure 7 displays an example of typical patterns arising in the Schelling's model. As it can be observed, the dynamics depends critically on the value of the tolerance parameter θ, defined above. More precisely, if θ is larger or smaller than $\theta_{c_1} = |V|/4$, $\theta_c = |V|/2$, and $\theta_{c_2} = 3|V|/4$. The initial state was chosen randomly with a binomial distribution, that is $S_k(t = 0)$ was $+1$ with probability $1/2$ and -1 with the same probability. Hence, the total magnetization is $M(t = 0) \approx 0$, and it is kept fixed during the evolution.

The simulation shown in Fig. 7, corresponds to a Schelling rule with a vicinity of $|V| = 20$ elements. Clearly three different cases can be distinguished, and at least three transition points, namely $\theta_{c_1} = |V|/4$, $\theta_c = |V|/2$, and $\theta_{c_2} = 3|V|/4$. For $1 < \theta \leq |V|/4$ (see Fig. 7a) one observes a non-segregated pattern, the states $S_k = \pm 1$ are swapping, more or less randomly in the system, without a formation of any kind of large scale structure. In a coarse graining scale, for instance, the scale of the vicinity, the coarse-grained magnetization, namely, $m = \frac{1}{|V|} \sum_{i \in V_k} S_i(t)$ is zero everywhere, as well as the energy.[7] In this situation, it is tempting to make an analogy with the Ising paramagnetic phase. For $|V|/4 < \theta \le |V|/2$, one observes how a segregation pattern arises (see Fig. 7b, c). More important the coarse-grained magnetization is locally non-zero, and the pattern presents domain walls, which are characteristic of a ferromagnetic phase in the Ising-like terminology. For $|V|/2 < \theta \le 3|V|/4$, one observes also segregation (see Fig. 7e), but the dynamics stops in a finite time. The final state is a quenched disordered phase for which one may conjecture an analogy with a "spin glass" phase, and the appearance of a kind of long-range order. The case $\theta = 3|V|/4$ in (see Fig. 7f) it is interesting because, although the are some islands of segregation, the system also recovers its original heterogeneity, with almost a null coarse-grained magnetization m.

3.4 Bootstrap Percolation

The spin dynamics for the case of Bootstrap percolation of Sect. 2.5 is always characterized by an energy decreasing principle, moreover because a +1 spin never flips to a −1, the magnetization is mandated to increase up to a constant value because of the impossibility to infect more individuals, or simply because the system has been fully percolated by the +1 spin states.

As said in Sect. 2.5, we shall consider a random initial state with a fraction p of the spins at the state $S_k = +1$ (that is, a fraction p of the population would be infected).

It is observed, that for a moderately large value of p, say $p \approx 1/2$, the system becomes unstable very fast, percolating the $S_k = +1$ state everywhere almost instantaneously.

However, as one decreases p, the system presents a well defined scenario. Figure 8 shows the typical evolution of a percolation pattern in time. More precisely, the system nucleates bubbles of infected states ($S_k = +1$) and two scenarios are possible, either these bubbles continues to grow or they stop (compare Fig. 8b, c). In analogy with the instability of a first order phase transition, it should exist a critical radius of nucleation that depends explicitly on p.

This critical radius of nucleation maybe estimated in the limit of large vicinity, in other words, in the range of validity of the mean field approximation. Let be p the fraction of infected sites initially distributed randomly in the system and a the radius of the vicinity ($\pi a^2 = |V|$). We shall add an infection bubble with a radius R (see Fig. 9a). A $S_k = -1$ state in the boundary of the infected circle will become infected if $\sum_k S_k(t) = (2p - 1)(\pi r^2 - A(R)) + A(R) > 0$, where $A(R)$ is the surface of the portion of the circle inside the infection bubble (see Fig. 9b). Therefore, the bubble will infect neighbors and will propagate into the system, if

---

[7]Notice that, as already said, the total magnetization is constant in the Schelling model. Therefore we cannot match the Schelling transitions observed here with the phase transition for the cases of the Glauber-Ising and the Q2R models.

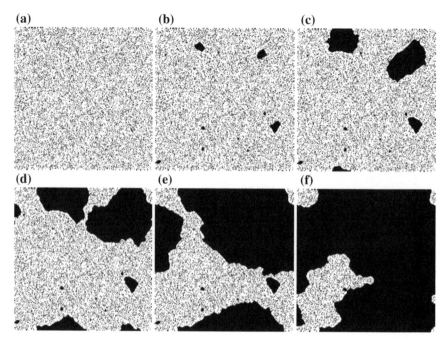

Fig. 8 Bootstrap percolation's patterns at six different time steps. The network is a square periodic lattice of $N = 256^2$ sites with a uniform vicinity of $|V| = 24$ sites. **a** Display the initial random state with an initial fraction 0.2 of $S_k = +1$ (that is, a given site is +1 with probability 0.2, and −1 with probability 0.8); In **b** one observes the nucleation of bubbles, which eventually would propagate the +1 state over the random phase; In **c** one observes that some infected bubbles have not reach the critical size and they do not propagate; however, in **d** big bubbles invade the system transforming the interface in a front propagation over the whole system (**e**) and (**f**)

Fig. 9 **a** Scheme for the mean field estimation of the critical radius of infection. The *gray* region represents the random initial data with a fraction p of +1. **b** Details of the geometry for the calculation of $A(R)$

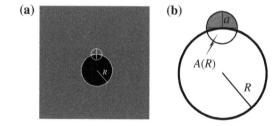

$$\frac{A(R)}{\pi a^2} > \frac{1 - 2p}{2(1 - p)}. \qquad (11)$$

The surface $A(R)$ follows from a direct geometrical calculation. In the large R/a limit, one gets

$$\frac{A(R)}{\pi a^2} \approx \frac{1}{2} - \frac{a}{3\pi R} + \mathcal{O}(R^{-3}),$$

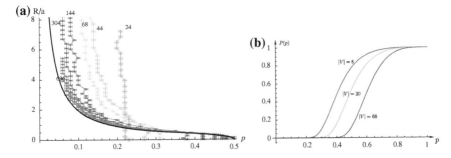

Fig. 10 **a** Critical radius of nucleation R/a as a function of p. As expected as $p \to 1/2$ the critical radius is zero, while as $p \to 0$ the critical radius diverges. The points correspond to the numerical simulations for different values of the vicinity size: $|V| = \{24, 44, 68, 144, 304, 696\}$ as indicated in the figure. **b** Estimation of the lower bound of the probability $P(p)$ of having a critical nucleation bubble of infected states, for $|V| = 8$, $|V| = 20$ and $|V| = 68$. One notices that this probability takes-off around a precise value of p

therefore, one concludes that the critical radius of nucleation scales as

$$\frac{R_c}{a} \approx \frac{2(1-p)}{3\pi p}.$$

Figure 10 shows a numerical study of the nucleation radius, for various vicinity sizes, $|V|$, as a function of p. Moreover the figure also presents the mean field estimation by an explicit geometrical calculation of the surface $A(R)$ and using the critical condition (11). One sees that the mean field approach matches perfectly with the data in the large $|V|$ limit.

However, a question remains open: what is the probability to obtain, ab-initio a bubble with a radius larger than R_c? This probability seems to be very small, because it is proportional to the probability to obtain πR_c^2 sates $+1$ all together, that is

$$P_{\text{bubble}} \approx p^{\pi R_c^2} = p^{|V|(R_c/a)^2} \sim p^{|V|\frac{4(1-p)^2}{9\pi^2 p^2}},$$

with R_c/a the function of p plotted in Fig. 10. Although, this probability $P(p)$ is quite small, it is a lower bound for the problem of Bootstrap percolation. If, initially, a bubble has a radius greater than $R_c(p)$, then the system percolates, and the nucleation bubble may not initially exist, but it may be built solely by the evolution, this provides a better estimation of the probability $P(p)$ of percolation.

4 Discussion

We have shown how different models amalgamate their underlying behavior under the common principle of the Ising-based models: Phase transitions, Bifurcations and Phase Diagrams and most important, the existence of a core principle, e.g., energy minimization which appears to be a robust feature of these models and which would require a deeper consideration.

It is a remarkable fact, however, how despite a continued interest over the last century, the Ising model continues to fascinate and amaze us, not only on it's original context, but also in some other areas of knowledge were it has been applied. The "paramagnetic-ferromagnetic" transition can be recovered in all models described here, with deeper consequences, for example, in the field of human behavior, specially social sciences. Here we can ask ourselves for example: can the sudden changes of opinion before an election or the choice of a product or racial segregation be related to the basic physics of the Ising model? Even more, the existence of an energy principle, something completely excluded and extraneous to the field of Social Sciences, seems to be the main thread behind, for studying and trying to understand human and social behavior. Certainly, delving deeper on this energy principle would require more attention and research. Finally we conclude by asking, Can we have some hope, in a near future and in the context of Social Science, of being able to develop predictive tools for studying and understanding better the human behavior?

Acknowledgments The authors acknowledge Eric Goles, Pablo Medina, Iván Rapaport, and Enrique Tirapegui for their partial contribution to some aspects of the present paper, as well as Aldo Marcareño, Andrea Repetto, Gonzalo Ruz and Romualdo Tabensky for fruitful discussions and valuable comments on different topics of this work. This work is supported by Núcleo Milenio Modelos de Crisis NS130017 CONICYT (Chile). FU. also thanks CONICYT-Chile under the Doctoral scholarship 21140319 and Fondequip AIC-34.

References

1. W. Lenz, Physikalische Zeitschrift **21**, 613–615 (1920)
2. E. Ising, Beitrag zur Theorie des Ferromagnetismus. Zeitschrift für Physik **31**, 253–258 (1925)
3. S.G. Brush, History of the Lenz-Ising Model Rev. Mod. Phys. **39**, 883 (1967)
4. R.J. Glauber, Time-dependent statistics of the Ising model. J. Math. Phys. **4**, 294–307 (1963). doi:10.1063/1.1703954
5. W.A. Brock, S.N. Durlauf, A formal model of theory choice in science. Econ. Theory **14**, 113–130 (1999)
6. W.A. Brock, S.N. Durlauf, Interactions-based models, ed. by J.J. Heckman, E. Leader. Handbook of Econometrics, **5**, Chapter 54, pp. 3297–3380 (2001)
7. J.-P. Bouchaud, Crises and collective socio-economic phenomena: simple models and challenges. J. Stat. Phys. **151**, 567–606 (2013)
8. G. Vichniac, Simulating physics with cellular automata. Physica D **10**, 96–116 (1984)
9. Y. Pomeau, Invariant in cellular automata. J. Phys. A: Math. Gen. **17**, L415–L418 (1984)
10. E. Goles, S. Rica, Irreversibility and spontaneous appearance of coherent behavior in reversible systems. Eur. Phys. J. D **62**, 127–137 (2011)
11. L. Onsager, Crystal statistics. I. A two-dimensional model with an order-disorder transition. Phys. Rev. **65**, 117–149 (1944)
12. C.N. Yang, The spontaneous magnetization of a two-dimensional Ising model. Phys. Rev. **85**, 808–816 (1952)
13. F. Urbina, S. Rica, E. Tirapegui, Coarse-graining and master equation in a reversible and conservative system. Discontin. Nonlinearity Complex. **4**(2), 199–208 (2015)
14. T.C. Schelling, Models of segregation. Am. Econ. Rev. **59**, 488–493 (1969)

15. T.C. Schelling, Dynamic models of segregation. J. Math. Sociol. **1**, 143–186 (1971)
16. T.C. Schelling, *Micromotives and macrobehavior* (W.W. Norton, New York, 2006)
17. N. Goles Domic, E. Goles, S. Rica, Dynamics and complexity of the Schelling segregation model. Phys. Rev. **E 83**, 056111 (2011)
18. J. Chalupa, P.L. Leath, G.R. Reich, Bootstrap percolation on a Bethe lattice. J. Phys. C Solid State Phys. **12**, L31–L35 (1979)
19. D. Stauffer, S. Solomon, Ising, Schelling and self-organising segregation. Eur. Phys. J. B **57**, 473 (2007)
20. A. Singh, D. Vainchtein, H. Weiss, Schelling's segregation model: parameters, scaling, and aggregation **21**, 341–366 (2009)
21. L. Gauvin, J. Vannimenus, J.-P. Nadal, Eur. Phys. J. B **70**, 293 (2009)
22. V. Cortez, P. Medina, E. Goles, R. Zarama, S. Rica, Attractors, statistics and fluctuations of the dynamics of the Schelling's model for social segregation. Eur. Phys. J. B **88**, 25 (2015)
23. J. Balogh, B. Bollobás, R. Morris, Majority bootstrap percolation on the hypercube. Comb. Probab. Comput. **18**, 17–51 (2009)
24. J. Balogh, B. Pittel, Bootstrap percolation on the random regular graph. Random Struct. Algorithms **30**, 257–286 (2007)

Nonequilibrium Trade-Investment Model for the Ranked Distribution of Gross Domestic Products Per-capita

R. Bustos-Guajardo and Cristian F. Moukarzel

Abstract The world distribution of Gross Domestic Products per capita (GDPpc) has been the subject of considerable debate for a long time. Basic properties such as the existence of convergence or divergence in time, and the broad features of its distribution, have not been agreed upon. In this work it is argued that the world distribution of GDPpc is well described by a multiplicative Trade-Investment model. We first derive analytically the typical value $g(r, t)$ of the ranked wealths for a system evolving under Trade (modeled as YS exchange) and Investment (modeled as RMN). The resulting analytical expressions are fitted to data for the GDPpc of up to 200 countries, in the period 1960–2013. Our results support the idea that biased international commerce, and not random noise, is the main cause for the observed divergence of GDPpc values in time.

1 Introduction

Knowledge of the mechanisms that shape wealth distributions has always been of great interest. For decades, the existence or not of concentration of wealth in the hands of a few has been debated [1–3]. Diverse positions on the matter are found in the literature, as well as several proposals to describe the shape of the wealth distributions observed in societies [4–8]. The large amount of economic data collected over the last years, and the development of new methods to analyze them, have provided excellent tools to test the different theories about wealth distributions.

One of the most popular economic indicators for the wealth of a country is the Gross Domestic Product per capita (GDPpc) g. Usually, g is interpreted as a rough measure of the average standard of living in a given country. Studies of the world

R. Bustos-Guajardo · C.F. Moukarzel (✉)
Departamento de Física Aplicada, CINVESTAV del IPN, 97310 Mérida,
Yucatán México
e-mail: cristian@mda.cinvestav.mx

R. Bustos-Guajardo
e-mail: bustosguajardor@gmail.com

© Springer International Publishing Switzerland 2016
M. Tlidi and M.G. Clerc (eds.), *Nonlinear Dynamics: Materials,
Theory and Experiments*, Springer Proceedings in Physics 173,
DOI 10.1007/978-3-319-24871-4_26

347

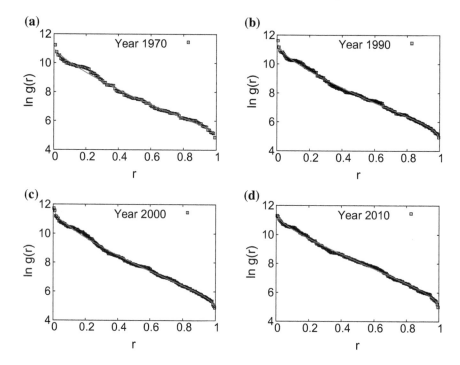

Fig. 1 Ranked values of GDP per capita for the years 1970, 1990, 2000 and 2010. *Solid lines* are fits of the data using (13)

distribution of g thus hint at how wealth is distributed among countries. The distribution of g has been suggested to display power law tails for the richest countries [9] but has also been modeled using lognormal distributions [5, 10]. Other studies [11–14] have pointed that the values of $\ln g(r)$, ordered by normalized rank $r = R/N$ (with R the absolute rank and N the number of countries), follow approximately a straight line as a function of r, except near the ends of the range [11] (see Fig. 1).

Following the analysis introduced in [11], a dynamical model, based on realistic micro-economic Trade-Investment (TI) rules, has recently been proposed [14] to describe the distribution of g. This model considers a population of equally able economic agents following a Trade-Investment (TI) dynamics [15, 16], and is constructed as follows. Investment represents the aggregated effect of local stochastic factors on the wealth of countries or individuals, such as natural phenomena, education, exploitation of natural resources, etc., and is modeled by equally distributed multiplicative noise acting independently on the "wealth" of each country. The second element determining the dynamics is international commerce, or trade [17–19], that shifts wealth conservatively among countries. Wealth exchanges are modeled as pairwise commercial interactions following Yard-Sale (YS) exchange rules [20, 21]. As will be seen in Sect. 2, whenever YS exchange systematically favor the rich agents, a wealth distribution appears where $\ln g(r)$ has a linear r-dependence.

This trait, combined with external multiplicative noise, gives rise to distributions that are very similar to those observed for the GDPpc of countries. Using our model to fit the world data, which is obtained from freely available databases, we can get useful information about the processes that determine the distribution of GDPpc in the world.

Given the micro-economic rules describing our TI dynamics, we focus on deriving analytically the ranked wealth distribution $g(r)$, i.e. the wealth as a function of the relative rank r. This is because our approach has the advantage of allowing the calculation of ranked distributions $g(r)$ in a relatively simple way. Notice that $g(r)$ and the probability density function $P(g)$ contain equivalent information, since one can write

$$r(g) = \int_g^\infty P(g')dg'. \tag{1}$$

The relative rank r equals one minus the cumulative distribution of wealth $Q(g)$. Of course, knowledge of $g(r)$ is enough to determine $P(g)$, since

$$P(g) = -\frac{1}{\partial g(r)/\partial r}. \tag{2}$$

An important advantage of studying $g(r)$ instead of $P(g)$ is the ease with which ranked distributions (instead of histograms), can be constructed from a small number of data points. Each country is a data point. The database we analyzed [22] has data for around 100–200 countries for each year. Additionally, ranked distributions $g(r)$ more clearly expose the details in the behavior of the tails of large and small values of g, which are barely noticeable when looking at $P(g)$.

Next section shows how analytic results can be derived for the ranked wealth of economic agents following the TI dynamics introduced above. Then, in Sect. 3 we fit these theoretical expressions to available data of GDPpc, for each year in the period between 1960 and 2013. Our model provides an acceptable description of the data, for all years and the entire range of ranks. Parameters obtained by fitting our model to the data give us information about the type, time-dependence, and relative importance of the main dynamical processes shaping the world distribution of g. Finally, Sect. 4 offers a discussion of our results.

2 Multiplicative Trade-Investment Dynamics

In this section we describe the micro-economic rules defining our model [14]. We consider a population of N economic agents, whose wealth evolves at discrete time intervals $\Delta t = 1$ through two independent processes: Random Multiplicative Noise (RMN) acting locally on each agent, and Yard-Sale (YS) exchange, stochastically shifting wealth between randomly chosen pairs of agents. The wealth of the agent with rank R at time t is denoted as $g(R, t)$. Ranks are assigned so as to have $g(1, t) > g(2, t) > \cdots > g(N, t)$.

The analytic expression used to describe the real world data of GDPpc is (13), which is derived in Sect. 2.4. In order to justify the derivation of this result, we first present an analysis of each component of the dynamics.

2.1 The Isolated Effect of Investments: The Lognormal Distribution

Investment represents the cumulative effect of local stochastic factors on the growth of wealth. These are modeled as Random Multiplicative Noise (RMN), the effect of which clearly does not conserve wealth. Under the effect of RMN, the evolution of wealth is given by

$$g(R, t + 1) = g(R, t) \cdot \mathscr{L}_I, \tag{3}$$

where the factor $\mathscr{L}_I$ is a random number drawn independently for each R and t. For simplicity, the distribution of $\mathscr{L}_I$ is assumed lognormal with parameters $\mu_I = \langle \ln \mathscr{L}_I \rangle$ and $\sigma_I^2 = \langle (\ln \mathscr{L}_I - \mu_I)^2 \rangle$. Both parameters are assumed the same for all agents, and can be in general time-dependent.

In the long run, the wealth distribution of agents following the dynamics in (3) tends to a lognormal distribution [23]. Using the definition of relative rank given in (1), the ranked distribution is found to be

$$\ln g(r, t) = \mu_A(t) + \sigma_A(t) \sqrt{2} \mathrm{erf}^{-1}(1 - 2r), \tag{4}$$

where erf^{-1} is the inverse error function and $r = R/N$ is the normalized rank. Parameters $\mu_A(t)$ and $\sigma_A^2(t)$ represent the cumulative effect in time of the drift μ_I and volatility σ_I^2. They are written as $\mu_A(t) = \int \mu_I dt$ and $\sigma_A^2(t) = \int \sigma_I^2 dt$, and are respectively the mean and the variance of $\ln g(r, t)$. For a multiplier $\mathscr{L}_I$ with time-independent distribution, both parameters grow linearly in time.

2.2 Stochastic Wealth Exchange: Yard-Sale Model

We next analyze the effect of wealth exchange under Yard-Sale (YS) rules. YS describes pairwise commercial interactions, where the value of the exchanged wealth is a random fraction of the fortune of the poorest of the two intervening agents [20, 21]. An interaction between agents with ranks R and R' is written as

$$g(R, t + 1) = \begin{cases} g(R, t) + \kappa g(R, t) & \text{if } R > R' \\ g(R, t) - \kappa g(R', t) & \text{if } R < R', \end{cases} \tag{5}$$

where κ is a random return with arbitrary distribution $\pi(\kappa)$. The only condition imposed on the return is $|\kappa| \leq 1$ (in each interaction the poor agent can win or lose no more than its total wealth). The upper line in (5) means that the poorest agent stochastically wins (if $\kappa > 0$) or loses (if $\kappa < 0$) a random fraction of its own wealth from/to the richest agent, while the lower line is wealth conservation.

The wealth distribution of agents following this exchange rule can be obtained by assuming the condition of Wealth Separation (WS) [14], which essentially means that $g(R, 0) \gg g(R', 0)$ whenever $R < R'$. The poor agent is always much poorer than the rich agent, and therefore the effect of the transaction on the rich agent can be neglected. Under the WS assumption, the term $-\kappa g(R', t)$ can be dropped from the second line in (5) and the dynamics can be approximated by a pure multiplicative process

$$g(R, t+1) \approx g(R, t) \cdot \left[1 + \theta(R - R')\kappa\right], \qquad (6)$$

where $\theta(R - R')$ is the Heaviside step function. Note that the random multiplicative factor in (6) now depends on the return κ and on the choice of interacting agents R and R'. Applying logarithms on both sides of (6), averaging over all choices of equally probable partner R' and over the return distribution $\pi(\kappa)$, which is the same for all pairs, the typical effect of YS exchanges after t timesteps is

$$\ln g(R, t) \sim \left(\frac{R - 1}{N - 1}\right) \phi_0 t, \qquad (7)$$

where $\phi_0 = \langle \ln(1 + \kappa) \rangle$ quantifies the average exchange strength for an interaction. Equation (7) can be generalized for time-dependent return distributions $\pi(\kappa, t)$, where the term $\phi_0 t$ is replaced by the cumulative $\int \phi_0(t')dt'$.

The return distribution $\pi(\kappa)$ determines the shape of the wealth distribution under YS rules. If $\phi_0 > 0$, wealth is redistributed between agents. The slope $\partial \ln g(R, t)/\partial R$ increases in time, producing a convergence of wealth towards a stationary wealth distribution. Equation (7) is not valid under these circumstances because the WS assumption fails in the long run. On the other hand, if $\phi_0 < 0$, wealth differences are amplified in time and wealth *condenses* onto a single agent [21]. In this nonequilibrium condensing state, (7) is valid for all times t, and we can rewrite it in terms of the relative rank r as:

$$\ln g(r, t) = \mu_B(t) - \alpha_B(t)(r - 1/2), \qquad (8)$$

with $\mu_B(t)$ the mean of $\ln g(r, t)$ and $\alpha_B(t) = \alpha_0 - \int \phi(t')dt'$ the cumulative effect of interactions. Wealth distributions described by (8) with decreasing values of the slope $-\alpha_B$ are the result of exchange interactions that systematically favor the richest agents.

2.3 Trade-Investment Dynamics with RMN and YS Exchange

In this section we introduce the complete TI dynamics, where now at each time step all
agents exchange part of their wealth and also are affected by external multiplicative
noise. Investment is described by (3). The condition of WS is assumed at all times,
so that YS exchange can be approximated by (6).

The full TI dynamics, combining the effect of one YS exchange and one RMN
acting on the wealth of the agents per timestep, is written as

$$\ln g(R, t+1) = \ln g(R, t) + \theta(R - R')\ln(1 + \kappa) + \ln \mathcal{L}_I. \qquad (9)$$

Thus, at each timestep, the wealth of an agent ranked R is multiplied by $(1 + \kappa)$ if
this agent interacts with a richer agent, and then it is unconditionally multiplied by
$\mathcal{L}_I$. Clearly, these multiplicative factors can be reordered without altering the end
result for $g(R, t)$. We therefore first consider the cumulative effect of all noises $\mathcal{L}_I$,
which produce a lognormal distribution of fortunes as given by (4). This distribution
satisfies the WS condition, and therefore we can now calculate the effect of exchanges
following the procedure described to obtain (7). The net effect of interactions is then
simply modifying the slope of $\ln g(r, t)$ by adding a term that is linear in r.

The final result [14], that combines the effect of simultaneously acting noise and
exchange, can be written as

$$\ln g(r, t) = \mu_C(t) + \sigma_C(t)\sqrt{2}\operatorname{erf}^{-1}(1 - 2r) - \alpha_C(t)(r - 1/2), \qquad (10)$$

where $\sigma_C^2(t) \sim \int \sigma_I^2 dt$ and $\alpha_C(t) = \alpha_0 - \int \phi(t')dt'$. The last term in (10) has been
rewritten to have zero average. Therefore, $\mu_C(t)$ is the average of $\ln g(r, t)$.

It is worth mentioning that, in deriving (10), we have omitted an additional source
of multiplicative noise, which comes from the stochastic nature of YS exchanges. In
previous work [14] we discussed a formal procedure to take into account the
effects of this extra noise source. Its net effect can be approximated by redefining
the noise intensity parameter σ_c. In the following, σ_c is the *total noise* parameter,
representing the combined effect of the external multiplicative noise and of noise due
to interactions. A numerical validation of (10) was presented in previous work [14].

2.4 TI Dynamics with RMN and Rank-Dependent Return Distributions

We now present the main theoretical result of this work, and the expression used to
describe the world distribution of GDPpc. This result is obtained by a generalization
of the methodology presented in previous sections. In order to construct the final
result, we consider more general return distributions, where now $\pi(\kappa)$ is not the
same for all pairs but depends on the ranks of the agents participating in the exchange

event. This can be interpreted as meaning that trade conditions depend on the wealth of the agents taking part in the exchange, which is a reasonable assumption.

The simplest case that goes beyond constant (i.e. rank-independent) return distributions is obtained by assuming a linear dependence on rank difference, as given by

$$\langle \ln(1+\kappa)\rangle_{(R,R')} = \phi_0 + \phi_1 \cdot \left(\frac{R-R'}{N-1}\right). \tag{11}$$

This amounts to the assumption that, for each pair of agents, only the difference between their ranks determines the average properties of their exchange interactions. Positive ϕ_1 represents increasing statistical advantage for increasingly poor agents in a pair. A negative ϕ_1 means that poorer agents are more strongly disfavored, the larger their rank difference with their commercial partner.

Following the procedure introduced in Sect. 2.2, we analyzed YS exchange with return distributions satisfying (11). The effect of this type of interactions once again is found after averaging (6) over the return distribution and over all partners R'. The result is

$$\ln g(r,t) \sim r\phi_0 t + \frac{1}{2}r^2\phi_1 t. \tag{12}$$

Finally, the wealth distribution of agents following a TI dynamics (9), with YS exchanges satisfying (11), will be given by [14]

$$\ln g(r,t) = \mu_D(t) + \sigma_D(t)\sqrt{2}\mathrm{erf}^{-1}(1-2r) - \alpha_D(t)(r-1/2) - \beta_D(t)(r^2-1/3) \tag{13}$$

where $\alpha_D(t)$ and $\beta_D(t)$ are the cumulatives of the interaction parameters ϕ_0 and ϕ_1. The noise intensity $\sigma_D(t)$ contains once again the combined effect of the random multiplicative and exchange noise, as $\sigma_C(t)$ in (10). Since all r-dependent terms in this expression have zero mean, $\mu_D(t)$ is simply the average of $\ln g(r,t)$.

Fits of (13) to real world data allow us to find the values of the relevant parameters and their time dependence. This is done in the next section.

3 Gross Domestic Product Per-capita

The data for Gross Domestic Product per capita provided by the World Bank [22] was fitted to our theoretical expression in (13), for each year between 1960 and 2013 (all values of g are expressed in constant 2005 US dollars). The quality of the fits obtained with (13) can be appreciated by examining Fig. 1.

For each year, we obtained an independent set of parameters μ_D, σ_D, α_D and β_D. Their values are plotted in Fig. 2 versus time. It is found that $\beta_D \neq 0$ for all years. This supports the convenience of considering rank-dependent return distributions, (11), for a better description of the data.

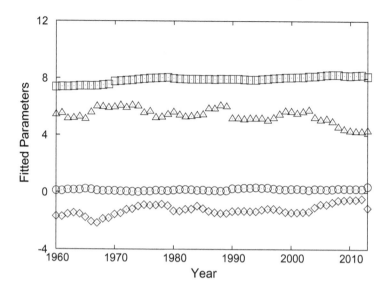

Fig. 2 Parameter values obtained by fitting (13) to the ranked distribution of GDPpc for each year: μ_D (*squares*), σ_D^2 (*circles*), α_D (*triangles*) and β_D (*diamonds*)

Squares in Fig. 2 represent the fitted values of μ_D for each year. This parameter tracks the overall growth of the world economies, since it is the logarithm of the geometric mean of g. Roughly speaking, a linear growth is seen in the values of μ_D. This represents a rise in the average standard of living of world inhabitants. However, μ_D only describes the mean of the distribution of $\ln g$, while the actual shape around this value is determined by the other parameters in (13).

The total noise σ_D^2 (circles in Fig. 2) determines the contribution of noise factors, as RMN and noise due exchanges, to the distribution of GDPpc. The fitted values are, in absolute value, small compared to the other parameters of our model. This implies that an important part of the divergence in the values of $\ln g(r)$ is due to the repulsive effect of commercial interactions, rather than being attributable to the dispersive effects of noise.

Parameters α_D and β_D (triangles and diamonds in Fig. 2, respectively) determine the contribution of commercial interactions or international trade to the distribution of g. Two main features can be observed in the time evolution of these parameters. First of all, β_D is negative in the whole period considered, so that the tail of poor countries curves downwards less strongly than the tail of rich countries curves upwards. The asymmetry of opposite tails of the distribution becomes weaker for recent years, as can be seen in Fig. 1. As a consequence, fitted values of β_D grows closer to zero in time. As seen in Fig. 2, β_D is found to grow in time, on average, implying ϕ_1 from (11) is negative. In other words, poor countries face more disadvantageous conditions the richer their commercial counterpart is. This effect might be responsible for making the tail of poor countries, which in early years is not too visible, increasingly bent downwards. Secondly, $\alpha_D > 0$, which means that $\phi_0 < 0$ in (11), so that wealth exchange favors the rich and thus promotes wealth concentration.

A more thorough analysis of these results can be found in [14], where the discontinuities observed in the time evolution of the fitted parameters are analyzed, as well as a comparison of (13) with simpler versions of the model.

4 Discussion

A theoretical description of the world distribution of Gross Domestic Products per capita was presented, comparing the data with the multiplicative Trade-Investment (TI) model described in Sect. 2. The basic micro-economic mechanisms that determine the distribution of GDPpc were assumed to be random noise (modeling investment), and conservative pairwise wealth exchanges resulting from trade between countries. Under the wealth separation approximation, which is compatible with the large inequality observed in the data for GDPpc, we solved analytically a dynamical model that takes into account the effects of Random Multiplicative Noise (RMN) and Yard-Sale exchange (YS). Trade was modeled considering rank-dependent return distributions, (11), representing commercial biases that depend on the relative wealth of the interacting agents.

Fits of our model to the data support the notion that exchange processes such as international commerce, and not random noise (as sometimes assumed), are the major forces shaping the distribution of GDPpc. It appears that the main reason of the divergence of the GDPpc is that trading conditions statistically favor richer countries. Our analytic results were derived under the assumption of trade promoting wealth concentration, so that our model suggests non-equilibrium wealth distributions. Exchanges are favoring the rich, and therefore wealth becomes increasingly concentrated in time. Section 3 shows that GDPpc behave in a way that is compatible with a non-stationary state, consistent with the assumptions of our model. We can test these conclusions comparing the data with a TI dynamics with a stationary state for the wealth distribution, as the classical Bouchaud and Mezard (BM) model [15]. The BM Model considers trade promoting redistribution of wealth, and the asymptotic behavior of the wealth distribution follows a power law for the very rich agents. However, as shown in [14], when using the BM the data is not described well.

We must mention that the data for GDPpc included in the available databases is not comprehensive. The number of countries described in each year is variable and generally increasing. The increment/decrement in number of countries presented in the databases can be due to the creation of new countries, the impossibility to measure the GDPpc for certain years or many other reasons. This may produce sudden changes in the distribution of GDPpc, which are not describable within our dynamical model. In spite of this, our model describes qualitatively well the real data for all considered years.

Our TI dynamics, with Yard-Sale exchange rules and RMN, seems to work as a good approximation to describe the world distribution of GDPpc. However, improvements to the model can be made. We could go beyond the mean-field approximation of wealth exchanges, analyzing the effect of the World Trade Web. Finally, distinct forms of return distributions can also be an alternative to describe more accurately the data. These possibilities will be explored in future work.

Acknowledgments The authors thank CGSTIC of CINVESTAV for computer time on cluster Xiuhcoatl. RBG acknowledges CONACYT for financial support.

References

1. T. Piketty, *Capital in the 21st Century* (Harvard University Press, 2014)
2. R.J. Barro, X. Sala-i-Martin, *Economic Growth*. 2nd Edn (The MIT Press, 2003)
3. O. Galor, Convergence? inferences from theoretical models. Econ. J. **106**(437), 1056–1069 (1996)
4. V. Pareto, *Cours d'Economie Politique* (Droz, Geneve, 1896)
5. H. Lopez, L. Serven, A normal relationship? Poverty, growth, and inequality. Policy Research Working Paper Series 3814 (The World Bank, 2006)
6. V.M. Yakovenko, J.B. Rosser, Colloquium: statistical mechanics of money, wealth, and income. Rev. Mod. Phys. **81**, 1703–1725 (2009)
7. S. Ispolatov, P.L. Krapivsky, S. Redner, Wealth distributions in asset exchange models. Eur. Phys. J. B **2**, 267–276 (1998)
8. A. Chatterjee, S. Yarlagadda, B.K. Chakrabarti (eds.), *Econophysics of Wealth Distributions* (Springer, Milan, 2005)
9. C. Di Guilmi, M. Gallegati, E. Gaffeo, Power law scaling in the world income distribution. Econ. Bull. **15**(6), 1–7 (2003)
10. See for example, http://en.wikipedia.org/wiki/International_inequality; A. Coad, The skewed world GDP distribution and the interdependence of national institutions. In: DIME Final Conference. http://final.dime-eu.org/files/Coad_C9.pdf (2011)
11. C.F. Moukarzel, Per-capita gdp and nonequilibrium wealth-concentration in a model for trade. J. Phys.: Conf. Ser. **475**, 012011 (2013)
12. B. Urosevic, H. Eugene Stanley, J. Shao, P.C. Ivanov, B. Podobnik, Zipf rank approach and cross-country convergence of incomes. Europhys. Lett. **94**(11), 48001 (2011)
13. R. Iwahashi, T. Machikita, A new empirical regularity in world income distribution dynamics, 1960–2001. Econ. Bull. **6**(19), 1–15 (2004)
14. R. Bustos-Guajardo, C.F. Moukarzel, World distribution of gross domestic product per-capita. J. Stat. Mech.-Theory Exp. P05023 (2015)
15. J.P. Bouchaud, M. Mezard, Wealth condensation in a simple model of economy. Phys. A **282**, 536–545 (2000)
16. N. Scafetta, S. Picozzi, B.J. West, A trade-investment model for distribution of wealth. Phys. D **193**, 338–352 (2004)
17. D. Garlaschelli, T. Di Matteo, T. Aste, G. Caldarelli, M.I. Loffredo, Interplay between topology and dynamics in the world trade web. Eur. Phys. J. B **57**, 159–164 (2007)
18. Diego Garlaschelli, Maria I. Loffredo, Structure and evolution of the world trade network. Phys. A: Stat. Mech. Appl., Elsevier **355**(1), 138–144 (2005)
19. M. Boguna, M.I. Loffredo, M.A. Serrano, D. Garlaschelli, The world trade web: structure, evolution and modeling. In: G. Caldarelli (ed.) Complex Networks, in Encyclopedia of Life Support Systems (EOLSS). Developed under the Auspices of the UNESCO (Eolss Publishers, Oxford, 2010)

20. C.F. Moukarzel, Multiplicative asset exchange with arbitrary return distributions. J. Stat. Mech.-Theory Exp. P08023 (2011)
21. C.F. Moukarzel, S. Goncalves, J.R. Iglesias, M. Rodriguez-Achach, R. Huerta-Quintanilla, Wealth condensation in a multiplicative random asset exchange model. Eur. Phys. J.-Spec. Top. **143**, 75–79 (2007)
22. The world bank database. http://data.worldbank.org/indicator/NY.GDP.PCAP.KD
23. S. Redner, Random multiplicative processes—an elementary tutorial. Am. J. Phys. **58**(3), 267 (1990)

Index

© Springer International Publishing Switzerland 2016
M. Tlidi and M.G. Clerc (eds.), *Nonlinear Dynamics: Materials,
Theory and Experiments*, Springer Proceedings in Physics 173,
DOI 10.1007/978-3-319-24871-4

CPSIA information can be obtained
at www.ICGtesting.com
Printed in the USA
LVHW02*1436040318
568593LV00001B/435/P